Re-Use and Recycling of Materials

Solid Waste Management
and Water Treatment

RIVER PUBLISHERS SERIES IN CHEMICAL, ENVIRONMENTAL, AND ENERGY ENGINEERING

Series Editors

ALIREZA BAZARGAN
NVCo and K.N. Toosi University of Technology
Iran

MEDANI P. BHANDARI
Akamai University, USA
Sumy State University, Ukraine
Atlantic State Legal Foundation, NY, USA

HANNA SHVINDINA
Sumy State University
Ukraine

Indexing: All books published in this series are submitted to the Web of Science Book Citation Index (BkCI), to SCOPUS, to CrossRef and to Google Scholar for evaluation and indexing.

The "River Publishers Series in Chemical, Environmental, and Energy Engineering" is a series of comprehensive academic and professional books which focus on Environmental and Energy Engineering subjects. The series focuses on topics ranging from theory to policy and technology to applications.

Books published in the series include research monographs, edited volumes, handbooks and textbooks. The books provide professionals, researchers, educators, and advanced students in the field with an invaluable insight into the latest research and developments.

Topics covered in the series include, but are by no means restricted to the following:

- Energy and Energy Policy
- Chemical Engineering
- Water Management
- Sustainable Development
- Climate Change Mitigation
- Environmental Engineering
- Environmental System Monitoring and Analysis
- Sustainability: Greening the World Economy

For a list of other books in this series, visit www.riverpublishers.com

Re-Use and Recycling of Materials
Solid Waste Management
and Water Treatment

Editors

Jibin K. P.

Mahatma Gandhi University, India

Nandakumar Kalarikkal

Mahatma Gandhi University, India

Sabu Thomas

Mahatma Gandhi University, India

Ange Nzihou

IMT Mines Albi CNRS, France

LONDON AND NEW YORK

Published 2019 by River Publishers
River Publishers
Alsbjergvej 10, 9260 Gistrup, Denmark
www.riverpublishers.com

Distributed exclusively by Routledge
4 Park Square, Milton Park, Abingdon, Oxon OX14 4RN
605 Third Avenue, New York, NY 10017, USA

First issued in paperback 2023

Re-Use and Recycling of Materials Solid Waste Management and Water Treatment / by Jibin K. P., Nandakumar Kalarikkal, Sabu Thomas, Ange Nzihou.

Routledge is an imprint of the Taylor & Francis Group, an informa business

Publisher's Note
The publisher has gone to great lengths to ensure the quality of this reprint but points out that some imperfections in the original copies may be apparent.

While every effort is made to provide dependable information, the publisher, authors, and editors cannot be held responsible for any errors or omissions.

ISBN 13: 978-87-7022-952-4 (pbk)
ISBN 13: 978-87-7022-058-3 (hbk)
ISBN 13: 978-1-003-33930-4 (ebk)

Contents

3   Autogenous Self-healing in Municipal Waste Incorporated Concretes   33

*Deniz Genc Tokgoz, Nesibe Gozde Ozerkan, and
Simon Joseph Antony*

MODULE 2: Water-Recycle and Reuse

8 Treatment of Whey Water from Food Processing Units Using Hybrid Methods **153**

*Britika Mazumdar, Gargi Biswas, Rajnarayan Saha,
and Susmita Dutta*

MODULE 3: Solid Waste Management – New Breakthrough

12 Photocatalytic Degradation of Plastic Polymer: A Review   225

Tarun Parangi and Manish Kumar Mishra

21 Recycling and Reuse of Metal Catalyst: Silica Immobilized Palladium Complex for C–C Coupling Reaction

Tahshina Begum, Sanjib Gogoi, and Pradip K. Gogoi

MODULE 4: RRR Strategy and Atmosphere

22 Investigation on Sound Absorber Performance, Insulation Property, and Dielectric Constant of Sugarcane Bagasse

Sakti Prasad Mishra, Punyatoya Mishra, and Ganeswar Nath

Foreword

Fernando Gomes
Associate Professor
Universidade Federal do Rio de Janeiro

Few times in life we have the chance to know such a promising young talent as Mr. Jibin, from who I am very proud to announce his first book as Editor, which is entitled **"Re-Use and Recycling of Materials Solid Waste Management and Water Treatment"**. This astonishing book deals with several critical subjects to our Society, which is responsible for the Anthropocene. Jibin and Senior editors (Prof. Ange Nizhou (IMT Mines Albi, France); Prof. Nandakumar Kalarikkal (MGU, India); and Prof. Sabu Thomas (MGU, India)) show us, through the 24 chapters of this book several ways to deal with our Environment impacts, starting by the removal of Arsenic (IV) from waters up to the sustainability of Waste Electrical and Electronic Equipment! So, without more delays, I recommend this book, and I hope you enjoy it as much as I am!

Preface

The recent Paris Climate Agreement should provide impetus to research and development in recycling, re-use and valorization, sustainable energy and materials from waste and biomass in view of the significant embedded demands towards safer environment. Likewise, with the growing trend towards renewables, there is a need to strengthen research and development as well as technologies to tackle the tremendous challenges linked with the sustainable management of resources.

In this context, this book gathers and disseminate cutting-edge research and experiences on the recycling, re-use and valorization (understood as the conversion of waste and biomass to energy, fuels and other useful materials), with particular focus on environmental indicators and sustainability goals. In this field strongly characterized by inter- and multidisciplinary, the quest for the recovery and generation of value in end-of-pipe processes demands cooperation by a large number of actors from various fields of engineering science including but not limited to bio-chemical, chemical, civil, environmental, materials, health and safety as well as social sciences. This is well illustrated in this book.

Among the topics developed in this issue, the reuse and recycling in a sustainable waste management approach is considered. Number of key issues such as waste electrical and electronic equipment (WEEE) which is one of the most notorious and the fastest growing waste streams in the world is discussed with a particular focus on the Indian case. In fact, a tremendous increase is observed annually of discarded quantities of WEEE arising high concerns from citizens. The objective is to treat WEEE as a means to recover valuable materials, including the metals, oil and gas product from the polymeric resin and glass fibre.

Polymers and plastics wastes are other families that are very abundant and represent a threat for the environment. They recycling is addressed in this book via couple of example dealing with photocatalytic degradation of plastic polymer, Thermo-mechanical process for recycling polystyrene waste, field evaluation of plastic mulch film for changes in its mechanical properties, fluoropolymer based tunable materials for emerging applications.

Other waste materials recycled in concrete are also discussed responding to the high demand for sustainable construction materials. In the same field dealing with wastes the book discussed the cases dealing with the thermochemical recycling of carbon based solid waste, waste papers as potential source cellulose nanofiber and nanocomposites. Recycling and reuse of metal catalyst (Silica immobilized palladium) is a presentation as an interesting case for recycling metal species. Agricultural waste recycling using microwave is another insightful case presented. Biowaste materials recycling and re-use are specifically addressed on study such as the ones on oxidation of lignin from wood dust to vanillin using ionic liquid medium and study of its antioxidant activity. Fabrication and characterization of hair keratin-chitosan based porous scaffolds for biomedical application. This field has gain a tremendous momentum over the last 10 years. A study on the investigation on sound absorber performance, insulation property and dielectric constant of sugarcane bagasse show also what could be profitably done with waste such as bagasse.

The treatment of wastewater for a better access to clean and/or potable water is among the highest priority worldwide. This issue is also addressed in this book with a critical review on wastewater treatment techniques for reuse of water in industries. Another study on the effect of local industrial waste additives on the Arsenic (V) removal is discussed. Treatment of whey water from food processing units using hybrid methods sounds a very interesting. In the other side, the Application of agro-residues based activated carbon as adsorbents for phenol sequestration from aqueous streams is discussed. Bioremediation which is one of the most abundant treatment of high strength post methanated distillery wastewater using constructed wetland technology is discussed. This topic is completed with a study on the reuse of magnetite (Fe_3O_4) nanoparticles in de-emulsification of emulsion effluents of steel-rolling mills.

Biochemical routes are another way to recycle and re-use waste materials. In light of declining oil prices, biofuel cannot be the only product obtained from biomass or wastes. The key feature is not what it may seem at the first instance, i.e., to develop new processes and products. Although the latter is essential for progress, there is first and foremost a need for a strong 'attractor' from the market side.

The biotechnology route is a promising way for bio-mass/wastes valorization. Industrial or white biotechnology uses microorganisms and isolated enzymes to generate production in different sectors including food and feed, chemistry, biofuels, chemical specialties, biomaterials and many others. Among other things, research in this area considers the optimization of the natural capacities of micro-organisms (bacteria, fungi, yeast, microalgae) and/or enzymes in biotransformation processes, the improved performance via process innovation and the genetic selection. In this book, the production of Hydrogen and Methane using dark fermentation is discussed.

The book also addresses the management of emissions and pollutants in air purification is highlighted.

Finally, in the above mentioned topics, this book reviews the ongoing research and highlights the direction where the R&D effort is trending from this point forward.

(Editors)

Jibin K. P.
Nandakumar Kalarikkal
Sabu Thomas
Ange Nzihou

List of Contributors

Adam Cenian, *Instytut Maszyn Przepływowych im R. Szewalskiego, Polskiej Akademii Nauk, E-mail: cenian@imp.gda.pl*

Ahmad K. Jassim, *Production and Metallurgical Engineering, Basra, Iraq, E-mail: ahmadkj1966@yahoo.com*

Amrita Kaurwar, *Indian Institute of Technology Jodhpur, Jodhpur*

Anand Plappally, *Indian Institute of Technology Jodhpur, Jodhpur, E-mail: anandk@iitj.ac.in*

Anubha Kaushik, *1. Department of Environmental Science and Engineering, Guru Jambheshwar University of Science and Technology, Hisar, India 2. University School of Environment Management, G.G.S. Indraprastha University, New Delhi, India*

Aparna Bhardwaj, *Department of Environmental Science and Engineering, Guru Jambheshwar University of Science and Technology, Hisar, India,*

Ashok N. Bhaskarwar, *Department of Chemical Engineering, Indian Institute of Technology Delhi, Hauz Khas, New Delhi, India, E-mail: ashoknbhaskarwar@yahoo.co.in*

Bartosz Hrycak, *Instytut Maszyn Przepływowych im R. Szewalskiego, Polskiej Akademii Nauk, E-mail: bartosz.hrycak@imp.gda.pl*

Biswajit Debnath, *1. Department of Chemical Engineering, Jadavpur University, Kolkata, India 2. Secretariat Member, International Society of Waste Management, Air and Water (ISWMAW), E-mail: biswajit.debnath.ju@gmail.com*

Britika Mazumdar, *Department of Earth and Environmental Studies, National Institute of Technology Durgapur, Durgapur, India*

Chander Prakash Kaushik, *1. Department of Environmental Science and Engineering, Guru Jambheshwar University of Science and Technology, Hisar, India*
2. Amity School of Earth and Environmental Sciences, Amity University Haryana, Gurgaon, India

Dariusz Czylkowski, *Instytut Maszyn Przepływowych im R. Szewalskiego, Polskiej Akademii Nauk, E-mail: dariusz.czylkowski@imp.gda.pl*

Deniz Genc Tokgoz, *Center for Advanced Materials, Qatar University, Doha, QATAR*

Ganeswar Nath, *Department of Physics, V S S University of Technology, Sambalpur, Odisha, India, E-mail: ganesh_nath99@yahoo.co.in*

Gargi Biswas, *Department of Chemical Engineering, National Institute of Technology Durgapur, Durgapur, India, E-mail: gargi129@gmail.com*

Gaweł Sołowski, *Instytut Maszyn Przepływowych im R. Szewalskiego, Polskiej Akademii Nauk, E-mail: gsolowski@imp.gda.pl*

Gyanashree Bora, *Department of Chemistry, Dibrugarh University, Dibrugarh, Assam, India*

Hyun-Chan Kim, *CRC for Nanocellulose Future Composite, Inha University, Incheon city, Republic of Korea*

Izabela Konkol, *Instytut Maszyn Przepływowych im R. Szewalskiego, Polskiej Akademii Nauk, E-mail: izabela.konkol@imp.gda.pl*

Jaehwan Kim, *CRC for Nanocellulose Future Composite, Inha University, Incheon city, Republic of Korea, E-mail: jaehwan@inha.ac.kr*

Juma Haydary, *Institute of Chemical and Environmental Engineering, Faculty of Chemical and Food Technology, Slovak University of Technology, Radlinského 9, Bratislava, E-mail: juma.haydary@stuba.sk*

Jung Ho Park, *CRC for Nanocellulose Future Composite, Inha University, Incheon city, Republic of Korea*

Jyotirekha G. Handique, *Department of Chemistry, Dibrugarh University, Dibrugarh, Assam, India, E-mail: jghandique@rediffmail.com*

Keshaw Ram Aadil, *Department of Biotechnology, National Institute of Technology Raipur, Raipur, Chhattisgarh, India, E-mail: kaadil7@gmail.com*

Krzysztof Pastuszak, *Katedra Algorytmów i Modelowania; Wydział Informatyki, Telekomunikacji i Informatyki Politechniki Gdańskiej, E-mail: krzpastu@pg.edu.pl*

Kulasekaran Jaidev, *Laboratory for Advanced Research in Polymeric Materials (LARPM), Central Institute of Plastics Engineering and Technology (CIPET), Bhubaneswar, Odisha, India*

Le Van Hai, *1. CRC for Nanocellulose Future Composite, Inha University, Incheon city, Republic of Korea
2. Department of Pulp and Paper Technology, Phutho College of Industry and Trade, Phutho, Vietnam, E-mail: levanhai121978@gmail.com*

Lovelesh Dave, *Indian Institute of Technology Jodhpur, Jodhpur*

Manish Kumar Mishra, *Department of Chemistry, Sardar Patel University Vallabh Vidyanagar, Gujarat, India, E-mail: manishorgch@gmail.com*

Mintu Job, *Department of Agricultural Engineering, Birsa Agricultural University, Kanke, Ranchi, E-mail: mintujob@rediffmail.com*

Mona Sharma, *Department of Environmental Science and Engineering, Guru Jambheshwar University of Science and Technology, Hisar, India, E-mail: drmonasharma1@gmail.com*

Nesibe Gozde Ozerkan, *Center for Advanced Materials, Qatar University, Doha, QATAR*

Omdeo K. Gohatre, *Laboratory for Advanced Research in Polymeric Materials (LARPM), Central Institute of Plastics Engineering and Technology (CIPET), Bhubaneswar, Odisha, India*

Parsanta Verma, *Department of Chemical Engineering, Indian Institute of Technology Delhi, Hauz Khas, New Delhi, India, E-mail: parsantaiitd@gmail.com*

Pradip K. Gogoi, *Department of Chemistry, Dibrugarh University, Dibrugarh, Assam, India*

Pratima Gupta, *Department of Biotechnology, National Institute of Technology Raipur, Raipur, Chhattisgarh, India, E-mail: pgupta.bt@nitrr.ac.in*

Pravin Kumar *Department of Mechanical Engineering, Delhi Technological University, Delhi, India, E-mail: pravin.dce@gmail.com*

Punyatoya Mishra, *Department of Physics, Parala Maharaja Engineering College, Berhampur, Odisha, India, E-mail: punyatoya.phy@gmail.com*

Pushpa Jha, *Sant longowal Institute of Engineering and Technology, India*

Rajesh Kumar Singh, *Management Development Institute, Gurgaon, Haryana, India, E-mail: rksdce@yahoo.com*

Rajnarayan Saha, *Department of Chemistry, National Institute of Technology Durgapur, Durgapur, India*

Ruth M. Muthoka, *CRC for Nanocellulose Future Composite, Inha University, Incheon city, Republic of Korea*

Sakti Prasad Mishra, *Department of Physics, V S S University of Technology, Sambalpur, Odisha, India, E-mail: saktimishra27@gmail.com*

Sandeep Gupta, *Indian Institute of Technology Jodhpur, Jodhpur*

Sanjay K. Nayak *Laboratory for Advanced Research in Polymeric Materials (LARPM), Central Institute of Plastics Engineering and Technology (CIPET), Bhubaneswar, Odisha, India*

Sanjib Banerjee, *Department of Chemistry, Indian Institute of Technology Bhilai, Raipur, Chhattisgarh, India,*
E-mail: sanjib.banerjee@iitbhilai.ac.in (S.B.)

Sanjib Gogoi, *Applied Organic Chemistry Group, CSTD, CSIR-NEIST, Jorhat*

Shanmugasundaram O. Lakshmanan, *Department of Textile Technology, K.S. Rangasamy College of Technology, Tiruchengode, India,*
E-mail: mailols@yahoo.com

Shashikant Shingdilwar, *Department of Chemistry, Indian Institute of Technology Bhilai, Raipur, Chhattisgarh, India*

Shri Sita Ram Bhakar *Department of Soil and Water Engineering, College of Technology and Engineering, Udaipur, Rajasthan*

Simon Joseph Antony *School of Chemical and Process Engineering, University of Leeds, Leeds LS2 9JT, UK, E-mail: S.J.Antony@leeds.ac.uk*

Sk Arif Mohammad, *Department of Chemistry, Indian Institute of Technology Bhilai, Raipur, Chhattisgarh, India*

Smita Mohanty, *Laboratory for Advanced Research in Polymeric Materials (LARPM), Central Institute of Plastics Engineering and Technology (CIPET) Bhubaneswar, Odisha, India*

Snehalata Ankaram, *Department of Zoology, Vasantrao Naik College of Arts, Commerce & Science, Aurangabad, Maharashtra, India,*
E-mail: asnehalata@yahoo.in

Sowjanya Makarla, *CVR College of Engineering, Hyderabad, Telangana,*
E-mail: madireddisowjanya@gmail.com

Sujatha Karuppiah, *Department of Physics, Vellalar College for Women, Erode, India*

Sunanda Roy, *CRC for Nanocellulose Future Composite, Inha University, Incheon city, Republic of Korea*

Sunil S. Suresh, *Laboratory for Advanced Research in Polymeric Materials (LARPM), Central Institute of Plastics Engineering and Technology (CIPET), Bhubaneswar, Odisha, India E-mail: sunilssuresh@gmail.com*

Susmita Dutta, *Department of Chemical Engineering, National Institute of Technology Durgapur, Durgapur, India, E-mail: susmita_che@yahoo.com*

Tahshina Begum, *Applied Organic Chemistry Group, CSTD, CSIR-NEIST, Jorhat, E-mail: tahshi.du@gmail.com*

Tarun Parangi, *Department of Chemistry, Sardar Patel University Vallabh Vidyanagar, Gujarat, India*

List of Figures

List of Tables

List of Abbreviations

2H-PFP	1,1,3,3,3-Pentafluoropropene
AFM	Atomic force microscopy
AIBN	2,2(-Azobisisobutyronitrile
ANOVA	Analysis of variance
AOP	Advanced oxidation process
AQI	Air quality index
ASR	Automobile shredder residue
ATM	Automated teller machine
ATR-FTIR	Attenuated total reflection Fourier transform infrared
ATRP	Atom transfer radical polymerization
BOD	Biological oxygen demand
BuVE	Butyl vinyl ether
$CaCO_3$	Calcium carbonate
CB	Conductance band
CD	Compact disk
CDPCD	Cyanomethyl 3,5-dimethyl-1H-pyrazole-1-carbodithioate
CEC	Cation exchange capacity
CGC	commercial grade carbon
CMS	Chloromethylstyrene
CNF	Cellulose nanofiber
COD	Chemical oxygen demand
COQ	Cost of quality
CPC	Civil Procedure Code
CPCB	Central Pollution Control Board
CPDB	2-cyano-2-propyl benzodithioate
CPTES	3-Chloropropyltriethoxysilane
CR	Crop residue
CRMP	Cobalt-mediated radical polymerization
CRT	Cathode ray tube
CSH	Calcium silicate hydrate
CS-TiO2	combustion-synthesized nano anatase titania

CTA	Chain transfer agent
CTFE	Chlorotrifluoroethylene
CuPc	Copper phthalocyanine
CW	Constructed wetland
CWM	Constructed wetland microcosm
DAT	Days after transplanting
DCV	Diesel commercial vehicle
DI	Deionized
DLS	Dynamic light scattering
DMA	Dimethyl acrylamide
DPPH	1, 1-Diphenyl-2-picryl hydrazyl
DSC	Differential scanning calorimetry
DSWMC	Qatar's domestic solid waste treatment center
DVD	Digital video disk
ECA	Ethyl cyanoacetate
ECM	Extracellular matrix
EDC.HCL	1-Ethyl-3-(3-dimethylaminopropyl)-3-ethylcarbodiimide hydrochloride
EDC	1-Ethyl-3-(3-dimethylaminopropyl)carbodiimide
EDS	Energy dispersive spectroscopy
EDX	Energy-dispersive X-ray
EEE	Electrical and electronic equipment
EFB	Empty fruit bunch
EHVE	2-Ethylhexyl vinyl ether
EM	Electromagnetic
EO	Ethylene oxide
EoL	End of Life
EPR	Extended producer responsibility
EU	European Union
FA	Fly ash
FATRIFE	2,2,2-Trifluoroethyl α-fluoroacrylate
FAV8	3,3,4,4,5,5,6,6,7,7,8,8,9,9,10,10,10-Heptadecafluoro-decyl vinyl ether
Fe(St)3	Ferric stearate
FePcCl16	Perchlorinated iron (II) phthalocyanine
FE-SEM	Field emission scanning electron microscopy
FTIR	Fourier transform infrared
FTIR	Fourier transform infrared spectrometer
GC	Gas chromatography

GC-MS	Gas chromatographic analysis coupled with mass spectrometry
GESAMP	The Joint Group of Experts on the Scientific Aspects of Marine Environmental Protection
GHG	Greenhouse gas
HD	High definition
HDPE	High density polyethylene
HFP	Hexafluoropropylene
HHV	Higher heating value
HMF	Hydroxymethyl furfural
HPCL	Hyper branched poly(3-caprolactone)
ICP-AES	Inductively coupled plasma-atomic emission spectrometry
ICT	IT and telecommunication equipments
IR	Infrared
ISW	Industrial solid waste
IT	Information technology
ITP	Iodine transfer polymerization
LB	Luria Bertani
LCA	Life cycle assessment
LCD	Liquid crystal display
LDPE	Low density polyethylene
LED	Light-emitting diode
LLDPE	Linear low-density polyethylene
LPG	Liquid petrol gas
MA	Methyl acrylate
MAA	Methacrylic acid
MAF	2-(Trifluoromethyl)acrylic acid
MAF-ester	Alkyl 2-trifluoromethacrylate
MAF-TBE	*tert*-Butyl 2-trifluoromethacrylate
MBR	Membrane bioreactor technology
MCC	Microcrystalline cellulose
MD	Machine direction
MFC	Microfibrillated cellulose
MIPs	Molecularity imprinted polymers
MMA	Methyl methacrylate
MMT	Million metric ton
Mn	Number average molecular weight
MNPs	Magnetite nanoparticles

MoEF	Ministry of Environment and Forests
MREW	Metal recovery from e-waste
MS	Mass spectrometry
MSW	Municipal solid waste
MTFMA	Methyl 2-trifluoromethylacrylate
MVE	Methyl vinyl ether
Mw	Weight average molecular weight
NB	Norbornene
NCC	Nanocrystals cellulose
NFC	Nanofibrillated cellulose
NGO	Non-governmental organization
NGT	National Green Tribunal
NHC	N-heterocyclic carbine
NMP	Nitroxide-mediated polymerization
NMR	Nuclear magnetic resonance
NP	Newsprint
NVP	*N*-vinylpyrrolidone
OD	optical density
OEM	Original equipment manufacturer
OMRP	Organometallic mediated radical polymerization
ONP	Old newsprint
OPW	Oxidized polyethylene wax
PA	Phosphonic acid
PAA	Poly(alkyl acrylates)
PAE	phthalate esters
PAM	Polyacralamide
PAM	Polyacrylamide
PAM-g-TiO2	Polyacrylamide grafted titania
PBFA	1,4-Phenylene *bis*(2-trifluoromethacrylate)
PBS	Phosphate buffered saline
PBS	Poly (butylene succinate)
PC	Portland cement
PCB	Printed circuit board
PDA	Photodiode array detection
PDS	Peroxydisulfate
PE	Polyethylene
PEG	Polyethylene glycol
PEMFCs	Polymer electrolyte membranes for fuel cells
PEO	Poly(ethylene oxide)

PET	Poly(ethylene terephthalate)
PFA	Phenyl 2-trifluoromethacrylate
PMDW	Post-methanated distillery wastewater
POM	Polyoxometalate
PP	Polypropylene
PPFR	Perfluoro-3-ethyl-2,4-dimethyl-3-pentyl persistent radical
PPRH	partially pyrolyzed rice husk
PPy/TiO2	Polypyrrole/titania
PS	Polystyrene
PS-g-TiO2	Polystyrene-grafted titania (embedding the into the PS)
PS-G-TiO2	Polystyrenre-grafted titania (grafting into the PS surface)
PU	Polyurethane
PVB	Poly(vinyl betrayal)
PVC	Poly(vinyl chloride)
PWS	paddy and wheat system
RAFT	Reversible addition fragmentation chain transfer
RAM	Radar absorbing material
RDF	Refused derived fuel
RDRP	Reversible deactivation radical polymerization
r-HDPE	Recycled high-density polyethylene
r-LDPE	Recycled LDPE
RO	Reverse osmosis
RT	Room temperature
SCN	Supply chain network
SD	Standard deviation
SDS	Sodium dodecyl sulfate
SEC	Size exclusion chromatography
SEM	Scanning electron microscope
SEP	Solid polymer electrolyte
SP	Superplasticizer
SR	Swelling ratio
St	Styrene
STHFA	4-(1,1,1,3,3,3-Hexafluoro-2-hydroxypropyl)-styrene
TAPE	*tert*-Amyl peroxy-2-ethylhexanoate
TBA	Thiobarbituric acid
TBPPi	tert-Butylperoxypivalate

TCD-GC	Thermal conductivity detector gas chromatography
TD	Transverse direction
TEA	Triethylamine
TEM	Transmission electron microscopy
TFE	Tetrafluoroethylene
TFEMA	2,2,2-Trifluoroethyl methacrylate
TFP	3,3,3-Trifluoropropene
Tg	Glass transition temperature
TG	Thermogravimeter
TGA	Thermogravimetric analysis
THF	Tetrahydrofurane
THP	Tetrahydropyrane
TKN	Total Kjeldahl nitrogen
TMP	Thermomechanical pulp
TOCNF	Tempo-oxidized CNF
TP	Total phosphate
TRAI	Telecom regulatory authority of India
UEEE	Used Electrical and Electronic Equipments
UNEP	United Nations Environment Program
UPLC	Ultra-high-performance liquid chromatography
UTM	Universal testing machine
UV	Ultraviolet
VAc	Vinyl acetate
VB	Valence band
VC	Vinylene carbonate
VCR	Videocassette recorder
VDF	Vinylidene fluoride
VE	Vinyl ether
VF	Vinyl fluoride
v-HDPE	Virgin HDPE granules
v-LDPE	Virgin low-density polyethylene
VSS	Volatile suspended solids
WB	waste biomass
WEEE	Waste Electrical and Electronic Equipments
WEEE	Waste electrical and electronic equipments
WHO	World Health Organization
WMS	WEEE management system
XRD	X-ray diffraction
XRF	X-ray fluorescence

MODULE 1

Sustainable Developments

1

Reuse and Recycling: An Approach for Sustainable Waste Management

Snehalata Ankaram

Department of Zoology, Vasantrao Naik College of Arts, Commerce & Science, Aurangabad, Maharashtra, India
E-mail: asnehalata@yahoo.in

Reuse and recycling of waste can be intercepted from disposal and dumping of waste at landfills. Best remedy for non-disposable and non-degradable materials is reusing and recycling directing back into consumer user chain. These are novel strategies of waste management and disposal issues. An approach to minimize consumption and extraction from natural resources in a more sustainable way by approving policy tools such as reuse and recycling.

Human refuse is exceedingly increasing score, unavoidable resulting from human activities posing a threat and imbalance in human health, social, economic, and environmental aspects. Waste handling and disposal a method along with integrated waste management is effectively concerned with three R's: reduce, reuse, and recycle. Biomass material is the key feed stock in the recycling process originating from both nature and human enterprises; agriculture, palm mills, sugarcane industries, paper industries, etc. Bioremediation of these wastes can be converted into value-added products through thermal and non-thermal processes with neutral carbon emission in environment. Energy is reclaimed from different sources by composting, pyrolysis, anaerobic digestion, biosorption, incineration, fermentation, etc. Energy can be recovered in various forms like electricity, oil, gas, briquettes, pellets, steam, compost, etc.

An uncountable number of feedstock for reuse and recycling involves waste from demolition sites, used tires, paper byproducts, plastics of different components with recycling codes, glass redirected for further use, preventing from landfill, minimizing emissions from landfill sites, and leaching in grounds.

There is an urge in shift of civilization from "use and throw" toward reuse and recycling era for "sustainable future".

1.1 Introduction

Biomass is an organic volatile matter and renewable energy source originating from living organisms. This volatile energy is used to produce heat, electricity, steam, biofuels, oils, etc. The technologies involved in conversion of this biomass resulting into energy yield, neutral carbon emission, declining pollution. Waste management by reusing and recycling biomass derived from different sources like agriculture, forest, and domestic organic waste can help reducing disposal and landfill of waste. It is renewable, widely available, carbon-neutral, and has the potential to provide significant employment in the rural areas. Green technology includes sustainable alternatives rather than depending on non-renewable vital sources. Recycling helps often in recovering the nutrients, metals, etc., from recyclates.

1.1.1 Biomass as Feedstock Source

The total biomass yield in India is 450 million tons per year, of which about 200 million tons is unused. India is credited with 32% of primary energy use by means of biomass likely with 20 GW of electricity generation from biomass residues (http://www.eai.in).

In India, about 350 million tons of agricultural waste is yielded every year possibly can generate 18,000 MW of power per year. Sugarcane industries being to a larger extent in India it is a chance with hope that sugarcane molasses can be used to generate electricity and also fertilizer for fields (http://www.hindustantimes.com). A successful experiment yielding 5- HMF biofuel extracted from sugarcane bagasse confirming cellulose and lignocelluloses as sources of sustainable options (Dutta et al., 2012).

Agriwaste such as sugarcane bagasse, wood scrapings, sawdust, coconut husk, oil mill waste, tree barks, rice husk, etc., are used as biosorbents in heavy metals' remediation from industrial wastewaster (Kadirvelu et al., 2001; Hegazi, 2013).

Woody biomass fly ash is reused as partial component in cement mixes as filler sand during concrete constructions (Berra et al., 2011). Increasing

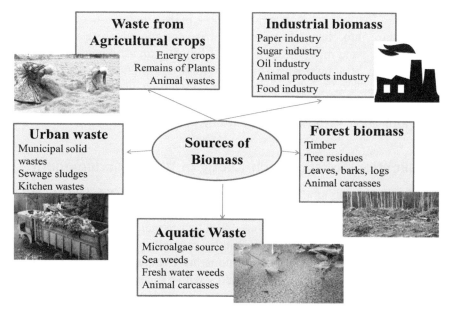

Figure 1.1 Different sources of biomass for biofuel production.

demands for fuels and power can be executed by minimizing dependence on fossil fuel sources through application of biomass.

The national biofuel policy of India structured in December 2009 based is on non-food feedstock for production of biofuels. Planting non-edible oilseeds like jatropha and karanjia are grown on waste lands devoid of agricultural practice. Biomass from *Jatropha curcas* is converted into biodiesel under The National Biodiesel Commission. Around 5% of ethanol derived from sugar molasses is mixed with petrol. Figure 1.1 illustrates the various sources of biomass naturally available in the environment. Most of these source are undepletable and an alternative for fossil fuels.

Biomass feedstocks employed for energy yield are sugarcane bagasse, rice husk, wheat husk, straw, cotton stalk, coconut shells and husk, soya husk, de-oiled cakes, coffee waste, jute wastes, groundnut shells, saw dust, etc. A non-food oriented crop algae, i.e., chlorella efficiently succeeded in the excess nutrient removal, viz., nitrogen and phosphorus from animal manure subjected to anaerobic digestion. Thus ascertained as a feedstock, algae a best choice for biomass production and an approach toward ecological method to balance and prevent disposal of effluent in environment (Chen et al., 2012).

Pyrolysis, a thermal decomposition process boosts up the bioconversion of swine manure into charcoal; a remedial base for animal waste management. The energy yield in this process is 160 kJ mol^{-1}, an ideal methodology for bioconversion of animal waste biomass (RoK et al., 2009). Conversion techniques like anaerobic digestion can be adopted to waste refuse with higher moisture while lower moisture thrash can be thermally converted by applying gasification, pyrolysis, or combustion (Williams et al., 2003).

1.2 Applications of Nanotechnology in Waste Management

Nanobiology is a widely practicing emerging technology in fields like health sciences, energy, crop and agriculture field, waste water treatment, etc. It has now applications in environmental zone and in waste management sector.

A step ahead in the field of solar energy research is seen by applying nanotechnology. Photosynthesis protein units are extracted from plants for conversion of solar energy to electrical energy with a potential of up to 1 year working capacity. Heavy multilayered photosystem I mobilized on cathode unit creates a photolytic effect generating a photocurrent of ~ 2 μA/cm^2 at moderate light intensities (Ciesielski et al., 2010).

Green manufacturing nanotechnology comprising reuse and recycling of fish and agricultural waste biomasses into useful products such as medicines, collagen rejuvenation, and cosmetic products for mankind by means of nanocrystals and nanofibrils was reported by (Morganti, 2013). An effective method of synthesizing iron nanoparticles was experimented by employing *Chlorella* sp. MM3 growth on brewery wastewater; a viable access toward safe environment. During the algal growth, all pollutants like total nitrogen, total phosphorus, and total organic carbon in wastewater were utilized, proceeding with formation of iron nanoparticle with the aid of algal particles (Subramaniyam et al., 2016).

1.3 Waste Recycling in India

Waste biomass rice husk in Bihar, India, proved to be an active ingredient for power generation in gasifiers. Bihar-based Husk Power Systems commenced by founders Gyanesh Pandey and Ratnesh Kumar with an innovative idea of simple designed gasifier operated by any trained person, combusting 50 kg of rice husk per hour and yielding 32 kW of power, providing off-grid power to 350 rural villages.

These power systems using rice husk with additives such as mustard stems, corncobs, grasses, and agricultural residue accomplished setting up other 75 biomass mini-plants across Bihar with a capacity of illuminating more than 12,000 rural households.

The sustainable impact on environment is with reduction of about 5800 tons of carbon dioxide emission for each megawatt of power generated and at individual mini-plants saving up to 125–150 tons of CO_2 per year confirming the best practice and alternative for fossil-fuel powered plants. Also, the refuse slag resulting after the process is diverted in making incense sticks, rubber, and manure, employing around 1200 women in byproduct recycling (Hanson, 2012; http://www.huskpowersystems.com/about-us/).

About 75% of coconut farming and production is lead by Asian countries like Indonesia, Philippines, and India with Indonesia the largest producing country. Coconut besides as a part of edible crop, it also benefits in yielding electricity, heat, fiberboards, organic manure, animal feeds, fuel additives with least emissions, and energy drinks. Figure 1.2 represents the flow chart of various means for waste management through three R's.

Coconut fruit is composed of 40% husk and 30% fiber consisting of cellulose, lignin, pyroligneous acid, gas, charcoal, tar, tannin, and briquette; a suitable contender for biomass fuel feedstock. Interesting is that coconut crop is a stable crop, grown throughout the year (Zafar, 2015). Hence, coconut waste can be diverted toward biofuel formation in an ecofrienldy way. Figure 1.3 depicts the different parts of coconut and its applications.

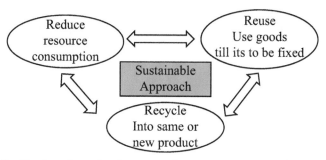

Figure 1.2 The flow chart of various means for waste management through three R's.

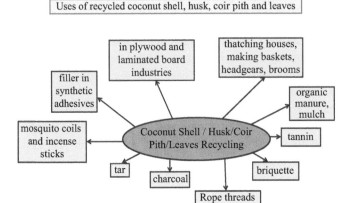

Figure 1.3   Different parts of coconut and its applications.

1.4 Advantages of Recycling

The pulverized and sieved coconut shell powder has gained the market value and proven to be a best substitute for bark powder, furfural, and peanut shell powder, due to easy processing, preventing fungal growth with greater water absorption property. Different mesh sizes ranging from 80 to 200 are manufactured depending on its use (http://coconutboard.nic.in/shelpwdr.htm).

Recycling by composting of waste biomass with mineral (mica) powder proved to be a promoting substitute for potassium. Isabgol straw and palmarosa distillation spent supplemented by mica waste were administered as organic raw material for composting process. Mica endowed compost displayed inclined potassium level compared with that of traditional compost (Basak, 2018).

Along with the additives such as fly ash, ground granulated blast-furnace slag, silica fume, and natural pozzolans, such as calcined shale, calcined clay, or metakaolin (www.ce.memphis.edu/1101/notes/concrete/PCA_manual/Chap03.pdf), ceramic waste is confirmed to be potent enhancer with cement augmenting the pozzolanic activity (Castillo, 2018). The study also revealed reusability of this ceramic waste in brick construction and improving the property of roof cover finishing.

Table 1.1 shows utilization of biowaste originated from coffee pulp, sugar bagasse, animal waste etc into the end product i.e energy production by various industries (Singh, 2013).

Table 1.1 Conversion of different waste biomass to energy in India and reported benefits (Singh, 2013)

Name of the Industry/ Agency	Feedstock/Waste	Conversion Route	Applications	Reported Benefits/Savings
Sakthi Sugars, Maharashtra	Sugarcane bagasse	Biomethanation	Heating	Reported IRR = 32% Biogas substituted for almost 87% of fuel oil consumption
K.M. Sugar Mills, Uttar Pradesh	Sugarcane bagasse	Biomethanation	Power plant (capacity 1 mW)	12,000 m^3 Biogas produced from 400 kL of spent wash per day
Pravara Nagar, Sugar factory, Maharashtra	Sugar factory press mud (75% organic matter; 29% solid content; 65% is volatile)	Biogas (having 60% methane)	Domestic fuel for cooking	Four biogas plants each having 85 m^3 Capacity is setup with MNRE financial assistance. Meeting cooking needs of 196 households for 4 h/day
Demo Plants by an NGO— appropriate Technology Institute (ARTI)	Sugarcane leaves left after harvesting	Oven and rotary kiln conversion to char and briquetting	Fuel for various applications	Plant output 100 kg char per day. Earning of Rs. 75,000 in 25 weeks of harvesting season
ASTRA, IIsc	Leafy biomass	Biogas	Fuel	It is claimed that 2/3 of the families of estimated 100 million rural household could be provided electricity if we use only 10% of around 1130 million ton leafy biomass waste available in India
AL-Kabeer Exports Pvt. Ltd., Medak (Andhra Pradesh)	Slaughter house waste (liquid and solid)	Two-stage digestion process for biogas production	Fuel	3000–4000 m^3 gas is produced, which saves furnace oil consumption worth Rs. 4 million per annum
Western Hatcheries, Ltd.	Poultry waste	UASB reactor for biogas production	Power plant (capacity 1.2 mW)	60 m^3 biogas is produced/ day from 200 TPD Poultry waste processing plant at Namakhal
MSW to Energy and Resources in Singupa town in Bellary District by Technology Informatics Design Endeavour (TIDe) and IIsc	Organic fraction of municipal waste	Plug flow biogas reactor	Not known	Data collected show that 1 kg waste gives 50–60 L of biogas. C/N ratio of compost is found to be 11.4

(Continued)

Table 1.1   (Continued)

Name of the Industry/Agency	Feedstock/Waste	Conversion Route	Applications	Reported Benefits/Savings
Coffee Board and Ministry of Commerce	Coffee pulping waste	Bioreactor for biogas conversion	Power generation	About 80 m^3 of biogas is produced from each ton of coffee parchment. The technology has been successfully demonstrated at 13 locations
TerI's Gurgaon campus at Gual Pahari	Fibrous and semi-solid organic wastes	Acidification and methanation process (biogas containing 70–75% methane, rest CO_2, and traces of H_2S and moisture)	Various uses	TERI claims it to be a useful way to turn wastes from food and fruit processing industries, hotels, pilgrim houses, hostels, housing colonies, community kitchens, vegetable markets, etc., into energy
Transport house, KSRTC, Bangalore	Canteen waste (rice straw, bagasse, paper shreds, garden cuttings, lawn mowing, vegetable peels, uneaten rice, plate and dish washings, and fruit and vegetable rejects)	Biomethanation	Fuel for food warming	The KSRTC plant can handle 25 kg of canteen rejects per day along with leaf litter, which produces 1.5 m^3 of biogas

1.4.1 Briquetting: A Suitable Option for Waste Management

An excellent example for waste into energy is replacing conventional coal with biomass briquettes. Briquettes are pressed small bricks under low pressure with densification yielding higher energy and dried to eliminate moisture, substitute for fuel source, formed of biomass origin derived from agricultural, paper, sugarcane bagasse, rice and coffee husk, saw-dust, cassava rhizomes, water hyacinth, groundnut and coconut hells, etc. (Lubwama and Yiga, 2017, 2018; Suttibak and Loengbudnark, 2018; http://www.renewableurja.com/).

The sustainable use of these biomass briquettes shows a calorific value of 3500 kcal/kg rather benign than to conventional coal of 5000 kcal/kg.

According to Rural Renewable Urja Solutions Pvt. Ltd., India, 1.30 kg of briquettes will replace 1 kg of coal, and 2.50 kg briquettes capable of replacing 1 kg of LPG in efficient smokeless stove discovered by Rural Renewable Urja Solutions Pvt. Ltd., India, has a tremendous scope for biomass source around 500 million metric tons per year including forest and agri contribution likely to be 18,000 MW. Preference can be given to bagasse as renewable source producing power of 7000 MW from sugar industries distributed throughout the country (http://www.ireda.gov.in).

1.5 Conclusion

National and enforcement/voluntary laws should be dealt depending on type of city and facilities available, obligatory waste segregation at generation site and transport facilities at recycling site, recycling claws should be assigned for types of waste. There should be an active participation of various stakeholders like citizens, leading authorities, local NGOs, and municipalities in recycling mechanism. Role of informal sectors including waste pickers, waste buyers, and recycling company contributing for waste reduction and 23% recycling rate is to be acknowledged (Sang-Arun, 2012). There should be availability and strengthening of timely subsidies, policy incentives, and safeguard system. Also, policies for agricultural waste resources should be designed. An organized and scientifically structured segregation of waste should be conducted. Plans and technology to gain zero waste and effective recycling are must. There is an urge of encouragement for active participation of all stakeholders (citizen, municipal authorities, NGOs, and rag pickers), awareness regarding resource depletion, environmental impact, and GHG emissions. There is a need of providing improved incentives to rag pickers.

The sustainable appliance of three R's can reduce the stress regarding the issues of storage, transport, and discarding the waste, thereby reducing the burden on land.

References

Basak BB. (2018). Recycling of waste biomass and mineral powder for preparation of potassium-enriched compost. Journal of Material Cycles and Waste Management. pp.1–7.

Berra M, De Casa G, Dell'Orso M, Galeotti L, Mangialardi T, Paolini AE, et al. (2011). Reuse of Woody Biomass Fly Ash in Cement-Based Materials: Leaching Tests. In: Insam H, Knapp B. (Eds). Recycling of Biomass Ashes. Springer, Berlin, Heidelberg.

Castillo RC. (2018). Use of Ceramic Waste as a Pozzolanic Addition on Cement. In: Martirena F, Favier A, Scrivener K. (Eds). Calcined Clays for Sustainable Concrete. RILEM Bookseries, vol. 16. Springer, Dordrecht.

Chen R, Li R, Deitz L, Liu Y, Stevenson RJ, Liao W. (2012). Freshwater algal cultivation with animal waste for nutrient removal and biomass production. Biomass and Bioenergy. Vol. 39:128–38.

Ciesielski PN, Hijazi FM, Scott AM, Faulkner CJ, Beard L, Emmett K, et al. (2010). Photosystem I – based biohybrid photoelectrochemical cells. Bioresource Technology. Vol. 101(9):3047–53.

Design and Control of Concrete Mixtures. EB001. Chapter 3 Fly Ash, Slag, Silica Fume, and Natura Pozzolans. p. 5772.

Dutta S, De S, Alam I, Abu-Omar MM, Saha B. (2012). Direct conversion of cellulose and lignocellulosic biomass into chemicals and biofuel with metal chloride catalysts. Journal of Catalysis. Vol. 288:8–15.

Hanson S. (2012). Energy for the Masses: Husk Power Helps Fuel India.

Hegazi HA. (2013). Removal of heavy metals from wastewater using agricultural and industrial wastes as adsorbents. HBRC Journal. Vol. 9(3):276–82.

http://www.hindustantimes.com/india/maharashtra-punjab-top-producers-of-green-energy-from-farm-waste/story-Ow46nfpSEgLy6OAe601PSN.html

http://www.renewableenergyworld.com/articles/2012/01/energy-for-the-mas ses-husk-powern helps-fuel-india.html. Originally published in ecomagi nation.

Kadirvelu K, Thamaraiselvi K, Namasivayam C. (2001). Removal of heavy metals from industrial wastewaters by adsorption onto activated carbon prepared from an agricultural solid waste. Bioresource Technology. Vol. 76(1):63–5.

Lubwama M, Yiga VA. (2017). Development of groundnut shells and bagasse briquettes as sustainable fuel sources for domestic cooking applications in Uganda. Renewable Energy. Vol. 111:532–42.

Lubwama M, Yiga VA. (2018). Characteristics of briquettes developed from rice and coffee husks for domestic cooking applications in Uganda. Renewable Energy. Vol. 118:43–55.

Morganti P. (2013). Saving the environment by nanotechnology and waste raw materials: use of Chitin Nanofibrils by EU research projects.2_art_Morgani 17/01/14 08:31 Pagina 89. Applied Cosmetology 31:89-96 (July/December 2013).

RoK KS, Cantrell B, Hunt PG, Ducey TF, Vanotti MB, Szogi AA. (2009). Thermochemical conversion of livestock wastes: carbonization of swine solids. Bioresource Technology. Vol. 100(22):5466–71.

Sang-Arun J. (2012). The 3Rs (reduce, reuse, recycle): an approach to sustainable solid waste management. BNDES seminar, 14–16 May 2012, Recife. IGES |http://www.iges.or.jp.

Singh J. (2013). Biomass in India. Akshay-urja. 45 vol. 6 (5 and 6):44–7.

Subramaniyam V, Subashchandra Bose SR, Ganeshkumar V, Thavamani P, Chen Z, Naidu R, et al. (2016). Cultivation of Chlorella on brewery wastewater and nano-particle biosynthesis by its biomass. Bioresource Technology. Vol. 211:698–703.

Suttibak S, Loengbudnark W. (2018). Production of charcoal briquettes from biomass for community use. IOP Conf. Ser.: Mater. Sci. Eng. 297012001. 8th TSME-International Conference on Mechanical Engineering (TSME-ICoME 2017). IOP Publishing. IOP Conf. Series: Materials Science and Engineering. 297. (2018) 012001. doi:10.1088/1757-899X/297/1/012001

Williams RB, Jenkins BM, Nguyen D. (2003). Solid waste conversion: a review and database of current and emerging technologies. Solid Waste Conversion; IWM-C0172. California Integrated Waste Management Board.

Zafar S. (2015). Energy Potential of Coconut Biomass. Bioenergy Consult.

2

Sustainability of WEEE Recycling in India

Biswajit Debnath[1,2]

[1]Department of Chemical Engineering, Jadavpur University,
Kolkata, India
[2]Secretariat Member, International Society of Waste Management,
Air and Water (ISWMAW)
E-mail: biswajit.debnath.ju@gmail.com

Waste Electrical and Electronic Equipments (WEEE) is one of the most notorious and the fastest growing waste streams in the world. It is growing exponentially every year. Globally, 41.8 million metric ton (MMT) of WEEE was generated in the year 2014 and India generates nearly 1.8 MMT WEEE every year. To combat the problem of WEEE, the new e-waste management rules 2016 have been enacted by the government of India. However, the implementation is not up to the mark. Various factors involving the economics as well as the social perspectives need to be addressed for a better WEEE management system (WMS) in India. The lion's share of the WEEE is handled by the informal sector. There are 178 Central Pollution Control Board (CPCB) registered WEEE recyclers equipped for dismantling, shredding, and separation of metals and non-metals in an environment friendly manner. The concept of reuse and refurbishing of Used Electrical and Electronic Equipments (UEEE) is very much in practice in India. Many are other different factors, both positive and negative are there that dictates the sustainability of WMS of India. The questions that arises are—what are the characteristics of WEEE recycling system in India? What are the issues and challenges pertaining to WEEE supply chain in India? What is the sustainability of this system? The present study focuses on answering these questions, providing current status from field studies, and addressing the issues and challenges. The study is expected to be beneficial for the stakeholders, policymakers, as well as the researchers in a holistic way.

2.1 Introduction

E-waste or electronic waste is a mini catastrophe and a threat to the modern developed society (Debnath et al., 2016). The definition of e-waste varies depending on the regulative body (Table 2.1). In a holistic way, it can be defined in the following manner—*any end-of-life electrical and/or electronic equipment and their accessories that have a power source or require a power source for functioning can be defined as e-waste* (Authors work). E-waste generation is increasing exponentially every year and there are some driving forces behind it (Perkins et al., 2014). The reason is the high rate of EEE obsolescence. Change in lifestyle, short innovation cycles, new models, and intentional complex design that shortens the lifespan are responsible for this high EEE obsolescence. Primarily, the electronics industry may be blamed for this purpose. The electronic industry is a very big industry which is expected to reach $400 billion in 2022 from $69.6 billion in 2012 [Corporate Catalyst (India) Pvt. Ltd. (2015)]. Technological advancement, short innovation cycles, witty business strategies, and complex designs are the reasons which reduce the lifespan of electronic devices (Debnath and Ghosh, 2017). As a result, the demand of new electronic devices increases. India is a big market for the electronic industry. For white goods (refrigerators, washing machines, TV, and air conditioners), only Indian market size is expected to become 2k billion INR in 2020 from 782 billion INR in 2014 with a CAGR

Table 2.1 Different definitions of WEEE/E-waste

Reference	Definition
European Directive 2002/96/EC	"Waste electrical and electronic equipment, including all components, subassemblies and consumables which are part of the product at the time of discarding". The Directive 75/442/EEC, Article I (a), defines as "waste" "any substance or object which the holder discards or is required to discard in compliance with the national legislative provisions."
Basel Action Network (www.ban.org)	"E-waste includes a wide and developing range of electronic appliances ranging from large household appliances, such as refrigerators, air-conditioners, cell phones, stereo systems, and consumable electronic items to computers discarded by their users."
OECD (www.oecd.org)	"Any household appliance consuming electricity and reaching its life cycle end."

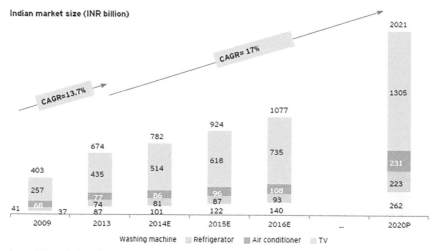

Indian market size (INR billion)

CAGR=13.7%

CAGR= 17%

Washing machine ■ Refrigerator ■ Air conditioner ■ TV

Source: JP Morgan, TechNavio, Spark Capital estimates, EY analysis

Figure 2.1   Indian white goods market.

of 17% (Figure 2.1) (FICCI, 2015). In 2016, nearly 44.4 million metric tons e-waste is produced globally which is equivalent to 4500 Eiffel towers (Baldé et al., 2017). It is expected to exceed 46 MMT in 2017 and 52.2 MMT by 2021 (Baldé et al., 2015, 2017). India generated 1.6 MMT of e-waste in 2014 which increased to 2 MMT in 2016 (Baldé et al., 2015, 2017). BRICS nation generated nearly 25% of the total WEEE generated in 2014 which have increased to nearly to 28% in 2016 (Baldé et al., 2015, 2017; Ghosh et al., 2016). India is also the second highest producer of e-waste among the BRICS nation only after China (Ghosh et al., 2016).

E-waste is a complex and heterogeneous material containing metals, polymers, cables and wires, PCBs, glass, hazardous materials, ceramics, concrete, woods, rubbers, etc. (Widmer et al., 2005; Ongondo et al., 2011). Figure 2.2 represents the composition of WEEE and it is quite clear that electronic items are getting lighter every year. This is primarily because the amounts of polymers are increasing and the metals are decreasing (Debnath and Ghosh, 2018). In India, e-waste recycling is primarily restricted to physical recycling and metal recovery. The presence of metal is the primary driving force which has helped the industry to foster (Debnath and Ghosh, 2018).

In India, the collection chain of e-waste is dominated by the informal sector (Ghosh et al., 2014b). In addition to that, the trans-boundary movement of e-waste is another big issue in India. India is a signatory to the

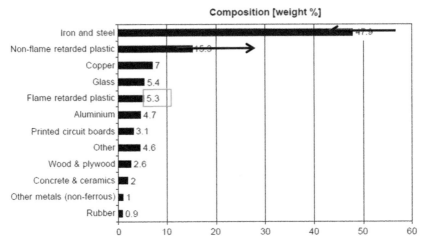

Figure 2.2   Composition of WEEE showing decrease in metals and increase in plastics.

Basel Convention and according to the convention India cannot import hazardous waste such as e-waste from other countries (Basel Action Network, 2002; UNEP, 2009). The hazardous waste rules (2016) of India also state that it is illegal to import e-waste from other countries [Hazardous and Other Wastes (Management and Transboundary Movement) Rules, 2016]. The e-waste management rule of India also focuses on EPR and says no to trans-boundary movement [E-Waste (Management) Rules, 2016]. However, some NGOs and port members confirm that e-waste is being imported in India either mixed with Scrap or in the form of charity. These are some loopholes in the rules itself (Ghosh et al., 2014a; Debnath et al., 2015).

E-waste management in India has been a very contemporary topic for the researchers. Previous literature focuses on technological aspects, policy, innovation, health hazards, and to some extent issues in the management system. Very few literatures have been found focusing on supply chain issues, field practices and sustainability aspects. The current study focuses on lifecycle of WEEE, supply chain issues, and sustainability aspects of WEEE management in India.

2.2 Methodology

This paper adopted the methodology of extensive literature survey and field visits. The literature survey is limited to journal publications, conference proceedings, and some reports. Search engines were explored with keywords including "E-waste management in India," "WEEE management in India," "E-waste supply chain," "WEEE supply chain," "Data security in e-waste," "E-waste recycling," etc. The relevant literatures have been cited and additional information from the associated cross-literatures has also been properly referred. Field visits were carried out with the intention to understand the practical situation. Informal sector visits were carried out during lean hours of the day and they were interviewed in a semi-structured way. Formal recycling unit was visited as scheduled. Based on the field study and the literature review, detailed analysis was carried out by brainstorming sessions. Based on the outcomes, the SCN of e-waste in India was formed. Thereafter, the sustainability analysis was carried out and some suggestive measures were proposed.

2.3 WEEE Lifecycle and Management in India

WEEE management in India is an interesting topic of research as there are many small quirks, twists, complexities, and heterogeneity involved all along the value chain. Undoubtedly, there are a number of environmental, social, economical, as well as political factors involved which makes the whole system inherently stochastic in nature. These factors are present all along the SCN which dictates the overall stability as well as the sustainability of the system.

2.3.1 Lifecycle of WEEE in India

It is important to understand the lifecycle of WEEE in India. Debnath et al. (2016) have discussed the lifecycle of WEEE and different phases of the lifecycle. Based on the aforementioned work, a structured framework of the life cycle of WEEE in India has been developed and shown in Figure 2.3. The figure captures the important stakeholders like OEMs, retailers, consumers, collectors, dismantlers, refurbishers, etc. The lifecycle path is described below.

The electronic item (e-item) reaches the consumers either via e-commerce or via the traditional retailers. After its intended use, the user decides to either replace it with a new one or to dispose it. If the user decides to exchange it via buy back scheme, sell it online, sell to personal contacts, or repair and reuse it, the e-item is termed UEEE. If it reaches its EoL, it is termed WEEE. In India, repairing and reusing e-items is a very common practice and hence it is really helpful in increasing the product lifecycle. The UEEE collected

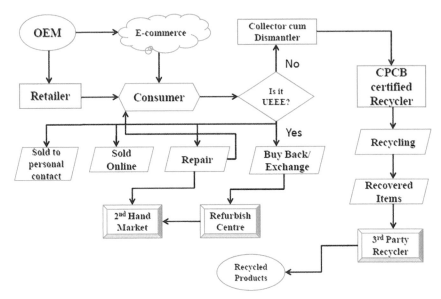

Figure 2.3   Lifecycle of WEEE in India.

by take back schemes or collected by collectors (either informal or formal) repairs and refurbishes the UEEE and sell them back in the second hand market. WEEE collected by the informal sector is dealt by them in a very crude way whereas the formal recyclers perform mechanical recycling and sell the recovered fraction to the respective recyclers.

2.3.2 WEEE Recycling Practices in India

In India, WEEE is recycled by two sectors, namely, informal and formal sectors. The informal sector is the unregistered and unauthorized sector that collects e-waste from doorsteps and recycles them in a very crude way and ends up polluting the environment which affects the flora and fauna. They belong to the Base of Pyramid of the society and their livelihood depends on the recycling of e-waste and other scraps (Wilson et al., 2006). The formal sector is the authorized sector that is making business out of e-waste in a comparatively environmentally benign way, registered and permitted to operate by the pollution control board (Ghosh et al., 2014b).

There are different recycling processes adopted by the informal as well as the formal sector. The informal sector uses very crude methods for recycling of WEEE. These methods are highly polluting and have poor efficiency.

They operate in several hotspots where they use very basic tools like hammer, chisel, etc., to dismantle the WEEE without any safety equipments (An et al., 2015). Some valuable materials are removed from the WEEE in a very primitive manner and the residual part is either burner or landfilled. Open burning of WEEE results in emission of toxic chemicals such as dioxins, polychlorinated biphenyls, polybrominated diephenyl ethers, other brominated flame retardants, etc. (Sthiannopkao and Wong, 2013). Generally, PCBs and wires are open burnt to recover valuable metals (Widmer et al., 2005). Acid bath or acid striping is another process used by the informal sector to stripe metals from electronic components. The resulting spent acid is filled with toxic chemicals and heavy metals which creates further environmental hazards as they are discharged in open land or running drains (Zhao et al., 2009).

The formal sector employs environmentally sustainable techniques to recycle WEEE. The formal recyclers primarily perform manual dismantling followed by mechanical dismantling (shredding, crushing, etc.) in their units (Debnath et al., 2018). Further separation of metals and non-metals is carried out using magnetic separation and eddy current separation. In this process, the major recovered fractions are metals, polymers, electronic components, and glass (Ghosh et al., 2014b). The metals reach fractions that are sent to the third-party recyclers or sister facilities who extract metals using pyro-metallurgical, hydro metallurgical, and combination of both techniques. The other fractions are recycled by respective third-party recyclers (Debnath et al., 2018).

2.4 Security Threats from WEEE in India

In recent days, when India is among the world's fastest growing economy and one of the largest consumers of electronics, most consumers are still unaware of how to dispose of their e-waste. Most Indians end up selling their e-waste to the informal sector, which poses severe threats to human (including children's) lives, with its improper and highly hazardous methods of extracting the trace amounts of precious metal from it and handling e-waste for profit (Debnath et al., 2019). Electronic gadgets have become a part of the style statement in today's dynamic world. In developed countries, where the informal sectors are non-existent, the electronic items are either sold after using it for someday, exchanged, or send to the recyclers. In developing countries, there is a big market for secondary and refurbished electronics (Hicks et al., 2005). There are very few people who are able to remove all of their data from their devices before channelizing it to any other nodal points. In India, it is quite evident that people are unaware of prevention techniques of their personal data stored in their electronic equipments. As a result, security threats may arise from those electronic equipments at their EoL (Roychowdhury et al., 2018). Whenever the used phones or any electronic equipment are to be handed over to any unknown source, all data prevention processes must be followed. People should remove all the data, break the link between the phone and the cloud server, and log out from all the user accounts so that no-one have access to the previous data. From the above ways of awareness, the amount of security threat can be minimized.

2.5 WEEE Supply Chain in India

The SCN of any organization is a very important element and the business sustainability depends on it. The SCN of WEEE India is interesting as well as complex (Debnath et al., 2014). Figure 2.4 represents the SCN of WEEE in India. The presence of the informal sector is an important characteristic of WEEE in India. The informal sector is responsible for collecting and recycling 95% of the total WEEE. However, the number is expected to be

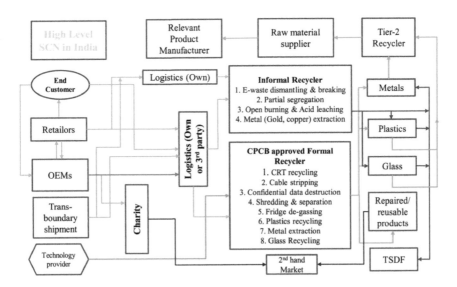

Figure 2.4 High-level SCN of WEEE in India.

less than 90% now as there has been a boost in formal recycling in the past few years (Ghosh et al., 2016). The informal ones are more popularly known as the *Kabadiwalas* in India. These people are door to door collectors of waste who collect recyclable dry waste along with e-waste in exchange of money. They sell the collected waste to big and medium scrap vendors. These vendors segregate the waste and channelize the waste to respective recyclers (Debnath et al., 2015).

Another noteworthy characteristic is the repairing and refurbishing practices which is helpful in extension of product lifecycle. In the forward supply chain, OEMs, end customers and retailers, and trans-boundary shipments are the provider of WEEE. WEEE enters the country by trans-boundary movement in the name of charity. Sometimes the WEEE is sold to the informal sectors. Then it is inspected for reusability. If it is found to be repairable, then it is sold in the second-hand market after repairing and refurbishing. The formal sector collects WEEE via auction, contractual basis, and individual disposal. The recycling operation is carried out within their unit. In the backward chain, there are tier-2 recyclers who received dismantle materials for further recycling (Ghosh et al., 2014a, 2016).

2.6 Case Studies

2.6.1 Case Study A

This organization is one of the oldest e-waste recycling plants in India with a capacity of 3650 tons per year. The unit is situated on 1.5 acres of land with 25,000 sq.ft. closed area and 60,000 sq.ft. of open area in a well-known industrial estate in Bangalore, Karnataka. The objective of this unit is to create an opportunity to transfer waste into socially and industrially beneficial raw materials like valuable metals, plastics, and glass using simple, cost-efficient, home-grown, environmental friendly technologies suitable to Indian conditions. Its products and services include (a) metal, plastic, and glass and different other recycled products, (b) gold recovery from PCB strips and components, and (c) silver recovery from silver-coated components. 100 employees are there 50% of which is female. The process line consists of manual dismantling, shredding, manual segregation, and density separation. In this unit, only mechanical dismantling and physical separation is carried out. Sister companies and other third-party industries are in contractual practice for metal recovery and other component recycling. Plastic and glass are reused and recycled in other companies. The glass is recycled by one

(a)   (b)

Figure 2.5   (a) and (b) Informal sector activities in Kolkata.

of the leading TV manufacturers which uses 30% of the waste glass and rest virgin glass for manufacturing the TV glass panel. Further details on this case study have been recently reported by Debnath et al. (2018).

2.6.2 Case Study B

This is a hotspot in Kolkata, West Bengal, situated near the Jodhpur Park Market area. There are more than eight shops in this area that buy e-waste (along with other dry recyclable waste) from the Kabadiwalas and segregate them in a number of categories. These people collect small- and medium-size WEEE. There are at least two middle men associated with each shop who help them to get customers for the waste. Each shop has some dedicated person (mostly children or aged person) who dismantle the WEEE (Figure 2.5(a) and (b)). Their primary target is recovering copper from coils, removal of electronic components from PCBs, and size reduction of PCBs. Screw drivers, chisels, and hammers are used for this purpose. The WEEE PCBs are sold to some person whose identity was not disclosed. Other fractions are channelized to respective recyclers.

2.7 Sustainability of WEEE Management System in India

Evaluation of sustainability of any system is an important exercise which requires consideration of three pillars of sustainability—environmental, economical, and social. Generalized guidelines for sustainability analysis have not been developed yet. Environmental, economical, and social aspects are explored in general (Ghosh et al., 2018). Previous studies have focused on determination of environmental impacts by means of LCA

(Ghodrat et al., 2017; Iannicelli-Zubiani et al., 2017). The economical and social parts have not been considered in those studies. Studies on social sustainability have been found in literature which focuses on informal sector only (Umair et al., 2015). The above-mentioned studies on environmental and social LCA are standalone ones and technology specific. Intervention in understanding the overall sustainability of the WMS has not been approached. Debnath et al. (2018) have established a generalized approach for sustainability evaluation. They have focused on MREW technologies for sustainability evaluation. Ghosh et al. (2018) have also followed a similar method but rather have taken a system approach in evaluating the sustainability. In this section, an attempt has been made to provide an overview of the sustainability of WMS in India.

2.7.1 Generalized Discussion on Environmental Sustainability

Environmental sustainability of waste management system of any country is dictated by some factors. These factors are allocated with some numerical values. These values are measure of magnitude of that particular factor on the environment (Debnath et al., 2018). An LCA exercise is perhaps the most assuring tool to describe the environmental sustainability (Debnath et al., 2018; Ghosh et al., 2018). However, in the current scenario due to lack of reliable data, LCA was not performed. A generalized idea on environmental sustainability has been presented keeping in mind the LCA perspectives.

1. Due to the informal recycling activities, enormous impact on environment is there. As described in previous sections, the methods used by the informal sector are very much crude and the resultant streams are harmful to the environment. Mechanical dismantling, open burning, and acid leaching are the methods which generate toxic flue gas and waste water containing heavy metals and other hazardous components. The possible contributions will be in the following impact categories— eutrophication, acidification, marine and freshwater ecotoxicity, human toxicity, global warming potential, photochemical oxidation, particulate matter formation potential, terrestrial acidification, and water depletion (Fugiel et al., 2017; Debnath et al., 2018).
2. Since the polymeric fraction is neglected in the informal sector, they often end up in the landfill or in open dumps. Such practices have severe environmental impact.
3. Due to the better collection efficiency of informal sector, more waste is collected. Also they work from door to door for waste collection.

In comparison to the formal sector, emissions due to transportation of e-waste are less compared to informal sector, because the WEEE needs to be collected from the users to the collection center and then from that center to the recycling facility. Certain amount of emissions is associated with this transportation via vehicles as they consume fuel. Global warming potential and particulate matter formation potential are the possible impact categories (Fugiel et al., 2017).

4. Mechanical recycling processes in the recycling facilities are energy intensive. Also during the processes, dust and dust laden fumes are generated. The potential impact categories are as follows:

   a. Pyro-metallurgical processes require high energy which implies high GHG emission. Human toxicity, climate change, marine eutrophication, freshwater eutrophication, fresh water ecotoxicity, and water depletion are the possible impact categories (Ghodrat et al., 2017).

   b. Hydro-metallurgical processes are energy efficient but they use several chemicals. As a result, waste water is generated containing spent acid and other substances which have significant water foot print. Eutrophication, acidification, global warming, abiotic depletion, human toxicity, marine and freshwater ecotoxicity, fresh water depletion, etc., are possible impact categories (Iannicelli-Zubiani et al., 2017; Debnath et al., 2018).

2.7.2 Generalized Discussion on Economic Sustainability

No business can foster without the economic sustainability. The economic sustainability of any WEEE business revolves around metals primarily. There are several other factors that actually help in understanding the real scenario. This section summarizes the factors that need to be taken into consideration for understanding of the economic sustainability of WMS of India. The points are stated below:

1. Due to transportation of WEEE from one node to another along the supply chain incurs additional supply chain cost.

2. High energy penalty of mechanical recycling and traditional pollution control equipments associated with it requires extra operation and capital cost.

3. Hydro-metallurgical processes require low energy but additional cost is there for waste water treatment. Sometimes spent acid and other chemicals are recycled back into the reactor which saves cost of virgin chemicals and reduces waste water loading.

4. Pyro-metallurgical processes have high energy penalty which means high operating cost.
5. Electrochemical refining also consumes huge amount of electricity.
6. Cost of setting up a WEEE recycling plant is very high. Other costs such as COQ, maintenance cost, legislative cost, and cost due to taxes are there.
7. Recent conference on e-waste raised the question of economical sustainability of metal recycling from WEEE. Polymer fraction of WEEE is often neglected and they end up in land fill sites. It is high time for recycling of WEEE polymers and generating revenues with the output products. It is possible to use these plastics for energy recovery, conversion into nanomaterials and even methanol. Interestingly, WEEE plastic recycling opens a new door for urban mining. Utilization of these resources will not only enhance the resource efficiency but also ensure circular economy.
8. Revenues generated by recycling and selling metals, polymer, and glass keep the business economically sustainable.

2.7.3 Generalized Discussion on Social Sustainability

Determination of social sustainability of system is a tedious and complex process. The behavior, mentality, education, and other social indicators change with geographical locations. Religion also plays a crucial role in social aspects (Ghosh et al., 2018). Performing a social LCA is the best way to determine the social sustainability (Debnath et al., 2018). Social LCA of WMS has not been a very contemporary topic as per the published literature. However, a study by Umair et al. (2015) provided some insight on informal WEEE recycling.

In India, the major portion of WEEE handled by the informal sector and some recycling is taking place. This is creating job opportunities for some people belonging to BOP level of the society. Giving recognition to the informal sector will be a good attempt. While the informal sector is still thriving the business, formal sectors are also making their efforts to make e-waste recycling a sustainable business. They are also expanding their capacities and creating job opportunities to more people. There were total 138 CPCB approved formal recyclers in India in 2014 and the number has gone up to 178 by 2016 (CPCB, 2014, 2016). The super networking ability of the informal sector can be utilized and merged with the eco-friendly processing ability of the formal sector. This attempt will be building a sustainable and synergetic system for WEEE management in India.

There has been an increasing trend in awareness regarding waste disposal among people and they are behaving in a much responsible manner. The bulk producers of WEEE are coming in the front and stepping up for a better option. This positive attitude has helped in implementation of EPR at a very grass root level. The amendment of the e-waste rules in March 2018 [E-Waste (Management) Amendment Rules, 2018] with the new targets for take back of WEEE under the umbrella of EPR is really being at the next step. There are more opportunities at this level to be explored for social sustainability. Working hours, child labor, health and safety, social security, equal opportunities, community engagement, local employment, public contribution to sustainable issues, contribution to economic development, governance, corruption, education level, awareness (United Nations, 2007; Umair et al., 2015; Debnath et al., 2018), etc. are a few key stakeholder and sub-stakeholder categories which should be considered to the least for further evaluation in a quantitative way.

2.8 Conclusions

There is a budding need of development of a sustainable WMS for India to deal with the monstrous amount of WEEE. The study has provided a generalized idea of WMS in India focusing on the lifecycle, processes, and the SCN. An important finding of the study is the security threat issue in the country which is often neglected or unspoken of. This is deemed to be a major issue in the country if not tackled properly. Two case studies have been presented—one in the informal sector and the other a formal recycling facility which gives the contrast of environment of both worlds. The generalized discussion on the sustainability of the WMS of India has identified several points, both positive and negative. These are crucial and need to be nurtured in a proper way to look forward to a sustainable future. For a developing country like India, more number of such studies are important which can provide local solutions to deal with the fluctuations as well as the major problems pertaining to the WEEE supply chain.

Acknowledgment

The author would like to acknowledge Prof. Ranjana Chowdhury, Chemical Engineering Department, Jadavpur University, and Prof. Sadhan Kumar Ghosh, Mechanical Engineering Department, Jadavpur University, for their help and immense support toward conducting research. International Society of Waste Management, Air and Water (ISWMAW) is acknowledged for providing partial funding towards research on WEEE management. TEQIP – III is gracefully acknowledged for providing partial funding to attend ICRM 2018 at Kerala. Last but not least, the author is thankful to Prof. Sabu Thomas for inviting to deliver the talk at ICRM 2018.

References

Agamuthu P. (2010). The role of informal sector for sustainable waste management. Waste Management and Research. 28:671–2.

An D, Yang Y, Chai X, Xi B, Dong L, Ren J. (2015). Mitigating pollution of hazardous materials from WEEE of China: portfolio selection for a sustainable future based on multi-criteria decision making. Resources, Conservation and Recycling. 105:198–210.

Baldé CP, Forti V, Gray V, Kuehr R, Stegmann P. (2017). The global e-waste monitor 2017: quantities, flows and resources. United Nations University (UNU), International Telecommunication Union (ITU) & International Solid Waste Association (ISWA), Bonn/Geneva/Vienna. ISBN Electronic Version: 978-92-808-9054-9.

Baldé CP, Wang F, Kuehr R, Huisman J. (2015). The global e-waste monitor– 2014. United Nation University.

Basel Action Network and Silicon Valley Toxics Coalition (SVTC) (2002) Exporting harm: The high-tech trashing of Asia.

Corporate Catalyst (India) Pvt. Ltd. (2015). A brief report on Electronics Industry in India. Available from: http://www.cci.in/pdfs/surveys-reports/Electronics-Industry-in-India.pdf, last accessed on 4th Jan, 2016.

CPCB. (2014). Available at: http://cpcb.nic.in/Ewaste_Registration_List.pdf (Accessed 9 March 2016).

CPCB. (2016). Available at: http://www.cpcb.nic.in/List_of_E-waste_Recycler_as_on_29.12.2016.pdf (Accessed 20 February, 2017).

Debnath B, Baidya R, Ghosh SK. (2015). Simultaneous analysis of WEEE management system focusing on the supply chain in India, UK and Switzerland. International Journal of Manufacturing and Industrial Engineering. 2(1):16–20.

Debnath B, Chowdhury R, Ghosh SK. (2018). Sustainability of metal recovery from E-waste. Frontiers of Environmental Science and Engineering. 12(6):2.

Debnath B, Das S, Das A. (2019). A study exploring security threats in waste mobile phones: a life cycle based approach. (Unpublished, under consideration in icset 2019).

Debnath B, Ghosh SK. (2017). E-waste recycling in India: a case study. The 32nd International Conference on Solid Waste Technology and Management (ICSW 2017), Philadelphia, PA, USA, 19–22 March, 2017, pp. 223–32.

Debnath B, Ghosh SK. (2018). Sustainable utilization of polymeric materials recovered from E-waste. International Conference on Current Trends in Materials Science and Engineering (CTMSE 2018), pp. 101, Hyderabad, India, 19–20 January, 2018.

Debnath B, Roychoudhuri R, Ghosh SK. (2016). E-waste management—a potential route to green computing. Procedia Environmental Sciences. 35:669–75.

ETC/RWM. European Topic Centre on Resource and Waste Management (Topic Centre of the European Environment Agency) part of the European Environment Information and Observation Network (EIONET); 2003. Available at: http://waste.eionet.eu.int/waste/6.

E-Waste (Management) Rules (2016) Available at: http://www.moef.nic.in/sites/default/files/notified%20ewaste%20rule%202015_1.pdf (accessed 25 August 2015).

E-Waste (Management) Amendment Rules (2018) Available at: http://envfor.nic.in/sites/default/files/e-%20waste%20amendment%20notification%202018184020.pdf (accessed 23 January 2019).

FICCI. (2015). Study on Indian electronics and consumer durables segment (AC, refrigerators, washing machines, TVs). E&Y. Retrieved from https://www.ey.com/Publication/vwLUAssets/EY-study-on-indian-electronics-and-consumer-durables/%24FILE/EY-study-on-indian-electronics-and-consumer-durables.pdf.

Fugiel A, Burchart-Korol D, Czaplicka-Kolarz K, Smoliński A. (2017). Environmental impact and damage categories caused by air pollution emissions from mining and quarrying sectors of European countries. Journal of Cleaner Production. 143:159–68.

Ghodrat M, Rhamdhani MA, Brooks G, Rashidi M, Samali B. (2017). A thermodynamic-based life cycle assessment of precious metal recycling out of waste printed circuit board through secondary copper smelting. Environmental Development. 24:36–49.

Ghosh A, Debnath B, Ghosh SK, Das B, Sarkar JP. (2018). Sustainability analysis of organic fraction of municipal solid waste conversion techniques for efficient resource recovery in India through case studies. Journal of Material Cycles and Waste Management. 1–17.

Ghosh SK, Singh N, Debnath B, De D, Baidya R, Biswas NT, et al. (2014a). E-waste supply chain management: findings from pilot studies in India, China, Taiwan (ROC) and the UK. The 9th International Conference on Waste Management and Technology, Beijing China, 29–31 October, 2014, pp. 1131–40.

Ghosh SK, Baidya R, Debnath B, Biswas NT, De D, Lokeswari M. (2014b). E-waste supply chain issues and challenges in India using QFD as analytical tool. In *ICCCM 2014:* Proceedings of International Conference on Computing, Communication and Manufacturing. pp. 287–91.

Ghosh SK, Debnath B, Baidya R, De D, Li J, Ghosh SK, et al. (2016). Waste electrical and electronic equipment management and Basel Convention compliance in Brazil, Russia, India, China and South Africa (BRICS) nations. Waste Management and Research. 34(8):693–707.

Hazardous and Other Wastes (Management and Transboundary Movement) Rules (2016) Available at: http://www.moef.nic.in/sites/default/files/HWM %20Rules%202015%20english%20version.pdf (accessed 30 August 2015).

Hazra J, Sarkar A, Sharma V. (2011). E-waste supply chain management in India: opportunities and challenges. Clean India Journal 7.

Hicks C, Dietmar R, Eugster M. (2005). The recycling and disposal of electrical and electronic waste in China—legislative and market responses. Environmental Impact Assessment Review. 25(5):459–71.

Iannicelli-Zubiani EM, Giani MI, Recanati F, Dotelli G, Puricelli S, Cristiani C. (2017). Environmental impacts of a hydrometallurgical process for electronic waste treatment: a life cycle assessment case study. Journal of Cleaner Production. 140:1204–16.

Ongondo FO, Williams ID, Cherrett TJ. (2011). How are WEEE doing? A global review of the management of electrical and electronic wastes. Waste Management. 31(4):714–30.

Perkins DN, Drisse MNB, Nxele T, Sly PD. (2014). E-waste: a global hazard. Annals of Global Health. 80(4):286–95.

Roychowdhury P, Alghazo JM, Debnath B, Ouda OKM, Chatterjee S. (2018). Security Threat Analysis and Prevention Techniques in Electronic Waste. In: Waste Management and Resource Efficiency. Springer, 2018. DOI: 10.1007/978.981.10.7290.1.85

Sthiannopkao S, Wong MH. (2013). Handling e-waste in developed and developing countries: Initiatives, practices, and consequences. Science of the Total Environment. 463:1147–53.

Umair S, Björklund A, Petersen EE. (2015). Social impact assessment of informal recycling of electronic ICT waste in Pakistan using UNEP SETAC guidelines. Resources, Conservation and Recycling. 95:46–57.

UNEP. (2009). Basel Convention on the control of transboundary movements of hazardous wastes and their disposal. United Nations Environment Programme (UNEP). Available at: http://www.basel.int/Portals/4/Basel%20 Convention/docs/text/BaselConventionText-e.pdf (accessed 15 February 2016).

United Nations Department of Economic. (2007). Indicators of sustainable development: Guidelines and methodologies. United Nations Publications.

Widmer R, Oswald-Krapf H, Sinha-Khetriwal D, Schnellmann M, Böni H. (2005). Global perspectives on e-waste. Environmental Impact Assessment Review. 25(5):436–58.

Wilson DC, Velis C, Cheeseman C. (2006). Role of informal sector recycling in waste management in developing countries. Habitat International. 30:797–808.

Zhao G, Wang Z, Zhou H, Zhao Q. (2009). Burdens of PBBs, PBDEs, and PCBs in tissues of the cancer patients in the e-waste disassembly sites in Zhejiang, China.

3

Autogenous Self-healing in Municipal Waste Incorporated Concretes

**Deniz Genc Tokgoz[1], Nesibe Gozde Ozerkan[1],
and Simon Joseph Antony[2],***

[1]Center for Advanced Materials, Qatar University, Doha, QATAR
[2]School of Chemical and Process Engineering, University of Leeds,
Leeds LS2 9JT, UK
E-mail: S.J.Antony@leeds.ac.uk
*Corresponding Author

For the sustainable development in the construction industry, there is a growing need for at least partially replacing the rapidly depleting natural resources,
and increasing re-utilization of conventional wastes. In this research work,
municipal polymeric wastes in various forms and fly ash from incineration
of municipal solid wastes are used together as secondary raw materials for
the preparation of concrete mixtures. The influence of various forms of
polyetylene (PE) substitution on the autogenous healing of concrete in terms
of its mechanical strength and permeability properties is determined using
suitable methods. Furthermore, the effects of autogenous healing phenomena on the microstructural characteristics of cement matrix are examined
by Fourier transform infrared spectrometer (FTIR) and thermogravimeter
(TG). The performances depended on the type of PE mixes and their
compositions.

3.1 Introduction

Concrete is the second most utilized material after water and it is the most
widely used construction material in the world (Aïtcin, 2000; Sabir et al.,
2001). The main components of concrete are cement, aggregates, and water.

Concrete is accepted as a versatile material which can be processed to meet specific requirements/needs and can be cast into a variety of shapes and sizes. Other properties of concrete such as resistance to water, fire, and cyclic loading, low maintenance requirement, and ability to utilize domestic and industrial wastes as composite concrete (Ozerkan et al., 2016; Tokgoz et al., 2016) are also important advantages of concrete over other construction materials. All of these advantages make concrete as the most popular and ubiquitous construction material in the world which can be used in constructing a variety of structures such as homes, schools, dams, ports, bridges, pavements, and highways.

One of the major issues of concrete structures is its sensitivity to crack formation (Mehta and Monteiro, 2006). There are many reasons for crack development in concrete such as hydration of cementitious materials, mechanical loading (static or cyclic), environmental conditions (e.g., freezing and thawing, alkali silica reactions, and sulfate attack), and volumetric instability (e.g., shrinkage and thermal deformation) (Yang et al., 2010). Once cracks have formed, aggressive liquids and gasses may penetrate into the matrix and cause substantial negative effects on the integrity and durability of the concrete as well as the strength, serviceability, and esthetics of the whole structure (Samaha and Hover, 1992; Van Tittelboom and De Belie, 2013; Tang et al., 2015a). The inspection and maintenance of cracks by conventional methods can be very challenging and costly (Li and Herbert, 2012) as most of the cracks are not visible or accessible. Hence, it is desirable to get information on the characteristics of cracks in concretes and to heal them where feasible.

Self-healing of cracks in cementitious composites is a complex phenomenon and has been studied intensively since 1990s (Dry, 1994; Edvardsen, 1999). Based on the mechanism of the healing, two approaches have been developed for self-healing in cementitious composites as autonomic healing and autogenous healing. In autonomic healing, some engineered agents, such as micro capsules (Boh and Sumiga, 2008; Mihashi and Nishiwaki, 2012) and hollow fibers (Dry, 1994; Joseph et al., 2010), containing chemical healing compounds are embedded in the cement matrix (Schlangen and Joseph, 2009). Once cracks formed, these agents are broken and they release the chemical compounds into the cracks for an immediate and efficient repair (Kessler et al., 2003; Joseph et al., 2010). The cost of autonomic healing increases if cracks grow and widen significantly (Van Tittelboom and De Belie, 2013).

Contrary to autonomic healing, autogenous healing in cementitious composites is relatively cheaper (Li and Herbert, 2012) because autogenous healing relies only on the self-healing properties resulting from the physical and/or chemical composition of the cementitious matrix itself (Rooij et al., 2013; Tang et al., 2015b). The major mechanisms of autogenous healing in the hardened concrete matrix are (i) formation of new CSH crystals by further hydration of the unhydrated cementitious particles; (ii) precipitation of $CaCO_3$; (iii) swelling of the CSH; and (iv) blocking of cracks by cement particles, hydration particles, loose particles, and any impurities (Edvardsen, 1999; ter Heide, 2005; Granger et al., 2007; Homma et al., 2009).

The overall contribution of the above-mentioned mechanisms to autogenous healing varies with the concrete age when cracking occur (Van Tittelboom and De Belie, 2013). For example, further hydration is the main mechanism for healing in young concrete (age $\leq$ 7 days) as it contains more unhydrated cementitious particles (Ferrara et al., 2014); whereas $CaCO_3$ precipitation becomes dominant mechanism at later ages (Hearn, 1998; Edvardsen, 1999; ter Heide, 2005). Autogenous healing occurs under certain circumstances. These are due to the presence of water, specific chemical ions (e.g., Ca^{2+}, CO_2), and small crack widths. Although the maximum crack width reported for autogenous healing varies in different studies, narrower crack widths (<100 μm) are more likely to be filled completely by healing products (ter Heide, 2005).

Due to its brittle nature, controlling the crack width in the concrete is difficult (Rooij et al., 2013). Application of fiber reinforcement provides a way to control the crack formation and propagation by improving the energy-absorbing capacity of the concrete (Li, 2003; Bunsell and Renard, 2005; Kamal et al., 2014); hence, relatively narrow crack widths can be achieved. These narrow cracks can be healed easily compared to wide cracks (Nishiwaki et al., 2012; Özbay et al., 2013; Ma et al., 2014).

Utilization of municipal solid wastes, i.e., plastic wastes (Naik et al., 1996; Siddique, 2008; Siddique et al., 2008; Saikia and de Brito, 2012; Sharma and Bansal, 2016), and their by-products, i.e., incineration ashes (Goh et al., 2003; Gao et al., 2008; Collepardi et al., 2012; Guo et al., 2014; Sua-Iam and Makul, 2015; Garcia-Lodeiro et al., 2016), have been applied to achieve sustainable development in the construction industry. This article aims to determine the influence of different type and form of PE substitutions on the self-healing behavior of concrete mixtures. Compressive loading was applied to generate micro cracks at 28 days water cured specimens and then loaded specimens were stored in water for 30 more days. At the end of self-healing period, the mechanical and permeability properties were determined. Microstructural characteristics of cement matrix were analyzed to determine the effect of self-healing process in all mixtures.

3.2 Experimental

3.2.1 Materials and Mixture Proportions

Portland cement (CEM I 42.5), incinerator FA, collected from the air pollution control system of DSWMC, and silica fume were used as binder. The chemical compositions were analyzed by a wavelength-dispersive XRF spectrometer. The percentage of loss on ignition at 750°C and specific gravity was determined according to ASTM C311 (ASTMC311/C311M-13, 2013). The insoluble residue and fineness was measured according to ASTM C114 (ASTMC114-15, 2015) and ASTM C204 (ASTMC204-11e1, 2011), respectively. The resulting chemical composition and physical properties are given in Table 3.1.

As for the natural aggregates, river sand with 4.75 mm maximum size was used as fine aggregate, and crushed limestone with a maximum size of 9.5 mm was used as coarse aggregate. The coarse and fine aggregates had

Table 3.1 Chemical composition and physical properties of PC, fly ash, and silica fume

Chemical Composition (% by Weight)	PC	FA	Silica Fume
CaO	64.95	45.0	1.05
SiO_2	21.92	1.89	89.5
Al_2O_3	4.32	0.784	0.32
Fe_2O_3	3.78	0.601	0.38
MgO	2.16	0.552	0.1
SO_3	2.08	8.67	0.1
Alkalies ($Na_2O + 0.658\ K_2O$)	0.68	18.3	–
Loss on ignition	1.00	1.9	2.3
Insoluble residue	0.68	1.06	1.0
Physical Properties			
Specific gravity	3.09	2.25	2.01
Blaine fineness (cm^2/g)	3527	–	–
Compressive Strength (MPa)			
2 days	21.4	–	–
7 days	28.9	12.5	13.2
28 days	40.2	18.2	–

specific gravities of 3.04 and 1.95 and water absorptions of 1.03 and 5.7%, respectively.

DRAMIX OL 6/16 straight cylindrical micro-steel fiber with a length of 6 mm and diameter of 0.16 mm was used as steel reinforcement. In addition to micro-steel fiber, r-HDPE fiber, which was collected and processed from Qatar's municipal waste stream, with a thickness of 0.10 mm and length of 3–10 mm was used. The mechanical properties of fibers along with their physical properties are presented in Table 3.2.

Recycled polyethylene granules, collected from a local plastic recycling plant, and neat polyethylene granules were used as partial natural coarse aggregate replacement. Neat low-density polyethylene granules (v-LDPE) and neat HDPE granules (v-HDPE) were in the form of spherical shape with an average diameter of 3.0±0.2 mm. r-LDPE and HDPE (r-HDPE) were in cylindrical shape with an average diameter and length of 3.0±0.5 mm for r-LDPE, and an average diameter of 4.0±0.2 mm and length of 3.5±0.5 mm for r-HDPE.

The experimental study consisted of seven different mixtures whose proportions are given in Table 3.3. Total fiber content kept constant at 2% by weight in all mixtures except Mix1 or namely Plain-F0 which means that there is no fiber reinforcement. Mix2 included only steel fiber reinforcement and labeled SF2. Mix3 was designed by including 1% steel fiber and 1% recycled HDPE fiber and named as SF1%-R-HDPE-F1%. In the design of remaining mixtures, only steel fiber was utilized as reinforcement and 10% by weight of natural coarse aggregate was partially substituted with neat and recycled PE granules. For all mixtures, a multi-carboxylate ether-based SP (EPSILONE HP 510) with a specific gravity of 1.11 was used. SP content kept constant in all mixtures, while water was gradually added to achieve slump values between 650 and 800 mm as requested by the European Federation of National Associations Representing Producers and Applicators

Table 3.2 Properties of the fibers

Fiber Type	Micro-steel	r-HDPE
Length (mm)	6	3.0–10.0
Diameter (μm)	160	100
Cross-sectional area (mm^2)	0.02	0.0078
Load at maximum load (N)	–	74.34
Tensile strength (MPa)	2.16	25.22
Elastic modulus (MPa)	210	672.29
% Total elongation at fracture	–	152.4

Table 3.3 Mixture proportions

Mix ID	Mix Design Label	W/B[a]	Water	PC	Silica Fume	FA	Aggregate (kg/m³)			SP	Fiber	
							Fine	Coarse	PE		Steel	PE
1	Plain-F0	0.49	196	320	40	40	974.2	800.1	–	12	–	–
2	SF2%	0.39	156	320	40	40	994.5	816.7	–	12	16.3	–
3	SF1%-R-HDPE-F1%	0.44	176	320	40	40	975.4	801	–	12	8	8
4	SF2%-V-LDPE10%	0.46	184	320	40	40	915.9	752.2	75.2	12	15	–
5	SF2%-V-HDPE10%	0.45	180	320	40	40	823.5	676.3	67.6	12	13.5	–
6	SF2%-R-LDPE10%	0.43	172	320	40	40	842.5	691.9	69.2	12	13.8	–
7	SF2%-R-HDPE10%	0.41	164	320	40	40	833.7	684.6	68.5	12	13.7	–

[a]B: binder (PC+ silica fume+ FA)

of Specialist Building Products for Concrete (EFNARC, 2002). Hence, water-to-binder ratios (W/B) were not constant but within the range of 0.39–0.49 as seen in Table 3.3.

3.2.2 Specimen Preparation and Initial Pre-loading

All solid ingredients were added into a mechanical mixer and mixed in dry state to obtain a homogenous distribution. Three quarters of mixing water mixed with the SP was added in the mixer and mixed for 2 min. Then, the remaining water was added gradually into the mixture to provide uniformity and mixed for a period of 2 more minutes. After completing the mixing procedure, slump flow diameters were measured. If the slump flow diameter was below 550 mm (EFNARC, 2002), then additional water was added and mixed until the final slump flow diameter was within the range of 550–800 mm.

From each mixture, Ø100 × 200 mm cylinder specimens were cast in one layer without any compaction. All specimens were demolded 24 h after casting and then kept in water for 28 days. The ultimate compressive strength at 28 days age was determined using three specimen from each mixture, and the remaining specimens were pre-loaded to 0 and 70% of their corresponding compressive strength. In the literature, application of higher loading rates could be found (Şahmaran et al., 2008; Yildirim et al., 2015); however, in this study, higher loading rates were avoided as the main aim was to create small cracks and similar damages in all mixtures. After pre-loading, all specimens were stored in water for an additional 30 days. At the end of the 30 days curing, all specimens were tested for permeability and compressive strength until failure.

3.2.3 Methods for Self-healing Evaluation

Water absorption test in accordance with ASTM C642 (ASTMC642-13, 2013) was performed on both 0 and 70% preloaded specimens at particular age (28+30 days) to determine the influence of healing on the permeability properties of all mixtures. To evaluate the regained strength after healing, compressive strength of the concrete specimens was determined at 28+30 days in accordance with ASTM C39 (ASTMC39/C39M-14a, 2014).

Microstructural characteristics have been determined by FTIR and TGA. Samples for these analyses were taken from the center of each specimen and then grinded into powder using an agate mortar and pestle. TGA analyses of the samples were performed using Perkin Elmer Pyris 6 TGA from 29 to 888°C at 10°C per minute rate under a dry nitrogen atmosphere purged at 20 mL per minute rate. For FTIR analyses, an ATR-FTIR spectroscopy (Nicolet 6700) with a nominal resolution of 1 cm^{-1} between 450 and 4000 cm^{-1} was used.

3.3 Results and Discussions

3.3.1 Compressive Strength

3.3.1.1 At 28 days

The compressive strength of each mixture at 28 days age is presented in Figure 3.1. As seen from the figure, the highest compressive strength was observed for Mix2 which includes only micro-steel fiber reinforcement, and the lowest compressive strength was observed for Mix4 which includes micro-steel fiber reinforcement plus v-LDPE granule as coarse aggregate replacement. The results revealed that PE granule incorporation in the mixtures (Mix4–Mix7) significantly reduced the compressive strength compared to Mix2. This result was consistent with the other studies in the literature (Batayneh et al., 2007; Ismail and Al-Hashmi, 2008; Hannawi et al., 2010; de Brito and Saikia, 2013; Ghernouti et al., 2015) which indicated the

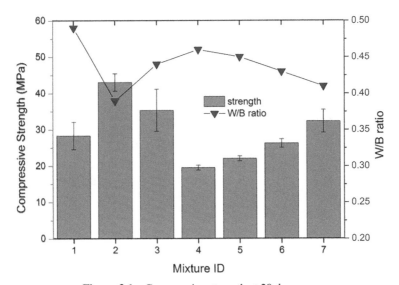

Figure 3.1   Compressive strength at 28 days age.

strength reduction with plastic incorporation into cementitious composites. This strength reduction usually increases with an increase in plastic content (Marzouk et al., 2007; Saikia and de Brito, 2014). In those studies, poor adhesion of plastic and the cement matrix is proposed as the main reason for the strength loss. Furthermore, porosity increase due to PE addition may also be responsible for the observed strength loss at some extent (Bignozzi and Sandrolini, 2006; Zhang et al., 2015; Tokgoz et al., 2016).

FTIR spectra of PEs given in Figures 3.2(a) and (b) can be useful for explaining the variation in observed strength loss with PE type. The characteristic peaks were observed at 2915–2848, 1462, and 719 cm^{-1} wavenumber for all PE types. r-LDPE showed weak peaks at 2350, 1741, 1240, and 876 cm^{-1} wavenumber. Weak peaks for r-HDPE were observed at 1721 and 1072 cm^{-1}. These weak peaks observed for r-LDPE and r-HDPE were probably due to oxidation reaction during recycling process and were assigned as C=O and –C–O groups. These functional groups in recycled PEs may interact and undergo reaction with cement matrix (via Van der Waals bonding and covalent/ionic bonding) (Larbi and Bijen, 1990; Madhu et al., 2014). This may result in stronger interfacial bonding between recycled PEs and the cement matrix compared to the neat PEs.

Neat HDPE (v-HDPE) has weak peaks at 3395 and 3192 cm^{-1}. These bands correspond to O–H stretching. The O–H group in the neat HDPE can interact with the inorganic binder (cement matrix) because of its hydrophilic nature (Liguori et al., 2014). This could explain better performance of the concrete incorporating neat HDPE (Mix5) compared to that of neat LDPE (Mix4).

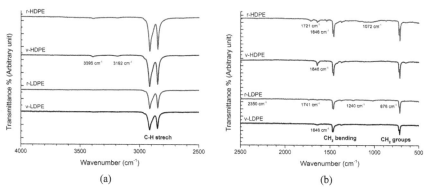

Figure 3.2 FTIR spectra of neat and recycled PEs (a) between 4000 and 2500 cm^{-1} wavenumber and (b) between 2500 and 500 cm^{-1} wavenumber.

In the current study, however, the main influencing factor for the observed strength reduction is probably water-to-binder ratio of mixtures (see Figure 3.1), though the type, shape, and size of PEs could also significantly influence the mechanical strength (Tokgoz et al., 2016).

3.3.1.2 At 28 + 30 days

In order to assess the influence of self-healing on mechanical strength, three pre-loaded and unloaded specimens from each mixture were tested for compressive strength at the end of 30 days healing period. As seen from Table 3.4, nearly all pre-loaded mixtures showed comparable or even higher compressive strength with respect to unloaded pairs (0% pre-loaded). T-test was applied to compare the compressive strength of healed specimens at 0 and 70% pre-loading levels and no statistically significant difference was found between 0 and 70% loading rate at 95% confidence level for all mixtures. This result confirms that self-healing process was occurred in all mixtures but at varying magnitude. According to Figure 3.3, which presents the percent change in strength after healing, plain concrete Mix1 indicated the highest self-healing capacity with 15%, while Mix4 indicated the lowest self-healing capacity with −6% change.

It can be concluded from Figure 3.3 that inclusion of fibers and granules decreased the self-healing efficiency of the concrete compared to the plain concrete (Mix1) although the amounts of cement, silica fume, and FA content were kept constant in each mixture. The possible reason for the observed lower healing performance in fiber and granule incorporated mixtures can be attributed to the wider crack widths. The poor adhesion of fibers/granules and the cement matrix and high porosity may probably cause wider crack formation in these mixtures when load is applied. Although healing products

Table 3.4   Mechanical properties of pre-loaded specimens after 30 days self-healing period

Mixture ID	Compressive Strength (MPa)	
	Pre-Load Level	
	0% (unloaded)	70%
Mix1	31.2 ± 2.5	36 ± 3.4
Mix2	45.4 ± 4.1	46.7 ± 3.7
Mix3	36.7 ± 2.3	38.3 ± 4.6
Mix4	21.4 ± 1.7	20.1 ± 0.4
Mix5	24.1 ± 3.1	23 ± 5.2
Mix6	27.4 ± 1	28 ± 1.8

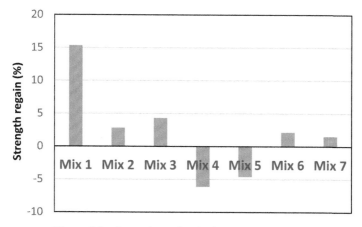

Figure 3.3 Strength regain of mixtures after healing.

can be formed, they may not be enough to completely close the wide cracks and may result in lower strength gain compared to unloaded pairs. Similar results regarding the better recovery of the mechanical strength in small crack widths were also found in the literature (Qian et al., 2009; Özbay et al., 2013; Qiu et al., 2016).

When Mix2 and Mix3 were compared, it was observed that the gained strength after healing was higher in Mix3 than Mix2. This difference was probably related with the higher numbers of fibers present in Mix3 compared to Mix2. It should be noted that the volume and so the number of PE fibers are higher compared to the volume and the number of steel fibers for an equal weight. As stated by Rooij et al. (2013), attachment of self-healing products may depend on the volume content of fibers. Therefore, the amount of fiber per volume may have a great influence on self-healing. Furthermore, as stated in FTIR analyses of PEs, C=O and –C–O functional groups in r-HDPE may promote precipitation of healing products such as $CaCO_3$ compared to the steel fibers (Homma, et al., 2009; Nishiwaki, et al., 2012).

Neat PE granule incorporated mixtures, namely Mix4 and Mix5, showed lower strength gain compared to Mix6 and Mix7 which included recycled PE granules. Functional groups in recycled PEs could probably facilitate formation and precipitation of $CaCO_3$ crystals during healing process and resulted higher strength gain compared to neat PEs.

3.3.2 Permeability

The permeability properties of the specimens after $28+30$ days curing from each mixture decided by their water absorption capacity are presented in Figure 3.4. As seen from the figure, water absorption capacity for all mixtures was below 10% even after 70% pre-loading. However, mixtures showed mixed response to self-healing in terms of water absorption. Water absorption decreased for Mix7 and Mix3, an increase was observed for Mix2 and Mix6, while no significant change was observed for Mix1, Mix4, and Mix5 when mechanical pre-loading was applied.

This indicates that there is no direct correlation between the water absorption and mechanical strength recovery for self-healed specimens. Similar results were observed by several authors investigating permeability properties. For example, Samaha and Hover (1992) reported that pre-loading levels of up to 70% of maximum strength do not lead to an increase in permeability. Rooij et al. (2013) highlighted the complexity of interpretation of water permeability in self-healed specimens as lower permeability does not necessarily describe self-healing in terms of recovery against mechanical action.

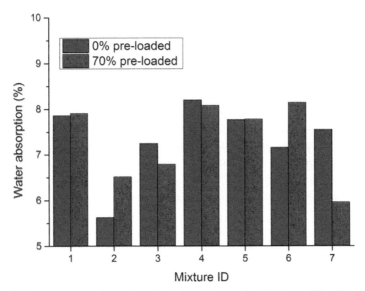

Figure 3.4   Water absorption of pre-loaded specimens after 30 days self-healing period.

3.3.3 Characterization of the Concrete Specimens

3.3.3.1 TGA analysis

Thermogravimetric curve of Mix1 after 30 days of healing period for 0% pre-loaded and 70% pre-loaded specimens are presented in Figure 3.5 as an example.

As seen from the figure, there is difference between the weight loss occurred in unloaded (0% pre-loaded) and loaded samples (70% pre-loaded). Three steps were identified in TG curve (Zhou, 2012; Biricik and Sarier, 2014; Karen Scrivener and Lothenbach, 2015). The first step is identified as dehydration of water molecules in hydrates such as CSH and ettringite which occurs within the room temperature to 200°C. The second step of thermal degradation is selected as dehydroxylation of free CH produced during curing which occurred between 350 and 550°C. The last step is due to decomposition of $CaCO_3$ which takes place within the range of 550–750°C. The mass loss occurred during TG analyses of each mixture is given in Table 3.5. Using the below stoichiometric equations, the amounts of portlandite $(Ca(OH)_2)$ and $CaCO_3$ were calculated for each mixture and the results are presented in Table 3.6.

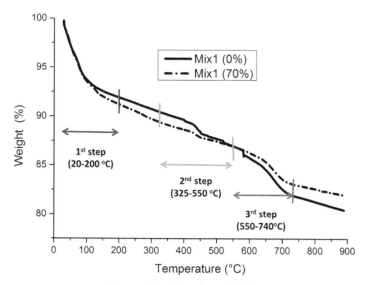

Figure 3.5   TG curve of Mix1.

Table 3.5   Percent mass loss in TGA

Mix Id	% Pre-Load	First Step (29–200°C)	Second Step (350–550°C)	Third Step (550–750°C)	% Residue (800°C)
1	0	7.8	3.2	5.2	81.3
	70	8.5	2.2	4	82.6
2	0	9.6	2.1	4.8	81.3
	70	9.6	2	4.4	81.6
3	0	10.1	2.8	4.8	79.5
	70	10.4	3.3	6.1	76.8
4	0	11.5	4.7	6.9	72.8
	70	10.1	3.1	5.1	78.5
5	0	10.2	2.3	4.4	80.3
	70	5.8	1.9	3.1	87.3
6	0	9.8	2.5	4	81
	70	10.5	3.5	5.8	76.8
7	0	10.7	3.2	6.5	76.1
	70	11.2	2.4	3.7	79.7

Table 3.6   Mass of $Ca(OH)_2$ and $CaCO_3$ calculated from the TG analyses for all mixtures under 0 and 70% pre-loading levels

Sample ID and Pre-Load Level	Mass % of $Ca(OH)_2$	Mass % of $CaCO_3$	Total Mass % of $Ca(OH)_2$ and $CaCO_3$
Mix1 (0%)	13.2	11.8	25
Mix1 (70%)	9	9.1	18.1
Mix2 (0%)	8.6	10.9	19.5
Mix2 (70%)	8.2	10	18.2
Mix3 (0%)	11.5	10.9	22.4
Mix3 (70%)	13.6	13.9	27.5
Mix4 (0%)	19.3	15.7	35
Mix4 (70%)	12.7	11.6	24.3
Mix5 (0%)	9.5	10	19.5
Mix5 (70%)	7.8	7	14.8
Mix6 (0%)	10.3	9.1	19.4
Mix6 (70%)	14.4	13.2	27.6
Mix7 (0%)	13.2	14.8	28
Mix7 (70%)	9.9	8.4	18.3

$$Ca(OH)_{2,measured} = WL_{Ca(OH)_2} \times m_{Ca(OH)_2}/m_{H_2O}$$
$$= WL_{Ca(OH)_2} \times \frac{74}{18} \tag{3.1}$$

$$CaCO_{3,measured} = WL_{CaCO_3} \times m_{CaCO_3}/m_{CO_2}$$

$$= WL_{CaCO_3} \times \frac{100}{44} \qquad (3.2)$$

where

WL; weight loss (%);
m; molecular weight (g/mol).

According to Table 3.5, all mixtures other than Mix4 and Mix5 indicated higher mass loss compared to their corresponding unloaded pair in the first step of TGA. If a sample is more hydrated, then higher mass loss should be observed in the first step of TGA for that particular sample. According to this, we can say that pre-loading improved the hydration process in Mix1, Mix2, Mix3, Mix6, and Mix7. This is in agreement with the results of compressive strength test (Table 3.4).

Table 3.6 provides the calculated mass percentages of Ca(OH)$_2$and CaCO$_3$ from all mixtures. The highest mass percentage of Ca(OH)$_2$ (19.3%) and CaCO$_3$ (15.7%) was calculated for Mix4 (unloaded), while the lowest [7.8% Ca(OH)$_2$ and 7% CaCO$_3$] was calculated for Mix5 after pre-loading. A direct correlation could not be evident between the amounts of Ca(OH)$_2$ and CaCO$_3$ and the observed strength.

Moreover, TGA results showed that significant amount of Ca(OH)$_2$ was present in both loaded and unloaded samples. This indicated the presence of very slow pozzolanic reactions, probably due to the incinerator FA used in this study. High proportion of calcium in the crystalline phase and presence of lower amount of silicates, aluminates, and iron-oxide in the FA can be given as reasons for the observed slow pozzolanic activity. Further details regarding the physical and chemical composition of FA can be found in Tokgoz et al. (2016).

3.3.3.2 FTIR analysis

FTIR spectra of the cement matrix from each mixture are shown in Figure 3.6. The band observed at 3640 cm^{-1} in the IR spectra corresponded to the O–H stretching of Ca(OH)$_2$ formed (Dutta et al., 1995; Oriol and Pera, 1995; Yousuf et al., 1995). This band diminished for all unloaded mixtures, indicating CSH bond formation. Except Mix5, the intensity of this band slightly increased for all loaded mixtures, implying ongoing Ca(OH)$_2$ formation by further hydration of unreacted cement particles.

The wideband appeared at 3405 cm^{-1} in the IR spectra was caused by the overlapping stretching vibrations (v1 and v3) of the –OH groups

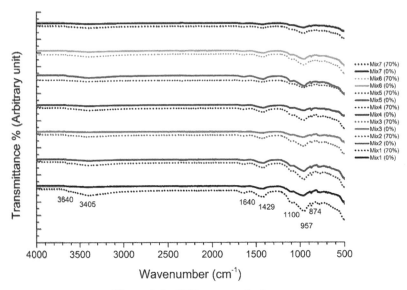

Figure 3.6 FTIR spectra of mixtures.

(Ping et al., 1999; Björnström et al., 2004). Except Mix5, this band became broader and smaller for all unloaded mixtures whereas the intensity of this band increases for loaded mixtures. This phenomenon indicated an increase in the amount of bonded –OH groups hence implying further hydration for loaded mixtures. The band at 1640 cm^{-1} in the IR spectra corresponded to the bending-in-plane vibrations (v2) of –OH groups of free water molecules. Similar to the stretching vibrations, the intensity of this band was lower for all unloaded mixtures, but Mix5.

All of the above findings indicated that further hydration was hindered in Mix5 after loading. This result is also consistent with TG analysis of Mix5 which indicated lowest mass loss during the first step of TG curve (Table 3.5) and its lower compressive strength after loading.

The bands observed at 1429 and 874 cm^{-1} were assigned as the stretching and bending-in-plane vibrations (v3 and v2) of the C–O bonds and indicating the presence of $CaCO_3$. The intensities for these bands were smaller for all unloaded mixtures than their corresponding loaded pairs, except Mix5. The intensity difference for these bands was most pronounced for Mix1 and implying formation of $CaCO_3$ as a result of healing process. This is in agreement with the compressive strength test results of Mix1 which indicated the highest healing efficiency with 15% strength regain.

The band at 1100 cm^{-1} was assigned to the stretching vibrations (v3) of the S–O bond of gypsum and ettringite (Fernández-Carrasco and Vázquez, 2009; Kontoleontos et al., 2012). The intensity of this band was higher for all loaded mixtures (except Mix5) compared to their unloaded pairs.

The strong band appeared at 957 cm^{-1} in the IR spectra corresponded to the stretching (v3) of Si–O bonds which accounts for the polymerization of the units of SiO_4^{4-} (Björnström et al., 2004). The intensity of this band was higher in all loaded mixtures, except Mix5, than their corresponding unloaded pairs. This indicates that the degree of polymerization of SiO_4^{4-} units was adversely affected with loading.

3.4 Conclusions

The experimental study regarding the influence of PEs incorporation on the self-healing properties of concrete mixtures was presented here. Addition of PEs reduced the compressive strength at varying magnitude as a result of different water-to-binder ratios, and PE types and sizes. Small cracks were obtained with 70% pre-load level in all the mixtures. Self-healing process occurred in all mixtures but at varying magnitude. Plain concrete (Mix1) showed highest strength gain after healing, while addition of PEs and steel fiber reduced the strength regain, probably, due to relatively wider cracks formation in these mixtures (Mix2–Mix7). An improvement in self-healing capacity was observed in Mix3 compared to Mix2. PE fibers in Mix3, probably, provided more nucleation sites for self-healing products compared to steel fibers alone (Mix2), and hence improved the self-healing capacity of Mix3. Recycled PE incorporated mixtures (Mix6 and Mix7) showed higher self-healing performance than those with neat PE (Mix4 and Mix5) due to the presence of C=O and –C–O functional groups in recycled PEs.

Water absorption of loaded and unloaded mixtures did not show a direct correlation with regained strength. However, all healed mixtures showed low water absorption properties like unloaded pairs.

For all mixtures, TGA and FTIR analyses revealed that there were significant differences in the structural chemical composition of loaded and unloaded concrete specimens. This highlighted that self-healing process influenced the chemical composition of the concrete mixtures. However, further research should be conducted on the mixtures having a constant water-to-binder ratio and microstructural analyses should be performed on the crack surfaces. Further methodologies to increase the binding of polymeric wastes in concretes could be explored.

Acknowledgments

This publication was made possible by NPRP grant number 6-1010-2-413 from the Qatar National Research Fund (a member of Qatar Foundation). The statements made herein are solely the responsibility of the authors.

List of Abbreviations

ATR-FTIR	Attenuated total reflection Fourier transform infrared
$CaCO_3$	Calcium carbonate
CSH	Calcium silicate hydrate
DSWMC	Qatar's domestic solid waste treatment center
FA	Fly ash
FTIR	Fourier transform infrared spectrometer
PC	Portland cement
PE	Polyethylene
r-HDPE	Recycled high-density polyethylene
r-LDPE	Recycled LDPE
SP	Superplasticizer
TG	Thermogravimeter
TGA	Thermogravimetric analysis
v-HDPE	Virgin HDPE granules
v-LDPE	Virgin low-density polyethylene
XRF	X-ray fluorescence

References

Aïtcin PC. (2000). Cements of yesterday and today–concrete of tomorrow. Cement and Concrete Research. 30:1349–59.

ASTMC39/C39M-14a. (2014). Standard Test Method for Compressive Strength of Cylindrical Concrete Specimens. ASTM International, West Conshohocken, PA.

ASTMC114-15. (2015). Standard Test Methods for Chemical Analysis of Hydraulic Cement. ASTM International, West Conshohocken, PA.

ASTMC204-11e1. (2011). Standard Test Methods for Fineness of Hydraulic Cement by Air-Permeability Apparatus. ASTM International, West Conshohocken, PA.

ASTMC311/C311M-13. (2013). Standard Test Methods for Sampling and Testing Fly Ash or Natural Pozzolans for Use in Portland-Cement Concrete. ASTM International, West Conshohocken, PA.

ASTMC642-13. (2013). Standard Test Method for Density, Absorption, and Voids in Hardened Concrete. ASTM International, West Conshohocken, PA.

Batayneh M, Marie I, Asi I. (2007). Use of selected waste materials in concrete mixes. Waste Management. 27:1870–6.

Bignozzi MC, Sandrolini F. (2006). Tyre rubber waste recycling in self-compacting concrete. Cement and Concrete Research. 36:735–9.

Biricik H, Sarier N. (2014). Comparative study of the characteristics of nano silica-, silica fume- and fly ash-incorporated cement mortars. Material Research. 17:570–82.

Björnström J, Martinelli A, Matic A, Börjesson L, Panas I. (2004). Accelerating effects of colloidal nano-silica for beneficial calcium-silicate-hydrate formation in cement. Chemical Physics Letters. 392:242–8.

Boh B, Sumiga B. (2008). Microencapsulation technology and its applications in building construction materials. Materials and Geoenvironment. 55:329–44.

Bunsell AR, Renard J. (2005). Fundamentals of Fibre Reinforced Composite Materials. Boston, MA/Philadelphia.

Collepardi S, Collepardi M, Iannis G, Quadrio Curzio A. (2012). SCC with ground bottom ash from municipal solid wastes incinerators. American Concrete Institute Special Publication. 453–64.

de Brito J, Saikia N. (2013). Chapter 7: Concrete with Recycled Aggregates in International Codes, 379–429.

Dry C. (1994). Matrix cracking repair and filling using active and passive modes for smart timed release of chemicals from fibers into cement matrices. Smart Materials and Structures. 3:118–23.

Dutta DK, Bordoloi D, Borthakur PC. (1995). Hydration of portland cement clinker in the presence of carbonaceous materials. Cement and Concrete Research. 25:1095–102.

Edvardsen C. (1999). Water permeability and autogenous healing of cracks in concrete. ACI Materials Journal 96:448–54.

EFNARC. (2002). Specification and Guidelines for Self-consolidating Concrete, European Federation of National Associations Representing producers and applicators of specialist building products for Concrete (EFNARC), Farnham, Surrey, United Kingdom.

Fernández-Carrasco L, Vázquez E. (2009). Reactions of fly ash with calcium aluminate cement and calcium sulphate. Fuel. 88:1533–8.

Ferrara L, Krelani V, Carsana M. (2014). A "fracture testing" based approach to assess crack healing of concrete with and without crystalline admixture. Construction and Building Materials. 68:535–51.

Gao X, Wang W, Ye T, Wang F, Lan Y. (2008). Utilization of washed MSWI fly ash as partial cement substitute with the addition of dithiocarbamic chelate. Journal of Environmental Management. 88:293–9.

Garcia-Lodeiro I, Carcelen-Taboada V, Fernández-Jiménez A, Palomo A. (2016). Manufacture of hybrid cements with fly ash and bottom ash from a municipal solid waste incinerator. Constrution and Building Materials. 105:218–26.

Ghernouti Y, Rabehi B, Bouziani T, Ghezraoui H, Makhloufi A. (2015). Fresh and hardened properties of self-compacting concrete containing plastic bag waste fibers (WFSCC). Construction and Building Materials. 82:89–100.

Goh CC, Show KY, Cheong HK. (2003). Municipal solid waste fly ash as a blended cement material. Journal of Materials in Civil Engineering. 15:513–23.

Granger S, Loukili A, Pijaudier-Cabot G, Chanvillard G. (2007). Experimental characterization of the self-healing of cracks in an ultra high performance cementitious material: mechanical tests and acoustic emission analysis. Cement and Concrete Research 37:519–27.

Guo X, Shi H, Hu W, Wu K. (2014). Durability and microstructure of CSA cement-based materials from MSWI fly ash. Cement and Concrete Composites. 46:26–31.

Hannawi K, Kamali-Bernard S, Prince W. (2010). Physical and mechanical properties of mortars containing PET and PC waste aggregates. Waste Management. 30:2312–20.

Hearn N. (1998). Self-sealing, autogenous healing and continued hydration: what is the difference? Materials and Structures 31:563–7.

Homma D, Mihashi H, Nishiwaki T. (2009). Self-healing capability of fibre reinforced cementitious composites. Journal of Advanced Concrete Technology. 7:217–28.

Ismail ZZ, Al-Hashmi EA. (2008). Use of waste plastic in concrete mixture as aggregate replacement. Waste Management. 28:2041–7.

Joseph C, Jefferson AD, Isaacs B, Lark R, Gardner D. (2010). Experimental investigation of adhesive-based self-healing of cementitious materials. Magazine of Concrete Research. 62:831–43.

Kamal MM, Safan MA, Etman ZA, Kasem BM. (2014). Mechanical properties of self-compacted fiber concrete mixes. HBRC Journal. 10:25–34.

Karen Scrivener RS, Lothenbach B. (2015). A Practical Guide to Microstructural Analysis of Cementitious Materials.

Kessler MR, Sottos NR, White SR. (2003). Self-healing structural composite materials. Composites Part A: Applied Science and Manufacturing. 34:743–53.

Kontoleontos F, Tsakiridis PE, Marinos A, Kaloidas V, Katsioti M. (2012). Influence of colloidal nanosilica on ultrafine cement hydration: physicochemical and microstructural characterization. Construction and Building Materials. 35:347–60.

Larbi JA, Bijen JM. (1990). Interaction of polymers with portland cement during hydration: a study of the chemistry of the pore solution of polymer-modified cement systems. Cement and Concrete Research. 20:139–47.

Li VC. (2003). On engineered cementitious composites (ECC) – a review of the material and its applications. Journal of Advanced Concrete Technology. 1:215–30.

Li VC, Herbert E. (2012). Robust self-healing concrete for sustainable infrastructure. Journal of Advanced Concrete Technology. 10:207–18.

Liguori B, Iucolano F, Capasso I, Lavorgna M, Verdolotti L. (2014). The effect of recycled plastic aggregate on chemico-physical and functional properties of composite mortars. Materials and Design. 57:578–84.

Ma H, Qian S, Zhang Z. (2014). Effect of self-healing on water permeability and mechanical property of Medium-Early-Strength Engineered Cementitious Composites. Construction and Building Materials. 68:92–101.

Madhu G, Bhunia H, Bajpai PK, Chaudhary V. (2014). Mechanical and morphological properties of high density polyethylene and polylactide blends. Journal of Polymer Engineering. 34:813–21.

Marzouk OY, Dheilly RM, Queneudec M. (2007). Valorization of post-consumer waste plastic in cementitious concrete composites. Waste Management. 27:310–18.

Mehta KP, Monteiro PJM. (2006). Concrete: Microstructure, Properties, and Materials, 3rd Edition. McGraw-Hill.

Mihashi H, Nishiwaki T. (2012). Development of engineered self-healing and self-repairing concrete-state-of-the-art report. Journal of Advanced Concrete Technology. 10:170–84.

Naik TR, Singh SS, Huber CO, Brodersen BS. (1996). Use of post-consumer waste plastics in cement-based composites. Cement and Concrete Research. 26:1489–92.

Nishiwaki T, Koda M, Yamada M, Mihashi H, Kikuta T. (2012). Experimental study on self-healing capability of FRCC using different types of synthetic fibers. Journal of Advanced Concrete Technology. 10:195–206.

Oriol M, Pera J. (1995). Pozzolanic activity of metakaolin under microwave treatment. Cement and Concrete Research. 25:265–70.

Ozerkan NG, Tokgoz DDG, Kowita OS, Antony SJ. (2016). Assessments of the microstructural and mechanical properties of hybrid fibrous self-consolidating concretes using ingredients of plastic wastes. Nature Environment and Pollution Technology. 15:1161–8.

Özbay E, Şahmaran M, Yücel HE, Erdem TK, Lachemi M, Li VC. (2013). Effect of sustained flexural loading on self-healing of engineered cementitious composites. Journal of Advanced Concrete Technology. 11:167–79.

Ping Y, Kirkpatrick RJ, Poe B, McMillan PF, Cong X. (1999). Structure of calcium silicate hydrate (C-S-H): near-, mid-, and far-infrared spectroscopy. Journal of the American Ceramic Society. 82:742–8.

Qian S, Zhou J, de Rooij MR, Schlangen E, Ye G, van Breugel K. (2009). Self-healing behavior of strain hardening cementitious composites incorporating local waste materials. Cement and Concrete Composites. 31:613–21.

Qiu J, Tan HS, Yang EH. (2016). Coupled effects of crack width, slag content, and conditioning alkalinity on autogenous healing of engineered cementitious composites. Cement and Concrete Composites. 73:203–12.

Rooij MD, Tittelboom KV, Belie ND, Schlangen E. (2013). Self-Healing Phenomena in Cement-Based Materials: State-of-the-Art Report of RILEM Technical Committee 221-SHC: Self-Healing Phenomena in Cement-Based Materials, RILEM State-of-the-Art Reports, 11.

Sabir B, Wild S, Bai J. (2001). Metakaolin and calcined clays as pozzolans for concrete: a review. Cement and Concrete Composites. 23:441–54.

Saikia N, de Brito J. (2012). Use of plastic waste as aggregate in cement mortar and concrete preparation: a review. Construction and Building Materials. 34:385–401.

Saikia N, de Brito J. (2014). Mechanical properties and abrasion behaviour of concrete containing shredded PET bottle waste as a partial substitution of natural aggregate. Construction and Building Materials. 52:236–44.

Samaha HR, Hover KC. (1992). Influence of microcracking on the mass transport properties of concrete. ACI Materials Journal. 89:416–24.

Schlangen E, Joseph C. (2009). Self-Healing Processes in Concrete. In: Self-Healing Materials. Wiley-VCH Verlag GmbH & Co. KGaA, pp. 141–82.

Sharma R, Bansal PP. (2016). Use of different forms of waste plastic in concrete - a review. Journal of Cleaner Production. 112:473–82.

Siddique R. (2008). Waste Materials and By-Products in Concrete.

Siddique R, Khatib J, Kaur I. (2008). Use of recycled plastic in concrete: a review. Waste Management. 28:1835–52.

Sua-Iam G, Makul N. (2015). Utilization of coal- and biomass-fired ash in the production of self-consolidating concrete: a literature review. Journal of Cleaner Production. 100:59–76.

Şahmaran M, Keskin SB, Ozerkan G, Yaman IO. (2008). Self-healing of mechanically-loaded self consolidating concretes with high volumes of fly ash. Cement and Concrete Composites. 30:872–9.

Tang P, Florea MVA, Spiesz P, Brouwers HJH. (2015a). Characteristics and application potential of municipal solid waste incineration (MSWI) bottom ashes from two waste-to-energy plants. Construction and Building Materials. 83:77–94.

Tang W, Kardani O, Cui H. (2015b). Robust evaluation of self-healing efficiency in cementitious materials - a review. Construction and Building Materials. 81:233–47.

ter Heide N. (2005). Crack healing in hydrating concrete. M.Sc. Dissertation, TU Delft University of Technology, Faculty of Civil Engineering and Geosciences.

Tokgoz DDG, Ozerkan NG, Kowita OS, Antony SJ. (2016). Strength and durability of composite concretes with municipal wastes. ACI Materials Journal. 113:669–78.

Van Tittelboom K, De Belie N. (2013). Self-healing in cementitious materials-a review. Materials. 6:2182–217.

Yang Z, Hollar J, He X, Shi X. (2010). Laboratory assessment of a self-healing cementitious composite. Transportation Research Record. 9–17.

Yildirim G, Keskin ÖK, Keskin SB, Şahmaran M, Lachemi M. (2015). A review of intrinsic self-healing capability of engineered cementitious composites: recovery of transport and mechanical properties. Construction and Building Materials. 101:10–21.

Yousuf M, Mollah A, Palta P, Hess TR, Vempati RK, Cocke DL. (1995). Chemical and physical effects of sodium lignosulfonate superplasticizer on the hydration of portland cement and solidification/stabilization consequences. Cement and Concrete Research. 25:671–82.

Zhang Z, Ma H, Qian S. (2015). Investigation on properties of ECC incorporating crumb rubber of different sizes. Journal of Advanced Concrete Technology. 13:241–51.

Zhou S, Lin Y, Zhao J, Zeng S, Zhou J. (2012). Utilization of the alkaline white mud as cement-based materials for the production of cement. Journal of Civil and Environmental Engineering. 2:1–7.

4

Burning the Crop Residues: A Major Environmental Problem in Delhi NCR

Pravin Kumar[1] and Rajesh Kumar Singh[2]

[1]Department of Mechanical Engineering, Delhi Technological University, Delhi, India
[2]Management Development Institute, Gurgaon, Haryana, India
E-mail: pravin.dce@gmail.com; rksdce@yahoo.com

The purpose of the paper is to highlight the issues related to the factors influencing the farmers of neighbouring states of Delhi to burn the crop residues (CRs), consequences on the health of the people, and alternative uses of the CRs other than the burning. Pre- and post-monsoon burning of the crops in Punjab, Haryana, Rajasthan, and Western Uttar Pradesh increases the particulate matter in Delhi NCR tremendously. During this time, the wind speed is also very slow which leads to air-lock in Delhi NCR. The environmental condition becomes so poor that the people cannot come out from their houses. The author has emphasized the existing alternative use as well as the new proposed technology development regarding the industrial uses of the CRs.

4.1 Introduction

Agriculture is a way of livelihood of a large section of the people of India. Approximately, 70% of the populations are dependent on the cultivation for the fulfilment of their basic requirements. The PWS (Paddy and wheat system) is the main crop produced in north-eastern states of India and generates a large volume of agricultural wastes in the form of husk, straw, and stubbles. The northwestern states especially Haryana and Punjab are the leading producers of rice and wheat in India. Rice and wheat are also produced in some parts of the western Uttar Pradesh and Rajasthan which pollute the environment of Delhi NCR. The farming dregs generated include mainly the cereal straws, stubbles, woody stems, cotton stalk and leaves, etc. A small part of the agricultural residue mainly leaves and stubbles is used as animal fodder, roofing and shedding of homes, cattle shed domestic usage fuel, and small-scale industries' raw material and fuel. Still, a bulky part of the stubbles and straw is not used properly and the dumping of huge quantity of farm yield is practically not feasible for the farmers.

According to Yadvinder et al. (2010), about 550 million tons (Mt) of CRs are produced per year in India. A large portion of the CRs (about 90–140 Mt per annum) is burnt on farms to clear the field for the next crop. Presently, more than 80% of the total rice straw produced annually is burnt by farmers just after the monsoon. During these 3–4 weeks in the month of October–November, Delhi NCR (National Capital Region) chokes on smoke due to the burning of the paddy straw in the neighbouring states. Burning the rice straw causes the gaseous emissions of 70% CO_2, 7% CO, 0.66% CH4, and 2.09% N_2O (Gupta et al., 2004). Burning CRs is a matter of serious concern not only for GHG (Green House Gases) emissions, but also for causing problems including pollution, health hazards, and loss of nutrients (Pathak et al., 2006). There are mainly two types of pollutants from the CRs burning: (i) gaseous air pollutants and (ii) particulate air pollutants. Gaseous air pollutants include SO_2, NO_X, O_3, CO_2, CO, etc., whereas the particulate air pollutants include PM10, PM2.5, and PM1.0. If the size of the suspended particles in the air is less than 10 μm, it is known as PM10. If it is less than 2.5 μm, it is known as PM2.5. Similarly, if the suspended particle size is very fine, i.e., less than 1 μm, it is known as PM1.0.

In this paper, the authors have highlighted the issue of CRs burning in the neighboring states of Delhi NCR, its impact on health, and alternative uses of CRs. The rest of the paper has been arranged as: Section 2 presents the literature review of the CRs burning in India and its impact on environment, Section 3 presents the CRs burning activities in NW India, Section 4 presents the factors enforcing the farmers to burn the CRs, Section 5 shows the consequences of the CRs burning, Section 6 represents the alternative uses and industrial applications of CRs, Section 7 discusses the legislation and government policy to prevent the burning of CRs, and Section 8 concludes the research work.

4.2 Literature Review

Crop residues burning is not a problem of India only but it is equally important for other countries like China, Russia, Thailand, Italy, and Pakistan. Many of research works have been done in the past on measurement of emissions due to CRs burning and other biomass burning and their environmental impact. In this section, the major works done in Indian context in the last one decade have been explored.

The recent study shows the importance of improved air quality in India, where approximately 600,000 annual premature deaths are due to outdoor air pollution ranking second only to China (Lelieveld et al., 2015; Ghude et al., 2016; WHO, 2016). PM2.5 exposure is responsible for the loss of average life expectancy by 3.4 years across the country and up to 6.4 years in Delhi (Ghude et al., 2016). In addition to the contributions of various other urban sources to air quality degradation, outdoor fires are a regional air pollution source dominated by fires in agricultural regions (Vadrevu et al., 2008). Particularly in northern India, fires are mostly from residue burning, which peaks in April to May (pre-monsoon) and October to November (post-monsoon), corresponding to burning after the wheat and rice harvests, respectively (Venkataraman et al., 2006; Vadrevu et al., 2011). The major works on the CRs burning and its effect on environment in the Indian context are summarized in Table 4.1.

4.3 Crop Residues Burning in North-Western States of India

Punjab and Haryana are the leading states in paddy and wheat production in India. Due to intensive and mechanized farming, approximately 70–80% of the paddy straw is burnt into the field which leads to the generation of SO_2, NO_X, and suspended particulate matters. The emissions from the burning of CRs in Punjab, Haryana, Rajasthan, and Western Uttar Pradesh choke Delhi NCR during pre- and post-monsoon. Figure 4.1 shows the air quality index of the Delhi NCR for pre- and post-monsoon. It has been observed that during the month of May (pre-monsoon) and October–November (post-monsoon), the air quality index becomes very poor. Therefore, this issue needs to be addressed properly.

Delhi's pre-monsoon and post-monsoon airsheds are characterized by high fire intensity and relatively weak northwesterly winds. The average post-monsoon Delhi airshed, extending northwest to Haryana and Punjab

Table 4.1 Reverences of the some of the major works on crop residue burning in Indian context

S. No.	Name of the Journal	References	Perspectives
1	Renewable and Sustainable Energy Reviews	Lohan et al. (2018)	Reviewed the amount of residue generation, its utilization *in situ* and *ex situ*, emphasize harmful effects of residue burning on human health, soil health and environment of northwest states of India especially in Punjab and Haryana
2	Renewable and Sustainable Energy Reviews	Vijay et al. (2018)	Studied the effects of agriculture crop residue burning (CRB) on aerosol properties and long-range transport over northern India during 09 and 17 November 2013, with the help of satellite measurements and model simulation data
3	Atmospheric Environment	Liu et al. (2018)	The influence of a single pollution source, outdoor biomass burning, on particulate matter (PM) concentrations, surface visibility, and aerosol optical depth (AOD) from 2007 to 2013 in three of the most populous Indian cities—Delhi, Bengaluru, and Pune have been examined
4	Environmental Pollution	Sharma et al. (2017)	Comprehensive measurements of AOD, PM, and black carbon (BC) mass concentrations have been carried out over Patiala, a semi-urban site in northwest India during October 2008 to September 2010
5	Journal of Cleaner Production	Yadav et al. (2017)	Discussed the sustainable and environmentally safer cropping systems with low global warming potential (GWP) and low energy requirement for rice fallow land of India

Table 4.1 (Continued)

S. No.	Name of the Journal	References	Perspectives
6	Atmospheric Environment	Gupta et al. (2016)	PM levels in the ambient air of three urban sites of strategic importance in Punjab were monitored from September 2013 to June 2014 covering two seasons of CRB episodes of rice and wheat, respectively
7	Renewable Energy	Sindhu et al. (2016)	The sugarcane crop residue is rich in cellulose and hemicelluloses. They emphasized on the production of bioethanol and other liquid transportation fuels from the sugarcane residue
8	Field Crops Research	Sidhu et al. (2015)	Review of the machines developed like happy seeder in last 10 years for sowing wheat into heavy rice residues in northwest India
9	Renewable and sustainable energy reviews	Hiloidhari et al. (2014)	Assessed crop residue biomass and subsequently bioenergy potential in all the 28 states of India using crop statistics and standard procedure. A total of 39 residues from 26 crops cultivated in India are considered for the study
10	Atmospheric Environment	Vadrevu et al. (2013)	Analyzed long-term trends (2003–2011) in CO retrievals and fire–CO relationships including CO profiles at nine different atmospheric levels
11	Atmospheric Environment	Mishra and Shibata (2012)	The present study deals with the spatial variability including the vertical structure of optical and microphysical properties of aerosols, during the CRB season (October and November) of 2009 over the Indo-Gangetic Basin

(*Continued*)

Table 4.1 (Continued)

S. No.	Name of the Journal	References	Perspectives
12	Journal of Atmospheric and Solar-Terrestrial Physics	Kharol et al. (2012)	Analyzed the variations in BC aerosol mass concentration over Patiala city, Punjab, India, during October/November 2008 associated with agriculture CRB activities
13	Science of The Total Environment	Agarwal et al. (2012)	Pulmonary function tests (PFTs) like force vital capacity (FVC), force expiratory volume in one second (FEV1), peak expiratory flow (PEF), and force expiratory flow between 25 and 75% of FVC (FEF25–75%) and oxygen saturation (SpO_2) level of 50 healthy inhabitants with respect to rice CRB were investigated for three rice cultivation periods from 2007 to 2009. The subjects were residents of five sampling sites selected in Patiala city
14	Agricultural Systems	Erenstein (2011)	Assessed the current crop residue management practices in Punjab and Haryana's rice–wheat, basmati–wheat, and non-rice–wheat cropping systems
15	Environmental Pollution	Vadrevu et al. (2011)	Characterization of the fire intensity, seasonality, variability, fire radiative energy (FRE), and AOD variations during the agricultural residue burning season using MODIS data
16	Atmospheric Environment	Singh et al. (2010)	India was analyzed for loss on ignition (LOI) and organic tarry matter (OTM) content in ambient air during CRB episodes and non-crop residue burning (NCRB) months in 2006–2007

Table 4.1   (Continued)

S. No.	Name of the Journal	References	Perspectives
17	Science of The Total Environment	Awasthi et al. (2010)	Variations in PFTs due to agriculture crop residue burning (ACRB) on children between the age group of 10–13 years and the young between 20 and 35 years are studied
18	Atmospheric Environment	Mittal et al. (2009)	A ground level study was deliberated to analyze the contribution of wheat and rice crop stubble burning practices on concentration levels of aerosol, SO_2, and NO_2 in ambient air at five different sites in and around Patiala city covering agricultural, commercial, and residential areas

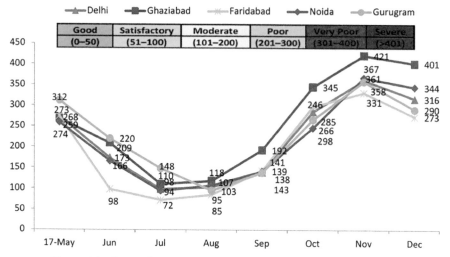

Figure 4.1   Pre- and post-monsoon air quality index 2017 in Delhi NCR.

Source: CPCB, December 2017).

and across the border to Pakistan, comprises a population of 63 million (39% in urban areas). The average pre-monsoon Delhi airshed extends less to the northwest and more to the south, covering only Haryana and southern Punjab. The average post-monsoon Delhi airshed comprises a smaller population of 39 million (48% in urban areas). In Delhi's pre-monsoon and post-monsoon airsheds and fire activity occurs primarily on cropland. On average, pre-monsoon and post-monsoon agricultural fires are associated with 99% of total FRP (Fire Radiative Power) within the seasonal airshed; in terms of area, the post-monsoon airshed averages 96% cropland, 4% urban and built-up, and 1% grassland, shrubland, and savanna, and the pre-monsoon airshed averages 83% cropland, 13% grassland, shrubland, and savanna, and 3% urban and built-up (Tianjia et al., 2018). Figure 4.2 shows the air quality trends of Delhi from 2009 to 2015.

Crop burning results in emissions of CO_2, CO, CH_4, N_2O, NO_x, NMHCS (Non-methane hydrocarbons), and aerosols. The emissions of CH_4, CO, N_2O, and NO_x had been estimated to be about 110, 2306, 2, and 84 Gg, respectively, from rice and wheat straw burning in India in the year 2000 (Gupta et al., 2004; Yadav et al., 2014). In Asia, paddy straw is a major CR. The approximate amount of CRs generation was 668 ton which could produce 187 gallons of bioethanol, if technology was available (Kim and Dale, 2004).

Incinerating fields is a process of uncontrolled combustion because of lack of oxygen. During a small time period (in weeks), a very large area of the farms in Punjab is set into the fire and availability of oxygen for complete combustion of the CRs becomes insufficient. Thus, it results into the emission

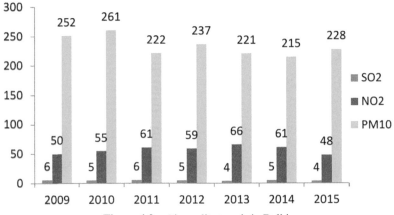

Figure 4.2   Air quality trends in Delhi.

Source: CPCB, December 2015.

of carbon dioxide (CO_2), carbon monoxide (CO), unburnt carbon (as well as traces of methane, i.e., CH_4), nitrogen oxides (NO_x), and comparatively less amount of sulfur dioxide (SO_2). According to Galanter et al. (2000) and Hays et al. (2005), the burning of 1 ton of paddy straw produces 3 kg particulate matter, 60 kg carbon monoxide, 1460 kg carbon dioxide, 199 kg ash, and 2 kg of sulfur dioxide. It also emits a huge quantity of particulates that are composed of a variety of organic and inorganic species. About 70% of the carbon in paddy straw released is as CO_2 upon the straw burning. Similarly, 7% carbon as CO, 0.66% carbon as CH_4, and 2.09% N as N_2O are released upon incineration.

Due to the light weight of the particulate matter (PM2.5) produced through the residue burning, it can stay in the air for a long time, causes smog, and travels hundreds of miles along with wind (Singh and Panigrahy, 2011; Jain, 2016). This is the reason that the Delhi NCR chokes during pre- and post-monsoon on residues burning in Punjab, Haryana, Rajasthan, and the farms in the border area of Pakistan. PM2.5 has greater stability timing in the air when contrasted to PM10 because of the balance between the downward force of gravity and aerodynamic drag force.

Figures 4.3 (a) and (b) show the variations in the level of PM2.5 and PM10, respectively, from 27th October 2017 to 30th November 2017 in Delhi NCR due to the burning of paddy straw in the neighboring states of Delhi. It has been observed that the maximum levels of PM2.5 and PM10 are ranging from 7th to 14th November 2017. One of the important reasons for the increased particulate matters is the burning of paddy straw at the large scale in Punjab and Haryana including western Uttar Pradesh and some parts of Rajasthan.

The total CRs produced in Punjab alone were 40.14 million tons (23 million tons of paddy straw and 17 million tons of wheat straw) (Singh et al., 2008; Chauhan, 2012; IARI, 2012; NAAS, 2012; Jain et al., 2014). More than 80% of paddy straw (18.4 million tons) and almost 50% wheat straw (8.5 million tons) produced in the state are being burnt in fields. Most of the paddy straw is burnt in the field except Basmati rice straw. On contrary, most of the wheat straw is used of fodder for the cattle and only small part is burnt in the field. In Haryana, 24.70 Mt of the CR is being burnt. Straw and husk of wheat and paddy alone contributed more than 85.6% in Punjab and 78.48% in Haryana, and rest contribution was from other crops such as cotton, mustard, and sugarcane crops as shown in Figure 4.4. The Punjab state produces 55.39 Mt CRs, among which, 22.32 Mt (40.17%) of the total residues have been found surplus with an average density of 4430 kg/ha (MNRE, 2009; Chauhan, 2011; NAAS, 2012).

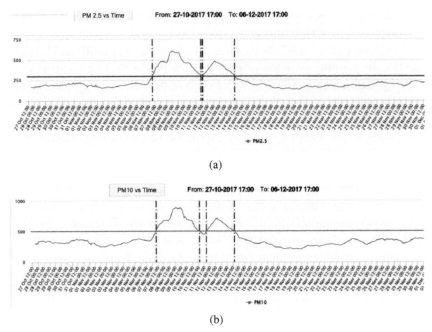

(a)

(b)

Figure 4.3   (a) Variations in the level of PM2.5 from 27th October to 30th November 2017 in Delhi NCR. (b) Variations in the level of PM10 from 27th October to 30th November 2017 in Delhi NCR.

Source: CPCBCCR, December 2017.

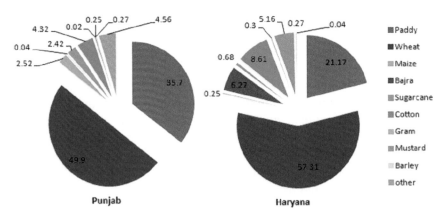

Figure 4.4 Crop-wise percentage of residue generation in Punjab and Haryana (MNRE, 2009; Chauhan, 2011, 2012; Jain et al., 2014; Sangeet, 2016).

In Punjab, the practice of mechanized harvesting has been increased in the past two to three decades which avoids the use of costly manual labours. However, the scattered, root-bound CRs are left behind on the farm. Using the combined harvesters, it is difficult to remove the scattered and root-bound CRs but this can help in removing the straw. The farmer finds that burning is the easy, fastest, and cheapest method to clear fields for the next planting (Gadde et al., 2009; Kumar et al., 2015a). An estimated 7–8 million tons of rice residue associated with post-monsoon agricultural harvesting are burned each year in Punjab, India (Kumar et al., 2015a).

Post-monsoon burning runs for several weeks and is associated with increased aerosol across the Indo-Gangetic Plains; however, pre-monsoon fires are not much dominant contributor to air quality degradation (Vadrevu et al., 2011; Singh and Kaskaoutis, 2014). Wheat and rice residue burning releases accumulation mode aerosols that mainly contribute to the PM2.5 fraction (Hays et al., 2005). Satellite-based studies of post-monsoon burning in the Indo-Gangetic Basin are shown in Figure 4.5 as elevated layers of aerosols from the surface to 4–4.5 km altitude, with peak concentrations below 1 km (Mishra and Shibata, 2012). Ground measurements in north-western India also show the increased aerosol, SO_2, and NO_2 concentrations during pre-monsoon and post-monsoon periods (Mittal et al., 2009).

Regarding the production of CRs, Uttar Pradesh (72 Mt) is chased by Punjab (45.6 Mt), West Bengal (37.3 Mt), Andhra Pradesh (33 Mt), and Haryana (24.7 Mt). Cereal CRs are generated maximum by paddy (53%) and

Figure 4.5 The image produced by NASA showing the cloud of fog due to stubble burning in NW states of India.

Source: IBT, November 10, 2017.

Figure 4.6   The farm area of Punjab and Haryana in hectares contributing the stubble burning.

Source: Hindustan Times, October 24, 2016.

then by wheat (33%) (Gadde et al., 2009). According to the estimation, paddy straw production was 22,289 Gg in excess in India annually and out of which, 13,915 Gg is burnt. Haryana and Punjab, the two states alone share 48% of the entire amount and burn the same in farms. Figure 4.6 shows the area of the farms in Punjab and Haryana contributing the stubble burning.

Primarily two types of residues from rice cultivation are produced-straw and husk. Although the technology of using rice husk is well established in

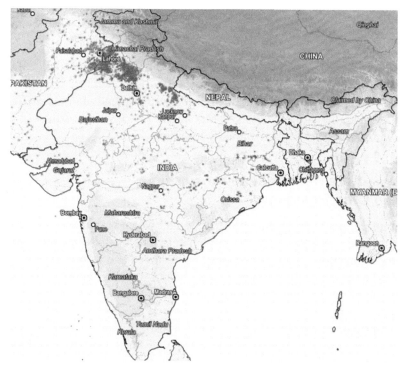

Figure 4.7 Satellite image between October 13 and 15 shows biomass fires in India.
Source: Down-to-earth, October 12, 2017.

many countries, paddy straw as of now is rarely used as a source of renewable energy. Husk procurement is easier than the straw because it is only available at the rice mills. In the case of a paddy straw collection, it is a tedious task and its availability is limited to harvest time. The logistics of the collection can be improved through baling but the necessary equipment is expensive and uneconomical for most rice farmers (Kumar et al., 2015a).

In India, the excess amount of residues (balance residue after domestic utilization) of cereal crops that amounts to 82 Mt is usually fired in the agricultural farm, in which rice and wheat contribute 44 and 24.5 Mt, respectively (IARI, 2012). We can see plumes of smoke rising from the fields which by the end of October or sooner becomes a thick blanket in the air over Haryana, Punjab, and western Uttar Pradesh extending up to the national capital (Sandhu, 2016). It is the season for paddy stubble burning in this region. According to another study, in India, 84 Tg of crop residues are burnt annually (Streets et al., 2003). In Figure 4.7, burning of rice straw at different locations in India is shown by red dots. It can be observed that the fires are dense in Punjab Haryana and western Uttar Pradesh.

Cereal crops (e.g., rice, wheat, maize, and millets) contribute 70% of the total CRs. Rice and wheat contribute 34% and 22% of the total CRs, respectively (MNRE, 2009). The surplus CRs (i.e., total residues generated minus the amount for another economic purpose) are typically burned on the field. On average, the surplus CRs (70 Mt) nearly (44.5 Mt rice straws and 24.5 Mt wheat straw) are burned annually in India.

4.4 The Factors Responsible for Burning of the Crop Residues

A number of alternative application of paddy and wheat residues have been suggested by the researchers but the implementations of these methods are far from the realization. The facilities must be localized and resources should be generated to motivate the farmers to think about the other applications of the CR. Some of the observed factors responsible for the burning of CRs are discussed below.

The scarcity of cheap labour for manual harvesting: Punjab and Haryana states have been shifted to the mechanized forming for the last two to three decades due to increasing labour cost and a trend of labour to shift from agriculture to industry due to high wage in industry compared to agriculture. Also, manual labour and animal power were not sufficient to cope with the workload of intensive agriculture. The contribution of manual labours in agricultural activities has been reduced from 62.67 to 35.96% from 1970–71 to 2012–13 in Punjab state (Lohan et al., 2015a). The mechanization has sharply increased due to the shortage of labour, increasing wage rates during harvesting season (Jat et al., 2009, 2011; Singh et al., 2011; Sidhu et al., 2015). In Punjab state, the manpower has substantially reduced from 7.5 to 0.69% due to increase in mechanical power from 17 to 76% and electrical power from 1.7 to 23.5 from 1960–61 to 2012–13 (Mittal et al., 2009).

Application of combine harvester at mass scale: The cropping intensity in Punjab has been increased due to increase in the availability of the farm power. The farm power availability has been increased from 0.37 to 5.68 kW/ha. Due to a substantial increase in the farm power, the cropping intensity and total food grain productivity have been increased from 112 to 196% and 668 to 3638 kg/ha, respectively, in Punjab state from 1960–61 to 2012–13 (Lohan et al., 2015a). The traditional method of harvesting paddy requires about 150–200 man-hour ha^{-1}. The introduction of combine harvester has eliminated the requirement of such a large manpower. The combine harvester performs a number of operations concurrently which saves both the time and the cost. The combine harvester cuts the crop, feeds the crop to the cylinder, threshes the grain from rear head, separates the grain from straw, cleans the grain, and handles the clean grain unit in one operation. The number of combine harvesters in the country increased dramatically approximately from 5000 in 1990–91 to 13,800 in 2012–13 (Lohan et al., 2015a; Kutcher and Malhi, 2010). Additionally, the combine harvester leaves a huge mass of residue (up to 9.0 t ha^{-1}) in the field after harvest and farmers are not furnished and prepared to manage the large mass of residues left in the field (Mittal et al., 2009). Thus, a farmer finds that it is easy and economical to burn the residue in the field to enable early sowing.

The small time gap between paddy harvesting and planting of wheat: The time available to prepare the field between paddy harvesting and planting of wheat in NW India is 7–10 and 15–20 days in basmati and coarse grain rice, respectively (Gupta, 2011). The combine harvester can cut the paddy crop at a certain height above the ground. It leaves two distinct straw components in the farm after harvesting: (i) the standing stubble residues and (ii) the windrows of loose CRs, big uneven heaped lines of straw. Approximately, a straw load of 7.5 t/ha remains in the field. Due to unavailability of the technologies for paddy residue management, farmers prefer burning the paddy residues in the field (Singh and Panigrahy, 2011; Jain et al., 2014).

Burning of weeds/pests and ash as nutrients for the land: The activities of burning weeds, insects, diseases, and pests with CRs result into control of unwanted grass and pests in the farm for the next crop (Gupta et al., 2011; Jain et al., 2014). Residual ash left behind burning behaves as a fertilizer which is a suitable source of potassium. It raises the short-term availability of specific nutrients (e.g., P and K) and lowers soil acidity, but leads to a loss of other nutrients (e.g., N, P and N, P, and S) and organic carbon (Singh et al., 2006; Jat et al., 2009; Mehta et al., 2013).

4.5 Consequences of Crop Residues Burning

The consequences of the CRs burning are very dangerous. Life expectancy of the people in Delhi NCR decreases. People are suffering from chronic diseases and respiratory problems. Some of the major consequences of the CR burning are discussed below.

Degradation of air quality and increasing the concentration of greenhouse gases: CR burning leads to the global warming due to various types of the emissions. Black carbon emission is the second leading contributor to current global warming, after carbon dioxide emissions (Ioannidou and Zabaniotou, 2007; Ramanathan and Carmichael, 2008). Burning of CR results into the emission of carbon dioxide (CO_2), carbon monoxide (CO), unburnt carbon (as well as traces of methane, i.e., CH4), nitrogen oxides (NO_x), and comparatively less amount of sulfur dioxide (SO_2).

Liberation of soot particles and causing smog in the environment: CR burning produces suspended particulate matter in the atmosphere. Due to the light weight of the particulate matter (PM2.5) produced through the residue burning, it stays in the air for a long time, causes smog, and travels hundreds of miles along with the wind. This is the reason that the Delhi NCR chokes during pre- and post-monsoon on residue burning in Punjab, Haryana, Rajasthan, and the farms in border area of Pakistan.

Health hazards to human, animals, and birds: The incineration of CRs causes severe impacts on human health, viz., aggravated chronic heart diseases and lung ailments, besides causing respiratory problems such as asthma, coughing; particularly affecting children, geriatrics, and pregnant women (Long et al., 1998; Treets et al., 2003; Mittal et al., 2009; Singh et al., 2009, 2011; Pathak et al., 2010).

Farm residues blazing emits a high magnitude of air pollutants like N_2O, CO_2, CH_4, CO, NH_3, SO_2, hydrocarbons, volatile organic compounds, and suspended particulate matter at a diverse pace which is observed in any grassland or forest fire because of separate composition of the farm residues and burning forms (Mittal et al., 2009; Zhang et al., 2011). These air pollutants cause adverse impacts on human health. They can cause chronic obstructive pulmonary diseases (COPD), pneumoconiosis, pulmonary tuberculosis, bronchitis, skin diseases, eye irritation, cataract, corneal opacity, and blindness.

Deterioration in soil health and fertility: Stubble and straw flaming also burn the nutrients present in the farm residues besides causing a huge pollution. The total quantity of carbon, 80–90% nitrogen, 25% of phosphorus, 20% of potassium, and 50% of sulfur present in various crop vanishes in the harmful gaseous forms and particulate matter causing the air pollution (Lefroy et al., 1994). One ton of paddy residues contains 6.1 kg N, 0.8 kg P, and 11.4 kg K

(Yang et al., 2003; Jat et al., 2013; Singh and Sidhu, 2014; Teodoro and Mendoza, 2016). Burning of paddy straw causes an intact loss of about 79.38 kg/ha N, 183.71 kg/ha P, and 108.86 kg/ha K (Jat et al., 2013).

It was estimated that burning of straw raised the soil temperature up to 33.8–42.2°C at 10 mm depth (Gupta et al., 2004). About 23–73% of nitrogen is lost; and the fungal and bacterial populations are decreased immediately up to 25 mm depth of soil. The burning of straw raised the temperature of the soil in the top 75 mm to such a high degree that the carbon–nitrogen equilibrium in soil changes rapidly (Singh et al., 2010; Kumar et al., 2015b).

4.6 Alternative Uses of Crop Residues

The paddy and wheat straw are very useful. Some of the alternative uses of paddy straw have been discussed in the following paragraphs. However, a large part of the wheat straw is used as food for the cattle and only small part of the wheat straw is burnt in the field.

Amalgamation of stubble in soil: According to Sidhu and Beri (2005), the best substitute for the burning of paddy residue is an amalgamation of stubbles in soil on the field site. It has been reported that the paddy residue incorporation in soil 3 weeks before cultivating considerably increases wheat yield in clay loam soils. It will also increase organic carbon in soil by 14–29% (Singh et al., 1996). In contrary, if the paddy residue is amalgamated instantaneously before sowing the wheat in rabi, then the crop production is reduced due to the arrest of inert nitrogen which adversely causes nitrogen deficit (Singh et al., 1996).

Crop stubble as animal fodder: One of the alternative utilizations of crop stubbles can be animal fodder. The northwestern states of India do not practice feeding cattle with a paddy straw because of its low feed value and the nutrients present in rice straw are not readily digestible to the livestock due to the presence of high silica and lignocellulosic content along with very low protein (2–7%). However, the residue of basmati variety of paddy is frequently consumed as animal fodder (Kumar et al., 2015b; Jain, 2016) because of its high palatability. The basmati paddy straw is fed to animals mixed with green fodder only in the dire scarcity of fodder (Kumar et al., 2013).

Wheat and paddy straw in mushroom farming: Paddy straw is a key ingredient to be utilized as a raw matter for mushroom cultivation in Punjab

(Chaudhary et al., 2009) but in general, farmers are using the wheat straw as raw material. For the production of button mushroom, some operations like the washing of straw and draining of excess water, cutting of straw, and preparation of bundles are necessary. A recent research conducted on paddy straw management (Roy and Kaur, 2016) revealed the estimated cost of these operations were $7 per quintal using paddy straw as a raw material rather it was $11 per quintal using wheat straw as a raw material.

Cattle shed and bed preparation: The paddy wastes used as bed material for cattle during winters has been a regular practice in few regions of India. The bed material of paddy helps improving milking capacity in terms of quality and quantity contributing to comfortable sleep of cattle warmth, udder health, and leg health. Moreover, the straw material leads to a hygienic, relaxed, greasy surrounding and it even prevents the chances of injury and lameness (Kumar et al., 2015b). The paddy straw used for bedding could be subsequently routine for composting. Each kilogram of straw absorbs about 2–3 kg of urine from the animal shed. Moreover, it can be composted by alternative methods on the farm itself.

Preparation of stubble compost fertilizer: Further, the stubble compost consists of approximately 2% of N_2, 1.5% of P, and 1.4–1.6% of K which improve crop yield by 4–9% (Sood, 2016).

Rice straw mulching: Rice straw could be used as mulch for other crops like wheat, maize, sugarcane, sunflower, soybean, potato, and chilli. It would improve crop yield in drylands as well as water stress conditions by conserving soil moisture, and it could also save about 7–40 cm of irrigation water (Sekhon et al., 2008; Yadvinder et al., 2010; Arora et al., 2011).

Mulching with CRs increases the least soil temperature in winter by reducing upward heat flux from soil and declines soil temperature during summer due to shading effect. The CRs play a significant role in betterment of soil acidity by releasing bases such as hydroxyls during the decomposition of CRs with higher C: N, and soil alkalinity through the application of residues from lower C: N crops such as legumes, oilseeds, and pulses (Pathak et al., 2011). The CRs also help in carbon sequestration in the soil (Jain et al., 2014). The CRs, particularly from wheat and rice crops, have a wide C: N ratio of 70:1 to 100:1. About 30–40% of C supplemented by CRs gets decomposed in about 2 months (Beri et al., 1992).

In situ incorporation: *In situ* incorporation enhances the decomposition of combine harvested residues to advance nutrients in the soil. Residue incorporation in the soil has many positive impacts on soil health attributes including enhancement in pH, organic carbon, infiltration rate, and water holding capacity (Gupta et al., 2004; Gangwar et al., 2006; Kumar et al., 2015b). It also increases hydraulic conductivity, CEC, and microbial biomass, enhances activities of enzymes such as dehydrogenase and alkaline phosphatase, reduces bulk density of soil by modifying soil structure and aggregate stability, surface crust formation, and water evaporation from the top few inches of soil, and prevents leaching of nutrients (Peter et al., 2014).

An increase in organic carbon increases bacteria and fungi in the soil. Beri et al. (1992) and Sidhu et al. (1995) reveal that soil treated with CRs held 5–10 times more aerobic bacteria and 1.5–11 times more fungi than soil from which residues were either burnt or removed. Due to increase in microbial population, the activity of soil enzymes responsible for the conversion of unavailable to an available form of nutrients also increases. It is reported that an addition 36 kg per hectare of nitrogen and 4.8 kg per hectare of phosphorous (6 g of nitrogen and 0.8 g of phosphorous per kg of paddy straw) leads to saving 15–20% of total fertilizer's use.

Back in soil: Composting is the decomposition of rice straw to enable recovery of portions of its nutrients and organic components if the feedstock materials have a high nitrogen content to obtain a better carbon to nitrogen ratio. Some of the factors that affect the composting are availability of oxygen, moisture content, pH value, temperature, and the carbon/nitrogen ratio. Rice straw takes longer time to decompose it may take a year.

Scientists have already developed a rapid composting technique to convert huge piles of rice straw into the organically rich soil. Generally, it takes about 45 days to prepare this rice straw compost. But the main problem with farmers is the small time gap between paddy harvesting and wheat sowing. They want quick solutions. That is the reason that the rice straw compost is not adopted in Punjab and Haryana.

Collection of residues for off-farm uses: The residue generated from the paddy–wheat cropping system can be laid too many uses but this is possible if the residue is carried out off the field. In some parts of northwestern India, it is gaining popularity in the wheat straw collection instead of rice because of its economical use for feeding animals. For removal and collection of straw after combine harvesting and using the residues for off-farm works, straw baler machines are very useful and commercially available. These balers, however, recover only 25–30% of potential straw yield after combining, depending upon the height of plant cut by combine harvester.

Domestic fuel: The rural population of Himachal Pradesh, Uttarakhand, and Jammu and Kashmir depends primarily on fuel-wood for fulfilling domestic needs/industrial fuel for cooking (combustion with dung cake/wood/ coal) (Jain et al., 2014; Kumar et al., 2015b). But now things have changed and the main reasons for un-preferred use of rice straw are the difficulty in procurement due to light weight and occupying more volume.

Combustion material for power plants: The technology has been developed to adopt the paddy straw as fuel in the power plant. The by-products are fly ash and bottom ash, which may be used in cement and/or brick manufacturing, and construction of roads. The paddy straw may be used in the form of bales directly in the furnace or in the form of shredded straw with pulverized coal. In the bales form, it is easy in handling and storage. In the shredded form of the paddy straw, thermal efficiency and heating value increase. Punjab Biomass Power, Ltd. is the first of the nine rice straw power plants coming up in Punjab. This type of plant near Ghanaur village in Patiala district is already functioning.

In biogas and biochar generation: The biomass of paddy residues is an efficient source of energy generated through anaerobic digestion, gasification, and pyrolysis technologies which offer an instant result for the decline of CO_2 concentration in the environment (Brar et al., 2000; Dhaliwal et al., 2011; Lohan et al., 2012). Using anaerobic digestion of 1 ton of paddy residue, 300 m^3 of biogas can be obtained (Koopman and Koppejan, 1997). The process generates suitable quality of gas consisting 55–60% methane and the spent slurry can be used as manure (IARI, 2012; Lohan and Sharma, 2012; Lohan et al., 2015b). One ton of paddy biomass can generate 300 kW h of electrical energy through gasification. During pyrolysis, about 50% of the carbon content in the CRs is immediately released as gas or volatile compounds, which could be effectively used as energy sources; the remaining carbon in biochar is highly recalcitrant and contributes to sequester C, once applied in soil (Lehmann et al., 2006).

Making pellets: The CR may be used as a fuel in the pellet form. The pellet mill is used to crush, press, compact, and form the straw, peanut shell, cob, cotton bar, soybean rod, weeds branches, leaves, sawdust, bark, and other solid wastes to prepare the pellets. Biomass pellet can be used for civil heating fuel and life fuel (Verma, 2014). This kind of fuel has high efficiency and is easy to store. It can also be used as the main fuel for the industrial boiler.

Raw materials for paper and pulp industry: Straw is a competitive, and alternative source of fiber for papermaking to reduce the pressures on forests. Rice straw can be used to make paper and various paper products (i.e., newsprint, copier paper, bond paper, etc.). A pulping technology could eliminate waste by turning rice straw into paper. The best method extracts cellulose from the straw to make paper and natural phenolic materials. Thus, the major portion of the burning paddy straw can be utilized as pulp for paper and cardboard.

Preparation of bioethanol from CRs: It is estimated that 250–350 L of ethanol could be produced from each metric ton of dry CRs. Considering that only 20% of world's rice straw is used for this purpose, this would lead to an annual ethanol production of 40 billion liters, which would be able to replace about 25 billion liters of fossil fuel-based gasoline (Jeffery et al., 2011). As a result, net GHG emissions could be reduced up to a 70 million ton of O_2 equivalent per year; however, large- and small-scale commercial machinery/technologies need to be developed in order to utilize this potential (Farrell et al., 2006; Marris, 2006; Prasertsan and Sajjakulnukit, 2006).

Packing materials: The compaction resistance and resiliency of rice straw make it a very good packing material. It saves the product from damages and gives the good cushioning effect.

Mixing with plastics: The paddy straw can also be used as reinforce material in plastics. A Chinese company has invented straw-based plastic. The plastics are made from rice and wheat talks and can be used in 3D printing, without sacrificing price or performance. The paddy straw is shredded into small pieces of 1.5–2 mm size and mixed with polypropylene, adding silane coupling agent and additives. The mixture is then extruded into granules using a twin screw extruder (Verma, 2014).

4.7 Government Initiatives and Legislative Policy to Stop Paddy and Wheat Straw Burning

India has strong legislation to control the pollution. Eleven major laws exist to control pollution in India and many forums for their implementation in various ways (Gathala et al., 2001). To prevent the burning of straw, Government invokes Section 144 of the code of civil procedure to ban the burning of paddy, but it is hardly implemented, and there is petite effort to sensitize farmers on the concern. As per the direction from

NGT (National Green Tribunal), a suitable coercive and penal action should be taken by the state government, including the launching of prosecution if persistent residue burnt by the defaulters (Kumar et al., 2015a). On the direction of High Court of Punjab and Haryana regarding the imposition of burning residue, the Government of Punjab imposed mandatory of SMS attachment on all the existing and new production of combine harvester.

In India, NGT forbids the tradition of straw and stubble flaming in highly polluted city New Delhi as well as its adjacent four states (Haryana, Rajasthan, Punjab, and Uttar Pradesh). Due to the mounting crisis combined with crop dregs burning in these states, numerous initiatives for its appropriate management have been approached. Government organizations and research centers are encouraging alternate utilization of straw and stubbles in lieu of blazing such as utilize farm dregs as animal fodder; utilization of stubbles in electricity generation; employment for mushroom farming, for quilt substance in cattle shed; utilization as bio-lubricant; paper and pulp production; biogas generation; and *in situ* amalgamations in soil (Kumar et al., 2015b). Some of the laws are in operation to regulate the pollution, viz., (i) Air Prevention and Control of Pollution Act, 1981; (ii) The Environment Protection Act, 1986; (iii) The Environment (Protection) Rules, 1986; (iv) The National Environment Tribunal Act, 1995; and (v) The National Environment Appellate Authority Act, 1997.

4.8 Conclusions

In this paper, the authors highlighted the issues of CR burning in the state of Punjab, Haryana, and Western Uttar Pradesh. Also, the conditions have been discussed in which the farmers have no option other than burning the CR. The contributions of Punjab and Haryana in the generation of NO_2, SO_2, PM2.5, and PM10 have been highlighted with the impact on the atmospheric environment of Delhi NCR. It has been observed that during post-monsoon in between November 7–15, 2017, the levels of PM2.5 and PM10 were very high. Visibility in entire Delhi NCR becomes very poor only due to the burning of rice straw in Punjab and Haryana during the peak season.

The consequences of the residue burning on the health of the human being including the wildlife, for this reason, have been highlighted. Some of the references have been observed in this study that shows the laboratory test on the health of the people of Patiala in Panjab and the adverse effect of the

CRs burning. It has been observed that the life expectancy of the people of Delhi NCR has also been decreased. Also, a large number of disease and deaths have been observed due to the pollution created by the CR burning. The farmers are forced to burn the residue due to lack of resources, lack of technology, unavailability of the manpower, and short duration between the harvesting of paddy and sowing of wheat in which the farmer cannot clear the farm quickly. Intense farming is one of the important reasons for the CR burning. On the basis of the study, some of the recommendations have been made as:

- The alternative uses of the rice and wheat straws must be popularized in the northwest states of India.
- A large number of power plants based on the CR as a fuel must be installed locally.
- The technology must be developed to convert the CR into the compost within a short duration of time without affecting the environment, i.e., land contamination, water pollution, and air pollution.
- The farmer should be given incentives for the alternative use of the CRs so that the monitory loss can be minimized.
- Farmers should be encouraged for diversification of the crops in place of fixed rotation of paddy and wheat production.
- The rice/wheat harvesters may be designed in such a way that minimum residues are leftover in the field just like cutting manually.
- Industrial application of the rice straw such as paper and pulp manufacturing, cardboard manufacturing, bioethanol, and biogas production must be promoted by the government.
- Strong implementation of the regulations related to prevention of the burning of CRs must be ensured.

References

Agarwal R, Awasthi A, Singh N, Gupta PK, Mittal SK. (2012). Effects of exposure to rice-crop residue burning smoke on pulmonary functions and oxygen saturation level of human beings in Patiala (India). Science of the Total Environment. 429:161–6.

Arora VK, Singh CB, Sidhu AS, Thind SS. (2011). Irrigation, tillage and mulching effects on soybean yield and water productivity in relation to soil texture. Agric. Water Management. 98:563–8.

Awasthi A, Singh N, Mittal S, Gupta PK, Agarwal R. (2010). Effects of agriculture crop residue burning on children and young on PFTs in North West India. Science of the Total Environment. 408(20):4440–5.

Beri V, Sidhu BS, Bhat AK, Singh BP. (1992). Nutrient balance and soil properties as affected by management of crop residues. In: Bajwaet MS, editor. Nutrient Management for Sustained Productivity. Proceedings of International Symposium, vol. II. Ludhiana, India: Department of Soil, Punjab Agricultural University; pp. 133–5.

Brar SS, Kumar S, Narang RS. (2000). Effect of moisture regime and nitrogen on decomposition of combine harvested rice residue and performance of succeeding wheat in rice-wheat system in Punjab. Indian Journal of Agronomy. 45:458–62.

Chauhan S. (2011). Biomass resources assessment for power generation: a case study from Haryana State, India. Biomass-Bioenergy. 34(9):1300–8.

Chauhan S. (2012). District wise agriculture biomass resource assessment for power generation: a case study from an Indian state, Punjab. Biomass-Bioenergy. 37:205–12.

Choudhary M, Dhanda S, Kapoor S, Soni G. (2009). Lignocellulolytic enzyme activities and substrate degradation by Volvariellavolvacea, The paddy straw mushroom/Chinese mushroom. Indian Journal of Agricultural Research. 43(3):223–6.

CPCB. (2011). Air Quality Monitoring, Emission Inventory and Source Apportionment Study for Indian Cities. Retrieved from http://cpcb.nic.in/FinalNationalSummary.pdf

CPCB. (December 2015). Air pollution in Delhi: an analytical study. Retrieved from http://www.cpcbenvis.nic.in/envis_newsletter/Air%20Quality%20of%20Delhi.pdf

CPCB. (December 2017). Ambient Air Quality Data of Delhi-NCR. Retrieved from http://cpcb.nic.in/openpdffile.php?id=TGF0ZXN0RmlsZS8xODNfMTUxNTU4ODM1OF9tZWRpYXBob3RvODM5LnBkZg==)

CPCBCCR. (December 2017) Central control room for air quality management—Delhi NCR. Retrieved from https://app.cpcbccr.com/ccr/#/dashboard-emergency-stats

Dhaliwal HS, Singh RP, Kaur H. (2011). Financial assessment of happy Seeder. Conservation Agriculture News Letters. 17:2011.

DownToEarth. (October 12, 2017). Intensity and extent of crop burning increases, and so is the level of air pollution. Retrieved from http://

www.downtoearth.org.in/news/pollution-scare-back-in-delhi-as-neighbo
urs-fail-to-prevent-crop-residue-burning-58867

Erenstein O. (2011). Cropping systems and crop residue management in the
Trans-Gangetic Plains: issues and challenges for conservation agriculture from village surveys. Agricultural Systems. 104(1):54–62.

Farrell AE, Plevin RJ, Turner BT, Jones AD, O'Hare M, Kammen DM.
(2006). Ethanol can contribute to energy and environmental goals.
Science. 311:506–8.

Gadde B, Bonnet S, Menke C, Garivait S. (2009). Air pollutant emissions
from rice straw open field burning in India, Thailand and the Philippines.
Environmental Pollution. 157:1554–8.

Galanter M, Levy H, Carmichael GR. (2000). Impacts of biomass burning
on tropospheric CO, NOx, and O_3. Journal of Geophysical Research.
105(D5):6633–53.

Gangwar KS, Singh KK, Sharma SK, Tomar OK. (2006). Alternative tillage
and crop residue management in wheat after rice in sandy loam soils of
Indo-Gangetic plains. Soil Research. 88:242–52.

Gathala M, Ladha JK, Balyan V, Saharawat YS, Kumar V, Sharma PK, et al.
(2001). Tillage and crop establishment affects sustainability of South
Asian rice–wheat system. American Society of Agronomy Journal.
103:1–10.

Ghude SD, Chate DM, Jena C, Beig G, Kumar R, Barth MC, et al.
(2016). Premature mortality in India due to PM2.5 and ozone exposure.
Geophysical Research Letters 43:4650–8.

Gupta PK, Sahai S, Singh N, Dixit CK, Singh DP, Sharma C. (2004). Residue
burning in rice-wheat cropping system: causes and implications. Current
Science India 87(12):1713–5.

Gupta R. (2011). Causes of emissions from agricultural residue burning
in North-West India; Evaluation of a technology policy response.
South Asian Network for Development and Environmental Economics
(SANDEE) PO Box 8975, EPC 1056, Kathmandu, Nepal.

Gupta S, Agarwal R, Mittal SK. (2016). Respiratory health concerns in
children at some strategic locations from high PM levels during crop
residue burning episodes. Atmospheric Environment. 137:127–34.

Hays MD, Fine PM, Geron CD, Kleeman MJ, Gullett BK. (2005). Open
burning of agricultural biomass; physical and chemical properties of
particle-phase emissions. Atmospheric Environment. 39:6747–64.

Hiloidhari M, Das D, Baruah DC. (2014). Bioenergy potential from crop residue biomass in India. Renewable and Sustainable Energy Reviews. 32:504–12.

Hindustan Times. (October 24, 2016). Delhi chokes on smoke from neighbouring states. Retrieved from http://www.hindustantimes.com/delhi-news/delhi-chokes-on-smoke-from-neighbouring-states/story-zAkXkfll e5MoUXLNYfZa0H.html

IARI. (2012). Crop residues management with conservation agriculture: potential, constraints and policy needs. Indian Agricultural Research Institute, New Delhi, pp. 7–32.

IBT. (November 10, 2017). Delhi air pollution: NASA images show stubble burning in neighbouring states worsened air quality. Retrieved from http://www.ibtimes.co.in/delhi-air-pollution-nasa-images-show-stubble-burning-neighbouring-states-worsened-air-quality-748862

Ioannidou O, Zabaniotou A. (2007). Agricultural residues as precursors for activated carbon production—a review. Renewable and Sustainable Energy Reviews. 11(9):1966–2005.

Jain AK. (2016). Residue Crop (Paddy Straw) Burning Shrouds NCR. In: Proceedings of the 2nd international seminar on utilization of non-conventional energy sources for sustainable development of rural areas, ISNCESR. Parthivi College of Engineering & Management, C.S.V.T. University, Bhilai, Chhattisgarh, India. 16th, 17th and 18th March.

Jain N, Pathak H, Bhatia A. (2014). Sustainable management of crop residues in India. Current Advances in Agricultural Science. 6:1–9.

Jat ML, Gathala MK, Ladha JK, Saharawat YS, Jat AS, Kumar V, et al. (2009). Evaluation of precision land leveling and double zero-till systems in the rice-wheat rotation: water use, productivity, profitability and soil physical properties. Soil Research. 105:112–21.

Jat ML, Kamboj BR, Sidhu HS, Singh M, Bana A, Bishnoi DK, et al. (2013). Operational manual for turbo happy seeder—Technology for managing crop residues with environmental stewardship. CIMMYT.

Jat ML, Saharawat YS, Gupta R. (2011). Conservation agriculture in cereal systems of South Asia: nutrient management perspectives. Karnataka Journal of Agricultural Science. 24:100–05.

Jeffery S, Verheijena FGA, van der Veldea M, Bastos AC. (2011). A quantitative review of the effects of biochar application to soils on crop productivity using meta-analysis. Agriculture, Ecosystems and Environment. 144:175–87.

Kharol SK, Badarinath KVS, Sharma AR, Mahalakshmi DV, Singh D, Prasad VK. (2012). Black carbon aerosol variations over Patiala city, Punjab, India—a study during agriculture crop residue burning period using ground measurements and satellite data. Journal of Atmospheric and Solar-Terrestrial Physics. 84:45–51.

Kim S, Dale BE. (2004). Global potential bioethanol production from wasted crops and crop residues. Biomass and Bioenergy. 26(4):361–75.

Koopman A, Koppejan J. (1997). Agricultural and forest residues generation, utilization and availability. Paper presented at the regional consultation on modern applications of biomass energy, Kuala Lumpur, Malaysia.

Kumar P, Kumar S, Joshi L. (2015a). Socioeconomic and environmental implications of agricultural residue burning: a Case Study of Punjab, India. Springer Open, p. 144.

Kumar P, Kumar S, Joshi L. (2015b). Alternative Uses of Crop Stubble, In Socioeconomic and Environmental Implications of Agricultural Residue Burning (pp. 69–89). Springer India.

Kumar V, Saharawat YS, Gathala MK, Jat AS, Singha SK, Chaudhary N, et al. (2013). Effect of different tillage and seeding methods on energy use efficiency and productivity of wheat in the Indo-Gangetic Plains. Field Crops Research. 142:1–8.

Kutcher H, Malhi SS. (2010) Residue burning and tillage effects on diseases and yield of barley (Hordeumvulgare) and canola (Brassica napus). Soil Research. 109(2):153–60.

Lefroy RD, Chaitep W, Blair GJ. (1994). Release of sulphur from rice residue under flooded and non-flooded soil conditions. Australian Journal of Agricultural Research. 45:657–67.

Lehmann J, Gaunt J, Rondo M. (2006). Bio-char sequestration in terrestrial ecosystems—a review. Mitigation and Adaption Strategies for Global Change. 11:403–27.

Lelieveld J, Evans JS, Fnais M, Giannadaki D, Pozzer A. (2015). The contribution of outdoor air pollution sources to premature mortality on a global scale. Nature. 525:367–71.

Liu T, Marlier ME, DeFries RS, Westervelt DM, Xia KR, Fiore AM, et al. (2018). Seasonal impact of regional outdoor biomass burning on air pollution in three Indian cities: Delhi, Bengaluru, and Pune. Atmospheric Environment. 172:83–92.

Lohan SK, Dixit J, Kumar R, Pandey Y, Khan J, Ishaq M, et al. (2015b). Biogas: a boon for sustainable energy development in India's cold climate. Renewable and Sustainable Energy Reviews. 43:95–101.

Lohan SK, Dixit J, Modasir S, Ishaq M. (2012). Resource potential and scope of utilization of renewable energy in Jammu and Kashmir, India. Renewable Energy. 39:24–9.

Lohan SK, Jat HS, Yadav AK, Sidhu HS, Jat ML, Choudhary M, et al. (2018). Burning issues of paddy residue management in north-west states of India. Renewable and Sustainable Energy Reviews. 81:693–706.

Lohan SK, Narang MK, Manes GS, Grover N. (2015a). Farm power availability for sustainable agriculture development in Punjab state of India. Agricultural Engineering International: CIGR Journal. 17(3):196–207.

Lohan SK, Sharma S. (2012). Present status of renewable energy resources in Jammu and Kashmir State of India. Renewable and Sustainable Energy Reviews. 16:3251–8.

Long W, Tate R, Neuman M, Manfreda J, Becker A, Anthonisen N. (1998). Respiratory symptoms in a susceptible population due to burning of agricultural residue. Chest. 113(2):351.

Marris E. (2006). Black is the new green. Nature. 442:624–6.

Mehta CR, Sharma S, Nair R, Singh KP. (2013). Impact of crop residue burning on environment and human health. Indian Farming. 63(4):24–35.

Mishra AK, Shibata T. (2012). Synergistic analyses of optical and microphysical properties of agricultural crop residue burning aerosols over the Indo-Gangetic Basin (IGB). Atmospheric Environment. 57:205–18.

Mittal SK, Singh N, Agarwal N, Awasthi A, Gupta PK. (2009). Ambient air quality during wheat and rice crop stubble burning episodes in Patiala. Atmospheric Environment. 43(2):238–44.

MNRE. (2009). Ministry of New and Renewable Energy Resources, Govt. of India, New Delhi Retrieved from www.mnre.gov.in/biomassrsources (accessed December 2012).

NAAS. (2012). Management of crop residues in the context of conservation agriculture. Policy Paper No. 58, National Academy of Agricultural Sciences, New Delhi, 58:12.

Pathak H, Bhatia A, Jain N. (2010). Inventory of greenhouse gas emission from agriculture. Report submitted to Ministry of Environment and Forests, Govt. of India.

Pathak H, Saharawat YS, Gathala M, Ladha JK. (2011). Impact of resource-conserving technologies on productivity and greenhouse gas emissions in the rice-wheat system. Greenhouse Gases: Science and Technology. 1(3):261–77.

Pathak H, Singh R, Bhatia A, Jain N. (2006). Recycling of rice straw to improve crop yield and soil fertility and reduce atmospheric pollution. Paddy Water Environment. 4(2):111–7.

Peter JK, Masih H, Kumar Y, Singh AK, Chaturvedi S. (2014). Organophosphate pesticide (Methyl Parathion) degrading bacteria isolated from rhizospheric soil of selected plants and optimization of growth conditions for degradation. International Journal of Research. 5:1–13.

Prasertsan S, Sajjakulnukit B. (2006). Biomass and biogas energy in Thailand: potential, opportunity and barriers. Renewable Energy. 5: 599–610.

Ramanathan V, Carmichael G. (2008). Global and regional climate changes due to black carbon. Nature Geoscience. 1(4):221–7.

Roy P, Kaur M. (2016). Economic analysis of selected paddy straw management techniques in Punjab and West Bengal. Indian Journal of Economics and Development. 12:467–71.

Sandhu K. (October 10, 2016). Fields on Fire: Burning Paddy Straw. The Indian Express, Kurukshetra, India.

Sangeet KR. (2016). Crop residue generation and management in Punjab state. Indian Journal of Economics and Development. 12(1):477–83.

Sekhon NK, Singh CB, Sidhu AS, Thind SS, Hira GS, Khurana DS. (2008). Effect of mulching, irrigation and fertilizer nitrogen levels on soil hydrothermal regime, water use and yield of hybrid chilli. Archieves of Agronomy and Soil Science. 54:163–74.

Sharma D, Srivastava AK, Ram K, Singh A, Singh D. (2017). Temporal variability in aerosol characteristics and its radiative properties over Patiala, northwestern part of India: impact of agricultural biomass burning emissions. Environmental Pollution. 231:1030–41.

Sidhu BS, Beri V, Gosal SK. (1995). Soil microbial health as affected by crop residue management. In: Proceedings of National Symposium on Developments in Soil Science, Ludhiana, India pp. 45–6. New Delhi, India; Indian Society of Soil Science; 2–5, November 1995.

Sidhu BS, Beri V. (2005). Experience with managing rice residues in intensive rice-wheat cropping system in Punjab. In: Abrol IP, Gupta RK, Malik RK. (eds.) Conservation Agriculture: Status and Prospects, pp. 55–63. Centre for Advancement of Sustainable Agriculture, National Agriculture Science Centre, New Delhi.

Sidhu HS, Singh S, Singh Y, Blackwell J, Lohan SK, Humphreys E, et al. (2015). Development and evaluation of the turbo happy seeder for sowing wheat into heavy rice residues in NW India. Field Crops Research. 184:201–12.

Sindhu R, Gnansounou E, Binod P, Pandey A. (2016). Bioconversion of sugarcane crop residue for value added products—an overview. Renewable Energy. 98:203–15.

Singh CP, Panigrahy S. (2011). Characterisation of residue burning from agricultural system in India using space based observations. Journal of the Indian Society of Remote Sensing. 39(3):423–9.

Singh J, Panesar BS, Sharma SK. (2008). Energy potential through agricultural biomass using geographical information system—a case study of Punjab. Biomass Bioenergy. 32:301–7.

Singh KK, Lohan SK, Jat AS, Rani T. (2006). New technologies of growing rice for higher production. Research Crops. 7(2):369–71.

Singh M, Sidhu HS, Singh Y, Blackwell J. (2011). Effect of rice straw management on crop yields and soil health in rice-wheat system. Conservation Agriculture News PACA. p. 18.

Singh RP, Kaskaoutis DG. (2014). Crop residue burning: a threat to South Asian air quality. EOS 95, 333–40.

Singh Y, Singh D, Tripathi RP (1996) Crop Residue Management in Rice-Wheat Cropping System. In: Abstracts of Poster Sessions 2nd International Crop Science Congress, p. 43. National Academy of Agricultural Sciences New Delhi, India, p. 26.

Singh Y, Gupta RK, Singh J, Singh G, Ladha JK. (2010). Placement effects on paddy residue decomposition and nutrient dynamics on two soil types during wheat cropping in paddy-wheat system in north western India. Nutrient Cycle in Agroecosystems. 88:471–80.

Singh Y, Humphreys E, Kukal SS, Singh B, Kaur A, Thaman S, et al. (2009). Crop performance in permanent raised bed rice–wheat cropping system in Punjab India. Field Crops Research. 110:1–20.

Singh Y, Sidhu HS. (2014). Management of cereal crop residues for sustainable rice-wheat production system in the Indo-Gangetic plains of India. Proceedings of the Indian National Science Academy. 80(1):95–114.

Singh Y, Singh B, Ladha JK, Khind CS, Khera TS, Bueno CS. (2004). Management effects on residue decomposition, crop production and soil fertility in a rice-wheat rotation in India. Soil Science Society of America Journal. 68:320–6.

Sood J. (2016). Not a waste until wasted. Down to Earth. Retrieved from http://www.downtoearth.org.in/content/not-waste-until-wasted (2013). Accessed on 15 Sept 2016.

Srinivas B, Rastogi N, Sarin MM, Singh A, Singh D. (2016). Mass absorption efficiency of light absorbing organic aerosols from source region of paddy-residue burning emissions in the Indo-Gangetic Plain. Atmospheric Environment. 125:360–70.

Streets DG, Bond TC, Carmichael GR, Fernandes SD, Fu Q, He D, et al. (2003). An inventory of gaseous and primary aerosol emissions in Asia in the year 2000. Journal of Geophysical Research: Atmospheres. 108(D21).

Teodoro C, Mendoza BC. (2016). A review of sustainability challenges of biomass for energy: focus in the Philippines. Journal of Agricultural Technology. 12(2):281–310.

Treets DG, Yarber KF, Woo JH, Carmichael GR. (2003). Biomass burning in Asia: annual and seasonal estimates and atmospheric emissions. Global Biogeochemical Cycles. 17:1099.

Vadrevu KP, Badarinath KVS, Anuradha E. (2008). Spatial patterns in vegetation fires in the Indian region. Environmental Monitoring and Assessment. 147:1–13.

Vadrevu KP, Ellicott E, Badarinath K. (2011). MODIS derived fire characteristics and aerosol optical depth variations during the agricultural residue burning season, North India. Environmental Pollution. 159(6):1560–9.

Vadrevu KP, Giglio L, Justice C. (2013). Satellite based analysis of fire–carbon monoxide relationships from forest and agricultural residue burning (2003–2011). Atmospheric Environment. 64:179–91.

Venkataraman C, Habib G, Kadamba D, Shrivastava M, Leon JF, Crouzille B, et al. (2006). Emissions from open biomass burning in India: integrating the inventory approach with high-resolution Moderate Resolution Imaging Spectroradiometer (MODIS) active-fire and land cover data. Global Biogeochemical Cycles. 20(2):1–12.

Verma SS. (2014). Technologies for stubble use. Journal of Agriculture and Life Sciences. 1(1):106–10.

Vijayakumar K, Safai PD, Devara PCS, Rao SVB, Jayasankar CK. (2016). Effects of agriculture crop residue burning on aerosol properties and long-range transport over northern India: A study using satellite data and model simulations. Atmospheric Research. 178:155–63.

WHO. (2016). Ambient Air Pollution: a Global Assessment of Exposure and Burden of Disease.

Yadav GS, Lal R, Meena RS, Datta M, Babu S, Das A, et al. (2017). Energy budgeting for designing sustainable and environmentally clean/safer cropping systems for rainfed rice fallow lands in India. Journal of Cleaner Production. 158:29–37.

Yadav M, Sharma MP, Prawasi R, Khichi R, Kumar P, Mandal VP, et al. (2014). Estimation of wheat/rice residue burning areas in major districts of Haryana, India, using remote sensing data. Journal of the Indian Society of Remote Sensing. 42(2):343–52.

Yadvinder S, Sidhu HS, Khanna PK, Kapoor S, Jain AK, Singh AK, et al. (2010). Options for effective utilization of crop residues. Directorate of Research Punjab Agricultural University, Ludhiana, India.

Yang SS, Liu CM, Lai CM, Liu YL. (2003). Estimation of methane and nitrous oxide emission from paddy fields and uplands during 1990–2000 in Taiwan. Chemosphere. 52:1295–305.

Zhang H, Hu D, Chen J, Ye X, Wang SX, Hao J, et al. (2011). Particle size distribution and polycyclic aromatic hydrocarbons emissions from agricultural crop residue burning. Environmental Science and Technology. 45:5477–82.

5

Waste Electrical and Electronic Equipments, Where Do We Stand and Where to Go: An Indian Scenario

Sunil S. Suresh*, Omdeo K. Gohatre, Kulasekaran Jaidev, Smita Mohanty, and Sanjay K. Nayak

Laboratory for Advanced Research in Polymeric Materials (LARPM)
Central Institute of Plastics Engineering and Technology (CIPET)
Bhubaneswar, Odisha, India
E-mail: sunilssuresh@gmail.com
*Corresponding Author

The present chapter critically analyzes the present scenario of waste electrical and electronic equipments (WEEE) generation, isolation, and recycling in India. Together with the population boom and the technological changes, usage of electrical and electronic equipments (EEE) has been escalated in India. Simultaneously, a significant amount of waste components are generating from products such as mobile phones, computers, and various home appliances. Current studies are indicating that WEEE generation in India depends upon domestic production and imports from the developed countries. This chapter significantly discusses the current WEEE generation of India in comparison with world scenario, various WEEE producers, and recycling procedures adopted. Besides, environmental aspects of WEEE and various research initiatives conducted by the Indian researchers are also reviewed based on the previous work analysis.

5.1 Introduction

In recent years, EEE market is enduring an exponential growth around the globe due to the enhancement in the usage of products like mobile phone, televisions, and computers including laptops, desktops, and notebooks. The electronic equipments like computers and mobile phones have become an integral part of humankind in their day-to-day life activity since those can provide better and faster exchange of information, efficiency in computing and entertainment purposes. Simultaneously, developments in the field of information and communication technology, population boom, and changes in the life style also result in the catastrophic rise in the utilization of EEE products by mankind. Moreover, the competition between the different manufactures of EEE, availability of the global market and consumer behavior to the gadgets, forces the manufacturers to reduce the cost of gadgets especially in the case of mobile phones and computer products. However, this condition accelerates the usage of electronic products (Osibanjo and Nnorom, 2007; Pérez-Belis et al., 2015; Ikhlayel, 2018; Suresh et al., 2018a).

Along with the increase in the number of electronic products, waste components developed from such products are also proliferated proportionally. WEEE or E-waste is the common terms which are used for indicating the waste materials or the products generated from the EoL EEE. Different organizations and researchers have been attempted to define WEEE in various definitions, as per directive put forwarded by EU, "WEEE means electrical or electronic equipments that is waste, including all components, sub-assemblies, and consumables that are part of product at the time of discarding waste" (Directive EC, 2012). Whereas an international organization step initiative (StEP-solving E-waste problem) indicates that "E-waste is a term used to cover items of all type of EEEs and its parts that have been discarded by the owner as waste without the intention of reuse" (StEP Initiative, 2014). In another terms, WEEEs are any electrical and electronic products (the equipments working under electrical current or electromagnetic field) discarded by its original consumers or manufactures after its valuable usage period, which includes various appliances such as washing machines, refrigerators, cooking ovens, air conditioners, and dryers, which are commonly considered as large household appliances; moreover, it also holds waste products from electronic gadgets like mobile phones, laptops, notepads, personal computers, and copying equipments those are exclusively used in IT and telecommunication filed.

5.2 Categories of WEEE

According to the EU directives, major EEE products are classified into 10 different categories as shown in Table 5.1 (Directive, 2003). Besides, Table 5.1 also conveys percentage of WEEE composition recovered in the EU (Ongondo et al., 2011). This directive also advices to consider the WEEE

Table 5.1 Categories of WEEE according to the EU directives (Directive, 2003; Ongondo et al., 2011)

Category Number	Category	Example of Products	Label	WEE Recovered in EU (%)
1	Large household appliances	Refrigerators, washing machines, freezers, cookers, dishwashers, microwave ovens, heaters, radiators, etc.	Large HH	49.07
2	Small household appliances	Vacuum cleaners, watches, grinders, toasters, fryers, kettles, hair cares, and trimmers	Small HH	7.01
3	IT and telecommunication equipments	Personal computers, flat screen monitors, copiers, faxes printers, telephone, and mice	ICT	16.27
4	Consumer equipment	Televisions, radios, video cameras, speakers, musical instruments VCRs, and CD and DVD players	CE	21.10
5	Lighting equipments	CFLs, incandescent lamps, and sodium lamp (however, this category rule out household luminaries	Lighting	2.40

(Continued)

Table 5.1   (Continued)

Category Number	Category	Example of Products	Label	WEE Recovered in EU (%)
6	Electrical and electronic (E&E) tools	Drills, milling, sewing machines, grinding, cutting machine, lawnmowers, and trimmers	E&E tools	3.52
7	Toys, leisure, and sports equipments	Exercising machines like tread mills, stretching machine, etc. Also includes various game equipments like computer games, video games, play stations, etc.	Toys	0.11
8	Medical devices	Medical devices, specialist waste (however, this category excludes implanted and infected products)	Medical equipments	0.12
9	Monitoring and control (M&C) instruments	Flow gages and various measuring equipments	M&C	0.21
10	Automatic dispensers	Hot and cold drinks dispersers, money dispensers (ATMs), hot and cold bottle dispersers	Dispensers	0.18

as a special waste and it must be collected and treated in a unique way for further processing and recycling. Among the categories, 1–4 is account for maximum share of WEEE generation around the globe. Currently, approximately 95 wt% of the waste generated falls under these categories (Directive, 2003), in which, large household and IT and communication appliances contribute around 42 and 33.9 wt%, respectively (Baldé et al., 2015); however, the amount of waste produced from the IT sector is gradually increasing day by day due to the growth and development of IT business.

5.3 Global Trends in Generation of WEEE

Globally, the generation of WEEE is continuously progressing at an alarming rate for the last several decades due to various reasons associated with advancement in the science and technology, reduction of lifecycle of EEE appliances, rapid growth of population, and changes in the economic growth. The day-to-day changes in the technologies such as software and hardware of the EEE products such as computers and mobile phones put pressure on the consumers to upgrade their gadgets to a new one even if it is in good working condition. Such consistent advent of new smart technologies resulted in rapid obsolescence of various electronic items in the last two decades. For example, the "digital transition" revolution of televisions into HD, which began in 2009, has caused large numbers of analog televisions to become obsolete even if it is functioning (Premalatha et al., 2014). Similarly, the displays used in the televisions and personal computers have changed from CRT to LCDs for sharper and clear images within a thin frame; as a consequence, huge amounts of CRT screen became obsolete (Socolof et al., 2005; Suresh et al., 2017a).

Besides, the obsolesce ratio of the EEE is increasing in the present context due to reduction in the lifespan of electronic devices; consequently, huge quantities of EEE components are coming to the waste yard. Various lifecycle studies over EEE products indicate that the lifespan of EEE is reducing annually, for instance, an average life of new generation computers is 3 years, which is 3–4 years less than the old generation computers (Dwivedy and Mittal, 2010a). Similarly, in the case of mobile phones, the life period is 1–2 years for the developed and developing countries. However, large household appliances like air conditioning units, washing machines, refrigerators, and electric heaters have a life duration of 12, 8, 10, and 20 years, respectively (Betts, 2008). Comparatively larger equipments have possessed a higher lifespan than the smaller equipments such as mobile phones and computers. As an outcome, a greater volume of such components either come to the recycling yard or export to developing countries such as India. Furthermore, the components generated during the repairing of EEE are also piling up extensively; however, most of those materials are ended up in municipal waste stream without any further consideration.

Globally, the quantity of abandoned EEE products is increasing from time to time; the quantity of WEEE in 2005 was 20 million tons, which consist of 8% of total municipal waste stream (Widmer et al., 2005). However, this digit has increased to 41.8 million metric ton (Mt) in the year of 2014, which includes 6.3 Mt of displays from various EEEs and 3.0 Mt of IT equipments such as mobile phones, personal computers, and so on. From this 41.8 Mt, only 6.5 Mt was recycled by the formal recyclers in proper manners, which are considered as 2.71% of the total WEEE produced (Baldé et al., 2015). The forecasting studies done by the various researchers indicate that the quantity of wastes will be increased to 50 Mt by the year of 2018 (Widmer et al., 2005; Baldé et al., 2015).

The developing and industrialized countries are generating a maximum amount of WEEE around the globe. In addition to the manufacturing of EEE, most of the countries such as United States of America (USA), China, and Russia are importing their WEEE and its components (after recovering valuable materials or as such) to poor or developing nations. The USA alone produced 2.63 million tons of WEEE in 2005, while in the year of 2009, the total E-waste generated in USA was increased into 3.19 million tons, which is 21.3% more than that of 2005. From the 3.19 million tons of WEEE, only 17.7% was recycled, or rest was landfilled, incinerated, or imported (Robinson, 2009; Electronics Take Back Coalition, 2012). However, in 2014, USA generated 11.7 Mt of E-waste, which can be represented as 12.2 kg/person (Baldé et al., 2015). Other than the USA, China is also generating a sizeable fraction of WEEE; nevertheless, China has a large number of informal recycling centers and those are doing primitive recycling techniques such as burning, melting, and acid recovery of the valuable metals, though such informal sectors do not focus on to the components such as plastics present in the WEEE. During the period of 2005, China has generated 2.5 million tons of E-waste and this figure was changed into 50 million units in the year of 2010 (Liu et al., 2006; Song et al., 2016). Similarly, Japan becomes the leading manufacturer of EEE products such as electronic items cameras and IT equipments even from the period of 1950s. In 2005, approximately 860 k tons of WEEE have generated in Japan from different home appliances, from it 80% of the products were recycled by manufactures. Conversely, remaining 20% were resold in the market of Japan or imported to other countries. Japan has initiated there WEEE recycle programs from the early periods of 1970s itself. Moreover, from the year of 2001, Japan commenced home appliances recycling law to reduce the burden of WEEE products particularly refrigerators, televisions, washing machines, and air conditioners over the environment. Further, this law was tailored on 2008, in order to incorporate LCD and plasma TVs and cloth dryers into the aforesaid category. This law is pressurized manufactures to take back their respective products from the consumer's inorder to recover or recycle components from it (Aizawa et al., 2008; Yoshida and Terazono, 2010; Baldé et al., 2015). WEEE is considered as the fastest growing waste stream in Brazil; it is estimated that about 14.12 Mt of WEEE products were generated in the year of 2014, representing 5.3–7.0 kg per capita/year. This study also signifies that half of the waste products produced in the Brazil are contributed by large HH appliances (Ghosh et al., 2016). In addition to it, country like Russia continues to generate E-waste. In 2014, around 1.478 Mt of E-waste was generated in Russia, which is composed of abandoned computers, televisions, mobile phones, and so on. Along with it, production of WEEE in various countries during the period of 2012–2014 is represented in Figure 5.1 (StEP Initiative, 2014; Baldé et al., 2015; Tansel, 2017). Moreover, a recent report by United Nation University (Baldé et al., 2015) indicates that the maximum amount of E-waste was generated by the Asian region (16 Mt) in the year of 2014, on considering per personal quantity of WEEE generated which is equivalent to 3.7 kg/person. In most of the countries, formal recycling centers

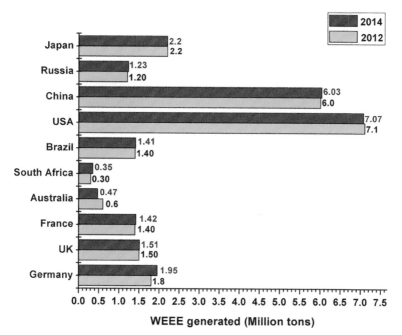

Figure 5.1 Comparison of WEEE generated in different countries for the years of 2012 and 2014 (StEP Initiative, 2014; Baldé et al., 2015; Tansel, 2017).

play an important role in reduction of WEEE generated; they are adopting technologically advanced setup to resolve the generated WEEE.

5.4 The Digital Revolution and Growth of EEE in India

In India, the initial phase of economical boom has started during the mid-1990s. In this period, large numbers of native industrial groups have been enthused into the manufacturing of EEE products. Along with it, various foreign-based EEE companies are also initiated production or import of their EEE products into the Indian market. Such situations pressurized different manufactures to reduce the cost of their products with incorporating improved technical features. Similarly, quality, cost effectiveness, improved purchasing capacity of individuals, and competition between the various EEE producers significantly resulted in the technical boom in the EEE sector, especially in the home appliances. Subsequently, in September 2014, government of India has launched a new policy called "make in India" which encourages various manufactures to develop their own products (including various EEE products) in India (Ministry of Electronics and Information Technology, 2016). Accordingly, various multinational companies like Foxconn, Lenovo, Micromax, Samsung, Xiaomi, etc., have been started manufacturing or initiated investment for production of their goods in India (Ministry of Electronics and Information Technology, 2016; Tansel, 2017). This movement is also facilitated transformation of India into a "global manufacturing hub."

Even if Asian countries are dominating the electronic market around the world, India has only small share of mere 0.7% of the global EEE production (Ministry of Electronics and Information Technology, 2016). However, this value will be overcome in the near future due to the various sensible programs and policies adopted by the government of India. For last several years, ministry of electronics and IT (MeitY) under the government of India has been regularly monitoring the growth of EEEs in India and the results are indicating that EEE productions are incessantly growing (Ministry of Electronics and Information Technology, 2016–2018). Similarly, as per MeitY new report in 2018, the production profile of EEEs is indicated in Table 5.2 (Ministry of Electronics and Information Technology, 2017). Table 5.2 signifies that production capacity of Indian EEE sector is growing annually with respect to its previous years.

In the same way, during the mid-1990s, there is a vital change happened in the ICT sector due to the development in internet technology and invention of mobile phone and portable computers (Ministry of Electronics and

Table 5.2 Annual growth of electronic sector with respect to market value in India (rupees in billion)

Electronic Items	2013–2014	2014–2015	2015–2016	2016–2017	2017–2018
Consumer electronics*	475.99	558.06	557.65	647.42	735.24
Industrial electronics	336.00	393.74	450.83	622.14	690.57
Computer hardware	174.84	186.91	198.85	203.82	214.01
Mobile phones	266.50	189.00	540.00	900.00	1320.00
Strategic electronics	138.00	157.00	180.55	207.60	235.62
Electronic components	321.02	397.23	453.83	520.99	583.51

Note: *Consumer electronics include various home EEE products such as refrigerators, washing machines, air conditioners, and micro ovens.

Information Technology, 2016; Borthakur and Govind, 2018). The changes in the EEE and ICT technologies are also reflected in amplification of various EEE products in the individuals and households. Among various EEE goods, ICT products such as mobile phones, computers, and laptops maximum share as compared with other appliances due to its wide usage and applicability. Moreover, introduction of smart phones has resulted in the mammoth growth in mobile users, due to connectivity, faster internet, and endless entertainment options. Further, India is in the second place while considering smart mobile phone usage around the world after China.

As per TRAI, approximately 113.26 million new mobile phone subscribers were added in the year of 2008 alone (TRAI, 2009). The usage of conventional desktops is minimized for the individual purposes in recent times due to the introduction of more efficient handheld portable devices. However, utilization of the computers and its related components is increased in commercial, industrial sectors, ICT, and government and non-governmental organizations those are expected to grow at a steady rate. Furthermore, advanced internet options triggered use of mobile phones and computers by the individuals. A study conducted by the IAMAI and IMRB indicated that internet users are continuously escalating in India; up to December 2017, India witnessed approximately 481 million internet users; however, by the end of June 2018; this figure forecasted into 500 million (IAMAI and KANTAR IMRB Report, 2016). Moreover, in order to improve the digital infrastructure and empowering ICT sector, the government of

India has launched a plan called "digital India" in July 2015 (Rahul, 2016). Additionally, as a part of digital India program, the government of India promotes cashless economy which includes various payments methods such as mobile phones money transfer, internet banking, debit or credit cards, e-wallets, etc. Among them, payment methods employing mobile phones are found to be fast growing in current situation due to ease of access (Watkins et al., 2012). Moreover, for last several years, there has been a significant improvement in the digital payments in India due to wide usage of mobile phones and internet. For instance, in the 2009–2010 duration, a total of 0.5 billion transactions occurred in India; however, this value shot up into approximately 19.40 billion transactions during the 2017–2018 period (up to March) (Ministry of electronics and information technology, 2018).

Similar to the ICT equipments, consumer electronics likes television, home theaters, and home appliances such as air conditioners, washing machines, and refrigerators are also rising persistently. The televisions with LCD and LED-based flat screens have shown substantial intensification in last several years because of changes in broadcast sector. The production of LCD/LED television was 14.5 million during the period of 2016–2017 but this was turned into 16.0 million on 2017–2018 (up to march) (Ministry of electronics and information technology, 2018). Nevertheless, as a result of technological development in the conventional, CRT-based TVs and DVDs are showing negative growth in recent times.

5.4.1 WEEE Generation—An Indian Scenario

The production and usage of the EEE products have been shown significant improvement in India as mentioned in the previous section. However, such products become obsolete after the completion of its successful life period. Amplified consumption of EEE, day to day changes in the technology, reduction in product quality and life period, economic development, and life style changes of individuals sequentially lead to the generation of huge amount of waste appliances. Various studies point out that penetration of ITC products like computers and mobile phones has huge impact on the production of WEEE in India (Dwivedy and Mittal, 2010a, b, 2012; Yedla, 2016). For example, because of technical changes, a large amount of CRT-based monitors are replaced with higher grade LCD or LED monitors. Conversely, such a situation leads to enhanced augmentation of CRT-based components in recycling yards. Even though various agencies are conducted studies on WEEE flow and generation in India, still there are reliable data on WEEE generation and its flow unavailable or ambiguous.

The quantity estimation of WEEE in India was primarily studied by international recourse group systems south Asia in 2005; according to them, the total generation of WEEE in India was 146,180.7 tons up to 2005 (Wath et al., 2011). Similar to it, CPCB has conducted another survey during the same period and approximated that 1.347 lakh Mt of WEEE products were generated (Borthakur and Sinha, 2013; Awasthi et al., 2016). The variation in the quantity of WEEE is associated inadequate data on accumulation of WEEE due to importing, informal recycle activities and also due to unawareness of customers about the quantity of their obsolete products in their hand. For instance, around 75% of the obsolete EEE products are arrested in store rooms of houses due to unawareness about the recycling procedure (Borthakur and Govind, 2017a). Another report estimated that a total of 332,979 metric tons waste components were generated domestically during the period of 2007, from the products like computers, mobile phones and TVs. Among them, the waste from TV sets has the shown maximum share of 275,000 metric tons with 82.6%, followed by waste computers 56,324 Mt and mobile phones 1655 metric tons. Similarly, a total of 50,000 metric tons of WEEE products are imported from the various countries across the globe during this period. Further, this study confirms that 4.9% of the products were recycled from the total 382,979 metric tons of WEEE products generated (Begum, 2013; Borthakur and Sinha, 2013). Besides, a study by Schluep et al. (2009) pointed out that approximately 823.6 K tons of various home appliances and IT products were released into the Indian market. However, around 439 K tons of waste were liberated from various home appliances and ITC products during the same period. Additionally, Manomaivibool (2009) estimated WEEE production in India is as 0.4 kg per capita during the year of 2007. The weight and quantity of the WEEE products have also significant role in production of waste. For example, the average weight of a waste computer is approximately 29.6 kg which is equivalent to weight of approximately 300 numbers of obsolete mobile phones (Wath et al., 2011). However, a large amount of mobile phones are discarded in per day around the globe. Weight composition of the different WEEE products is represented in Figure 5.2(b) (Wath et al., 2011).

The production and increased demand of mobile phones and computers are forced to enhance the generation of WEEE in India. In the present situation, India is positioning fifth as the WEEE generator nation and the volume of WEEE is increasing gradually with a rate of approximately 25% per annum (Garlapati, 2016). Similarly, data are also indicating that India has generated 1.85 metric tons during the period of 2016 (Garlapati, 2016). Pathak and Srivastava (2017) estimated current WEEE generation and with

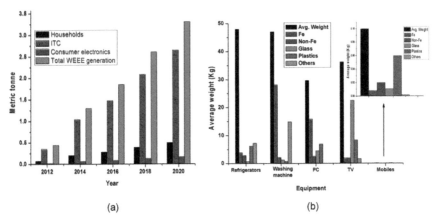

Figure 5.2   (a) WEEE generation from different categories of EEE (note: 2018 and 2020 are forecasted values). (b) Weight composition of the different WEEE products.

the help of mathematical model, they have also forecasted it to the future years (Figure 5.2(a)). It can also be seen from Figure 5.2 that the generation of WEEE from each category is rising annually. The production of WEEE has shown 493.2% augmentation in the year 2018 as on comparison with 2012 scenario. Nonetheless, total generation of WEEE during this period (2012–2018) is merely depended on the ICT products. The ICT products have been shown an increment of 497.1% during 2018 and it will be raised to 645.7% in 2020 (forecasted value) as compared to the 2012 situation. Along with ICT products, household products are also contributed a major share into the WEEE stream during the period of 2012–2018. According to Dwivedy and Mittal (2010a), obsolete computers including desktops, notebooks, and laptops and television together generated 31 million units, which is responsible for the 30% of the total generated WEEE during the period of 2007–2011. At present, the life span of the computers is reduced from 10 years to an average of 2–3 years which is accelerating total waste generation. For instance, the utilization of the computer in India during 2007 was 20 million; conversely, 2.2 million of the computers become out dated in the same period, which is approximately 11% of the total computers. Nowadays, obsolesce ratio of the computers is increasing compared to the period of 1990s. Further, it has been indicated that around 30,000–40,000 computers become out dated annually from different IT sectors of Bangalore alone (Needhidasan et al., 2014; Borthakur and Govind, 2017b). Compared to the other waste products, recycling of computer waste is extremely complicated due to the presence of diverse materials and various hazardous components (Suresh et al., 2017b, 2018b).

Moreover, various states in India are producing different quantities of WEEE into the waste stream (Figure 5.3) which depends upon the population density, technological awareness of the individuals, and IT sectors present in the state. Considering the various regions present in India, southern region is accountable for the largest production of WEEE with a value of 28.70%, followed by the north region (25.10%), and west region (24.10%). Among the different regions, northeast is liable for lower generation of the WEEE product with 2.35%. However, on coming to the specific state, Maharashtra holds 13.9% of total share of WEEE produced in India followed by Tamilnadu (9.3%) and Andhra Pradesh and Telugana (8.3%) together. Along with it, Lakshadweep is considered as the lower generator of WEEE (0.005%) among all the state/union territory in India. This is due to the fact that the penetration of computer equipments and other IT products is very lower in this region (Secretariat, 2011; Gupta and Kumar, 2014). A recent study estimated the WEEE generated in economically important cities in

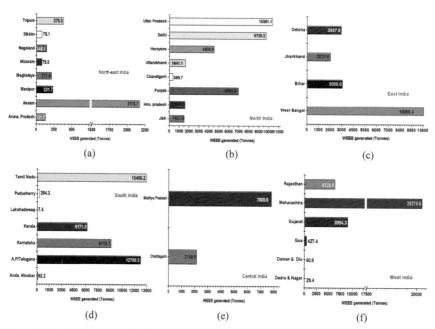

Figure 5.3 WEEE generation in various regions of India [(a) Northeast India, (b) Northern India, (c) Eastern India, (d) Southern India, (e) Central India, and (f) Western India].

India, which has been indicated that Mumbai has the first rank in production of WEEE (12 Mt) followed by Delhi-NCR and Bangalore with 98,000 and 92,000 tons, respectively (Network, 2005).

5.4.2 How much of WEEE Importing into India?

As indicated in the aforementioned section, the quantity of the domestic generation of WEEE is growing due to continuous removal of EEE products after its successive life cycle. In addition to it, India is also suffering from illegal imports of WEEE from other developed countries. The Basel Convention came into effect in 1992, for preventing trans-boundary movement of hazardous waste and its disposal from developed countries to developing or poorer countries (Song et al., 2017). Even if a huge amount of WEEE is still exporting into the Asian countries like India, Pakistan, and China from industrialized countries which is clearly indicating violating of Basel Convention policies. The developed countries like the USA are shipping their 50–80% of the generated domestic WEEE waste into the countries such as China and India without considering its recycling potential (Puckett et al., 2002). The reason behind the trading is related with the economical benefit and minimizing the environmental pollution and manpower during its recycling and recovering of the materials. Moreover, the labor cost of the recovering of materials from WEEE products is comparatively lower in developing countries. For instance, in the USA, recyclers have to contribute 20$ for recycling of a single personal computer; however, the same can be recycled in India with only of 2$; consequently, a significant amount of WEEE has imported into India (Chatterjee and Kumar, 2009).

In current WEEE stream, computer waste is the most trading goods into the Indian recycling market due to its higher penetration rate in the global market. Generally, the seaports like Dubai and Singapore act as a transit points for WEEE from the developing countries; at this place, most of the waste computers are labeled "used working computers" in order to make loopholes in the existing Indian regulations or reduce customs duty (Link, 2012a). In India, hazardous waste rules, 2008, permitted the import of hazardous waste such as WEEE if it is for reuse, recycling, and reprocessing procedure; however, importing of such waste is not allowed for the disposal procedure (Secretariat, 2011). Besides, reports are indicating that other than the domestic production, approximately 50,000 metric tons of WEEE products are importing into the India annually (Pinto, 2008). Further, it was calculated that out of the WEEE imported to India, 80% is coming from the USA and remaining 20% imported from EU (Skinner et al., 2010). Furthermore, more than 70% of the WEEE collected by the various recycling centers in Delhi (capital of India) was previously shipped or discarded from the developed nations (Goodship and Stevels, 2012).

5.4.3 Producers of WEEE in India

The generation of waste from the EEE components is beginning from the manufacture and it is extended up to the final recyclers. The hierarchy of WEEE production in India is proposed by the MoEF under the government of India. According to MoEF, there are three types of WEEE generators in India (Figure 5.4), which are described as below (MoEF, 2008; Borthakur and Sinha, 2013).

5.4.3.1 First level: primary WEEE producers

The primary waste generators are including various manufacturers and merchants of the EEE products. Considerable quantities of the waste products are generated by the manufactures during the production period because of failure in machinery, quality issues, and malfunction of their products. Even if those are not used by the consumers, they are considered as waste and most of the companies are giving it to the recyclers for further component recovery. Besides, some of the EEE products become broken or get damaged while transporting into the manufacture shops. Similarly, some of the consumers are replacing or exchanging their EEE products or its components with retailers during the guarantee or warranty period. The authorized and unauthorized servicing centers of EEE products also play an important role in generation of WEEE. However, such products or components are further moved into the scrap dealers or recyclers (Borthakur and Singh, 2012; Borthakur and Sinha, 2013).

5.4.3.2 Second level: secondary WEEE producers

In the same way, various public and private sectors, governmental organizations, and individual households are regarded as the secondary level WEEE creators. In which, IT and its associated sectors are the major contributors of WEEE among secondary producers. This is due to the fact that the average life of the equipments employed in the IT sectors is considerably less compared with other sectors like banking and commercial sectors. For example, an average life of the computers in the IT sectors is reduced to 4–5 years due to the hasty technological changes such as up-gradation of both software

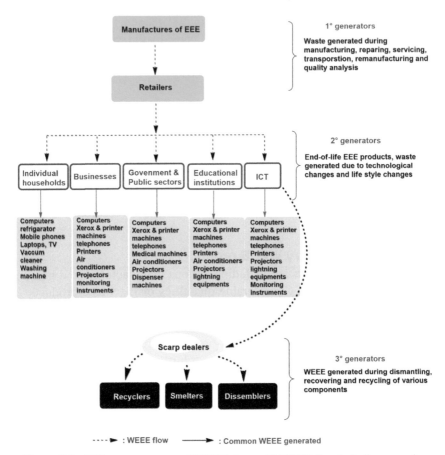

Figure 5.4   Different generators of WEEE along with WEEE flow in Indian scenario.

and hardware. The cities like Bengaluru, Pune, Chennai, and Hyderabad are responsible for the maximum generation of WEEE, as those are the prime IT hubs in India (Borthakur, 2014; Borthakur and Govind, 2017a). In addition, individual household is adding 20% of waste products into the WEEE stream. Wherein, obsolete products like mobile phones and computers are concerned as the major WEEE come out from the individual households. Additionally, other home appliances like washing machines, electrical ovens, television, electrical mixers, and refrigerators are also rejecting by the individuals in substantial quantity once its life cycle is completed, albeit such products are having much higher average life period compared to former products. The people belong to the middle and high income group societies are leading in generation of WEEE compared to the lower income groups in India. Nonetheless, the WEEE products generated by the households are pursuing different fates such as dumped along with the municipal waste, selling it to the informal recyclers, or kept it in store room incessantly. Likewise, governmental and non-governmental offices, banking, and commercial sectors are also accumulating a reasonable amount of WEEE into the waste yard. Unlike the other non-governmental sectors, obsolete products generated in the governmental organizations are moved into the hand of formal recyclers, since entire reselling procedures are followed by apposite auction procedure.

5.4.3.3 Third level: tertiary WEEE producers

Finally, various scarp dealers including dissemblers, smelters, and recyclers have significant role in generation of WEEE. The informal recycling centers are considering majority of the generated or imported WEEE in India contrast with the formal recycling centers. Most of the WEEE products have been collected, retained, recovered, and dismantled by the scrap dealers. Rather than producing WEEE, these groups are accountable for recycling and recovering of the materials from the waste components. However, during recovery of valuable metals such as gold, copper, silver, tin, etc., a huge amount of secondary components like plastics and composites are avoided by them; consequently, they were accumulated and become threat to the environment (Suresh et al., 2018b).

5.4.4 Disposal and Recycling Practices of WEEE Products Adopted in India

In India, damaged or EoL EEEs are handled in various approaches such as reuse, servicing, incineration landfill, and recycling. Generally, ordinary people will look for the servicing or reuse of the larger equipments like TV, refrigerators, washing machines, and micro ovens as compared with the smaller equipments like mobile phones. In addition, second hand usage of EEE products is also widely adopted in India after refurbishment of products. The formal and informal sectors around the country also play a vital role in elimination of WEEE generated.

5.4.4.1 Informal recycling and formal recycling

Huge quantity of WEEE generated in India is generally ended up into the hands of informal recyclers due to loopholes in the governmental laws. It has been indicated that above 90% of the WEEE components are dealt by the unauthorized recycling centers (Kumar et al., 2017). However, such domains do not follow governmental rules and regulations for the recycling and recovering of materials from WEEE. Moreover, informal sectors generally give their prime importance to the metals rather than plastics since those have higher economical values. For example, precious metals such as gold copper and tin present in the PCBs are recovered by the chemical leaching or smelting technique. Similarly, copper present in the wires and cables is recovered either manually or burning techniques. However, the leftovers formed after the isolation of metals are either dumped in the landfills or incinerated. However, such acts harmfully affect the health of both environment and humans. In addition, some informal recyclers have been shown interest over the isolation of engineering grade polymers such as ABS, HIPS, PC, or blends of polymers from structural components of WEEE. During this procedure, some of the recyclers isolate the flame retardant and non-flame retardant polymers using a simple density technique. Further, polymers are isolated, granulated, and resold to the pellet manufacturing units (Link, 2012b, 2014; Marconi et al., 2018). The cost of the isolated polymers lies in between 2 and 200 INR/kg which depends upon the category and quality of the materials (Link, 2012b).

The cities like Bangalore, Delhi, and Hyderabad have become prime important centers for WEEE recycling activities in India because of fast and emerging growth of IT sectors in these regions. Bangalore city contains large numbers of WEEE informal recycling centers in the country which contains 150 WEEE informal WEEE recyclers and 250 scrap sellers. Besides, through the recycling activities, Bangalore alone produces approximately 1000 tons of plastics, 300 tons of Pb, 43 tons of Ni, 350 tons of Cu, and 0.23 tons of Hg per annum from the WEEE components. Additionally, places like Mondoli industrial area in Delhi also contain a huge number of mid-scale and household sized industries for recycling of WEEE. Moreover, such places are engaged with recycling and recovery of materials from components like PCBs, CRTs, cables, structural parts, batteries, etc. Besides, it is also estimated that starting from initial waste collection, recovery, isolation, and final sales of the materials, approximately 700–1000 individuals are occupied in informal recycling procedure. Generally, local and poor individuals are working in daily income in informal recycling domains without wearing any personal safety equipments (Link, 2012b; Needhidasan et al., 2014; Perkins et al., 2014; Pradhan and Kumar, 2014; Reddy, 2015).

Compared with the informal recycling sectors, formal recycling sectors are following environmental friendly segregation, isolation, and recycling of the components such as metals, plastics, composites, and refractories from the WEEE. In formal sectors, mainly three types of operational techniques such as automated, semi-automated, and manual operations are adopted for the isolation of various materials. According to the CPCB, presently 178 formal recycling/dismantling units are registered across the country. Among the total formal sectors, 57 numbers of formal hubs belong to Karnataka followed by Maharashtra and Uttar Pradesh with 32 and 22 numbers, respectively. Additionally, CPCB reports that altogether these formal recycling centers can generate 43,8085.62 metric ton recycled material annually. Among the total 178 formal sectors, 27 units have higher capacity which can generate greater than 5000 Mt recycled materials per annum. However, ~51% of the formal recyclers registered in the CPCB are only having small annual recycling capacity (up to 1000 Mt/annum). Uttar Pradesh has the highest recycling capacity WEEE products of 86,130 Mt/annum followed by Rajasthan with 68,670 Mt/annum (Central Pollution Control Board, 2014; Awasthi and Li, 2017; Awasthi et al., 2018).

The formal recyclers follow strict collection and transportation procedure (in a closed vehicle) of WEEE from the source to the recycling center. Once the WEEE product reaches the recycling yard, they are isolated according to the product type and size. Subsequently, metals, plastics, and other components are isolated with fully or semi-automated instruments; during this stage, the products containing toxic components are segregated accordingly. Most of the formal recyclers are equipped with various innovative indigenous instruments in order to reduce the cost of processing. The plastics isolated from the WEEE components are randomly analyzed with various chemical and instrumental techniques. Further, they grounded into granular forms, if any metallic portions entrapped in the plastic, they will be recovered with the help of density method or gravity separation method during this stage. The recovered plastics and metals are further sent to the appropriate industries for the auxiliary applications. Similarly, PCBs and CRTs which contain various toxic elements get special attention in formal recycling centers and they are recycled according to the environmental friendly manner (Chatterjee and Kumar, 2009; Guptha and Shekar, 2009; Begum, 2013). The wastes developed after the recycling technique of various components may be used for the recovery of energy or send it to the landfilling. However, various reports are indicating that only fewer amount of WEEE is coming to the formal centers for recycling procedure. The general procedure adopted in the formal and non-formal recycling hubs is indicated in Figure 5.5.

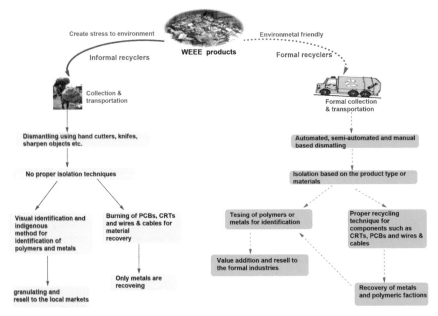

Figure 5.5 Process flow in the formal and informal recycling centers.

5.4.5 Environmental Aspects of WEEE Recycling and Disposal

The disposal and recycling of the WEEE products are still a question mark to the manufactures and consumers since those are included with various toxic chemical components. The manufacturers of the EEE products incorporate additives like flame retardants (brominated and phosphorus) and heavy metals into their products to suppress ignition and smooth functioning of the electronic components, respectively (Suresh et al., 2018b). As a result, during the manufacturing, disposal, and recycling of electronic products, a significant amount of toxic elements are liberated into the environment. The researchers have indicated the possibility of emission of dust particles that contain heavy metals and flame retardants during dismantling analysis of WEEE components (Sepúlveda et al., 2010). This kind of particle may travel a long distance from the source of production and re-deposit and contaminates water or soil systems. Similarly, burning of plastic products leads to the production of anthropogenic pollutants and extremely toxic products like polyhalogenated dioxins and furans. Studies are indicating that incineration of polymers like PVC in the presence of Cu (which is common conductive medium in PVC wires and cables) directs to the formation of dioximes

(Lett et al., 1992; Suresh et al., 2017c). Similarly, cleaning and landfilling of the WEEE products also contributed to the leaching of the chemicals incorporated in the polymer matrix. Additionally, several research works have indicated that the people living nearby the recycling centers and workers of the recycling hubs have high exposure of toxic chemical components. Hence, a considerable amount of toxic chemicals persist in the human body and body fluids like blood and milk (Sjödin et al., 2003; Julander et al., 2005; Schecter et al., 2006).

5.4.6 Recycling-based Research Initiated in India and its Major Outcomes

Similar to the global trend, Indian academicians and industrialist are paying their attention on recovering and recycling of components from WEEE products due to its heterogeneous material composition and wide availability. A study conducted by Borthakur and Govind (2017b) indicated that for a period of 30 years (1986–2016), 4201 numbers of research publications are published around the world. However, during this period, Indian research centers compiled only 276 scientific documents, which is 6.6% of the total generated documents (Borthakur and Govind, 2017b). Similar to this, Dey and Jana (2014) performed a study on patents documented in India in the field of recycling technologies. This reports pointed out that around 135 numbers of patents were applied into the Indian patent office during the period of 1995–2013, in which 54% is filled from various Indian industries or individuals. Both of the aforementioned reports were indicating that numbers of publication and filling up of the patents have been growing gradually as research institutes and industrialists have shown their interest over waste generated from the EEE products.

In the present scenario, many of the Indian researchers are engaged in research correlated with waste polymers, metals, refractories, and composite products. Some of the major research related with WEEE and its outcomes are listed in Table 5.3. The recovery and recycling of polymers from WEEE are considered as the "fruitful area" in current circumstances, since polymers carries 20–35% of the total WEEE. Along with it, different polymers such as ABS, PC, PVC, PMMA, PS, and their blend such as PC/ABS are commonly used in structural parts of the EEEs; consequently, those components are usually available from WEEE stream. Moreover, the plastic components recovered from the WEEE have mechanical and thermal properties less than or equivalent to its virgin

Table 5.3   Some of the major research related with recycling of WEEE components by Indian researchers

Author/ Year	Research Area	Component Selected	Major Outcomes
Debnath et al. (2016)	Recycling and reusability	Integrated circuit (ICs), PCBs	This work examines effective recovery and recycling options of ICs present in the PCBs which are collected from different ITC products. This work concludes that a proper recovering and recycling of ICs could be helpful for sustainable WEEE management
Biswal et al. (2015)	Value addition of plastics using WEEE component	Non-metallic fraction of PBCs	Developed composite materials with non-metallic fractions of PCBs and polypropylene. As compared with the respective virgin material developed composites have shown improvement in the corresponding mechanical properties. The authors believe that the developed composite could be useful for the production of automotive parts
Anandhan and Bhowmick (2013)	Polymer recycling and value addition	Computer scarps	Developed thermoplastic elastomeric blends employing ABS recovered from computer structural parts and waste nitrile rubber. After modification, generated blends have shown improved mechanical properties and swelling resistance in lubricating oil compared to its counterparts
Suresh et al. (2018c)	Value addition of plastics	Computer and data cable waste	PVC and PMMA components previously employed in the data cables and light guidance panels, respectively, were recovered and used for the preparation of recycled blend. Further authors indicated that functionalized NBR/NBR mixture can considerably

Table 5.3 (Continued)

Author/ Year	Research Area	Component Selected	Major Outcomes
			improve the mechanical and thermal properties of the recycled blend composed of PVC and PMMA. This study paves new method for value addition of recycled blends
Verma et al. (2016)	Recovery of components	PCBs	This work indicated an economical approach for recovery of brominated epoxy resins, copper foils, glass fiber, and metallic contents using dimethylformamide (DMF). Interestingly, DMF can initiate breakage of inter-hydrogen bonds of brominated epoxy resins and thereby leads to dissolution at higher temperature. Further, it is possible to regenerate DMF from the system and which could be useful for further applications
Bhakar et al. (2015)	Life cycle assessment of WEEE components	LCD, LED, and CRT screens	Investigated environmental impact of waste monitors like LCDs, LEDs, and CRTs which was previously employed in EEE products. From this study, they have concluded that LEDs have lower impact on environmental compared to LCDs and CRTs
Tiwary et al. (2017)	Recyclability of WEEE	PCBs	Authors have applied low-temperature ball milling technique for generation of nanomaterials having sizes of 20–150 nm from PCBs without any chemical treatment. Similarly, by this method, it is also able to separate components such as metals, oxides, and polymers from PCBs

(Continued)

Table 5.3 (Continued)

Author/ Year	Research Area	Component Selected	Major Outcomes
Sarath et al. (2018)	Value addition of polymeric components	Structural casings parts and touch pad recovered from mobile phone	The mechanical, thermal, and thermo-mechanical properties of polymers such as PC/ABS and mixed plastics components have been improved via co-recycling with silicon rubber recovered from mobile phone. Additionally, this work signifies that chemical modification of silicon rubber with GMA can further improve mechanical properties of polymers compared with non-modified silicon rubber
Suresh et al. (2018b)	Recovery and recycling analysis	Light guidance panels from LEDs	The polymer employed in the light guidance panels identified with various physical and chemical analyses. PMMA is used as a light guidance panel in LEDs due to exceptional clarity and mechanical properties. Further, the thermal and mechanical properties of recovered PMMA are compared with virgin PMMA. This work shows even after the mechanical recycling recovered PMMA has significant mechanical and thermal properties; therefore, the same could be useful for the production of other industrial products
Senophiyah-Mary et al. (2018)	Recovering components from PCBs	PCBs	Developed an economical method for removal of epoxy coatings present in the PCBs. The authors describe that when PCBs are soaked with the 5N NaOH for 8 h and reacted in bath sonicator for 5 min, it could be possible to break chemical bond and peel off epoxy coating from PCBs. Further, the chemical which employed in for the work can be potential to reutilize since they are less contaminated

counterparts (Suresh et al., 2017c,d). In general, recyclers are encountering problems such as degradation in the polymeric chains and reduction of mechanical properties after recycling procedure. However, such situations can be controlled by proper value addition of polymeric using external agents (Suresh et al., 2018a,b). Similar to the polymers, PCBs isolated from the waste products are also an interesting material for recycling analysis since those are rich in components like refractories, metals, and polymers. Most of the research works are focused only on the isolation of the metals rather than polymers; albeit, some of the research indicated the possibility of recovered polymeric fraction for vale addition of other polymers.

5.5 Conclusion and Future Perspectives

The current work indicates that WEEE generation in India is dangerously accelerating in the recent period. During the entire study, it has been noticed that numerous documents are available on estimation and quantification of WEEE in India; however, most of them are conflicting with each other. This indicates that there are no appropriate systems available for the quantification of the WEEE products generated. Along with the domestic generation, a huge amount of WEEE products are importing to India through using illegal pathways mostly via seaports. In order to trace out such illegal imports, the authorities must keenly observe and returned such illegal items to its originated place. Moreover, WEEE flow in India is in diverted manner; informal traders and recyclers handle 95% of the waste products generated/imported. However, recycling procedure adopted by them is highly unsafe to the workers and also harmful to the environment. The proper knowledge about the handling of WEEE materials in a safe manner should be educated to the poor people those who are working in such an environment. The burning of wires and cables and PCBs for isolation of the metals has to be banned since they act as a catalyst to air pollution. In order to control WEEE and improve the recycling status, both informal and formal recyclers act together. It is necessary to identify and categorize different informal sectors present in India. In which those are showing better performance can be either converted into a formal recycler or integrated with a formal recycler.

The flow of WEEE in India is primarily controlled by various manufacturers, retailers, business sector, public and non-public sectors, ITC sectors, and individual households. Furthermore, products such as computers, mobile phones, and home appliances are contributing the maximum share of WEEE due its lower average life period. Therefore, manufactures required to look for improving the product quality along with they have to offer free or cost effective services the appliances which are damaged or outdated. The program such as "EPR" is effectively implemented by European countries but in India it is still on initial stages. The complete implementation of EPR could lead to the better collection and recycling of WEEE components. Moreover, many of the studies are indicating that value-added polymers have comparable thermo-mechanical properties with its virgin counter parts. Therefore, manufactures have to employ such materials in manufacturing of EEE products or lower grade products.

References

Aizawa H, Yoshida H, Sakai, SI. (2008). Current results and future perspectives for Japanese recycling of home electrical appliances. Resources Conservation and Recycling. 52:1399–410. DOI: 10.1016/j.resconrec. 2008.07.013

Anandhan S, Bhowmick AK. (2013). Thermoplastic vulcanizates from post consumer computer plastics/nitrile rubber blends by dynamic vulcanization. Journal of Material Cycles and Waste Management. 15:300–9. DOI: 10.1007/s10163-012-0112-7

Awasthi AK, Zeng X, Li J. (2016). Comparative examining and analysis of E-waste recycling in typical developing and developed countries. Procedia Environmental Sciences. 35:676–80. DOI: 10. 1016/j.proenv.2016.07.065

Awasthi AK, Li J. (2017). Management of electrical and electronic waste: a comparative evaluation of China and India. Renewable and Sustainable Energy Reviews. 76:434–47. DOI: 10.1016/j.rser.2017.02.067

Awasthi AK, Wang M, Wang Z, Awasthi MK, Li J. (2018). E-waste management in India: a mini-review. Waste Management and Research. 36(5):408–14. DOI: 10. 1177/0734242X18767038

Baldé CP, Wang F, Kuehr R, Huisman J. (2015). The global E-waste monitor – 2014, United Nations, University; IAS – SCYCLE, Bonn, Germany. Available at https://i.unu.edu/media/unu.edu/news/52624/UN U-1stGlobal-E-Waste-Monitor-2014-small.pdf [Assessed June 3, 2018]

Betts K. Producing usable materials from E-waste. (2008). Environmental Science and Technology. 42:6782–3. DOI: 10.1021/es801954d

Biswal M, Jada N, Mohanty S, Nayak SK. (2015). Recovery and utilisation of non-metallic fraction from waste printed circuit boards in polypropylene composites. Plastic and Rubber Components. 44:314–21. DOI: 10.1179/1743289815Y.0000000021

Borthakur A. (2014). Generation and management of electronic waste in the city of Pune, India. Bulletin of Science Technology and Society. 34:43–52. DOI: 10.1177/0270467614541242

Borthakur A, Govind M. (2017a). How well are we managing E-waste in India: evidences from the city of Bangalore. Energy, Ecology and Environment. 2:225–35. DOI: 10.1007/s40974-017-0060-0

Borthakur A, Govind M. (2017b). Emerging trends in consumers' E-waste disposal behaviour and awareness: a worldwide overview with special focus on India. Resources, Conservation and Recycling. 117:102–13. DOI: 10.1016/j.resconrec.2016.11.011

Borthakur A, Govind M. (2018). Management of the challenges of electronic waste in India: an analysis. Waste Resouce Management. 171:14–20. DOI: 10.1680/jwarm.17.00035

Borthakur A, Singh P. (2012). Electronic waste in India: problems and policies. International Journal of Environmental Science. 3:353–62. DOI: 10.6088/ijes.2012030131033

Borthakur A, Sinha K. (2013). Generation of electronic waste in India: current scenario, dilemmas and stakeholders. African Journal of Environmental Science and Technology. 7:899–910. DOI: 10.5897/AJEST2013.1505

Begum DKJA. (2013). Electronic waste (E-waste) management in India: a review. IOSR Journal of Humanity and Social Science. 10:46–57. Available at: https://pdfs.semanticscholar.org/8cda/238ee8d21bce264737283 08a6ce7e6db186f.pdf [Assessed 12 June 2018].

Bhakar V, Agur A, Digalwar AK, Sangwan KS. (2015). Life cycle assessment of CRT, LCD and LED monitors. Procedia CIRP. 29:432–7. DOI: 10.1016/j.procir.2015.02.003

Central Pollution Control Board. (2014). List of Registered E-Waste Dismantler/Recyclerin the country, Central Pollution Control Board, New Delhi. Available at: http://cpcb.nic.in/List_of_E-waste_Recycler_as_on_29.12.2016.pdf [Accessed July 20, 2017].

Chatterjee S, Kumar K. (2009). Effective electronic waste management and recycling process involving formal and non-formal sectors. International Journal of Physical Science. 4:893–905. Available at: http://www.academicjournals.org/journal/IJPS/article-full-text-pdf/1CA355921575 [Assessed 09 July 2018].

Debnath B, Roychowdhury P, Kundu R. (2016). Electronic Components (EC) reuse and recycling–a new approach towards WEEE management. Procedia Environmental Sciences. 35:656–68. DOI: 0. 1016/j.proenv.2016.07.060

Dey S, Jana T. (2014). E–waste Recycling Technology Patents filed in India-An Analysis. Journal of Intellectual Property Rights. 19:315–24. Available at: http://nopr.niscair.res.in/bitstream/123456789/29505/1/JIPR%2019%285%29%20315-324.pdf [Assessed July 01, 2018].

Directive. (2003) Directive of the European Parliament and of the Council of 4 July 2012 on waste electrical and electronic equipment, WEEE 2002/96. Available at: http://eur-lex.europa.eu/legal-content/EN/TXT/?uri=CELEX:32002L0096 [assessed May 30, 2018].

Directive EC. (2012). Directive 2012/19/EU of the European Parliament and of the Council of 4 July 2012 on waste electrical and electronic equipment, WEEE. Official Journal of the European Union L. 197:38–71.

Dwivedy M, Mittal RK. (2010a). Future trends in computer waste generation in India. Waste Management. 30:2265–77. DOI: 10.1016/j.wasman. 2010.06.025

Dwivedy M, Mittal RK. (2010b). Estimation of future outflows of E-waste in India. Waste Management. 30:483–91. DOI: 10.1016/j.wasman.2009. 09.024

Dwivedy M, Mittal RK. (2012). An investigation into E-waste flows in India. Journal of Cleaner Production. 37:229–42. DOI: 10.1016/j.jclepro.2012. 07.017

Electronics Take Back Coalition. (2012). Facts and figures on E-waste and recycling. Available at: http://www.electronicstakeback.com/wpcontent/uploads/Facts and Figures [Assessed May 25, 2018].

Garlapati, Vijay Kumar. (2016). E-waste in India and developed countries: management, recycling, business and biotechnological initiatives. Renewable and Sustainable Energy Reviews. 54:874–81. DOI: 10. 1016/j.rser.2015.10.106

Ghosh SK, Debnath B, Baidya R, De D, Li J, Ghosh SK, et al. (2016). Waste electrical and electronic equipment management and Basel Convention compliance in Brazil, Russia, India, China and South Africa (BRICS) nations. Waste Management and Research. 34:693–707. DOI: 10. 1177/0734242X16652956

Goodship V, Stevels A. (Eds.). (2012). Waste electrical and electronic equipment (WEEE) handbook. Elsevier. Woodhead publishing.

Gupta V, Kumar A. (2014). E-waste status and management in India. Journal of Information Engineering and Applications. 4:41–8.

Guptha CN, Shekar GL. (2009). Electronic waste management system in Bangalore–a review. JK Journal of Management and Technology. 1(1):11–24.

IAMAI and KANTAR IMRB Report. (2016). Internet in India 2016. Available at: http://bestmediainfo.com/wp-content/uploads/2017/03/Internet-in-India-2016.pdf [Assessed 22 June 2018].

Ikhlayel M. (2018) An integrated approach to establish E-waste management systems for developing countries. Journal of Cleaner Production. 170:119–30. DOI: 10.1016/j.jclepro.2017.09.137

Julander A, Westberg H, Engwall M, Van Bavel B. (2005). Distribution of brominated flame retardants in different dust fractions in air from an electronics recycling facility. Science of the Total Environment. 350:151–60. DOI: 10.1016/j.scitotenv.2005.01.015

Komissarov VA. (2012). WEEE is the most growing waste flow in the world. Situation with WEEE management in Russia and other countries, Russia, Moscow. Available at: http://ac.gov.ru/files/content/2535/komissarov-va-pdf.pdf [Accessed May 20 2018].

Kumar R, Verma A, Rajput N. (2017). E-waste management in India: formal vs informal sectors. Available at: http://www.ijramr.com/sites/default/files/issues-pdf/1424.pdf [Assessed 23 July 2018].

Link T. (2012a). E-WASTE IN INDIA: System failure imminent–take action NOW. Available at: http://www.toxicslink.org/docs/06040_repsumry.pdf [Assessed July 01, 2018].

Link T. (2012b). Empa, "Improving plastic management in Delhi. A report on WEEE plastic recycling," Delhi, India. Available at: http://toxicslink.org/docs/Improving_Plastic_management_in_delhi.pdf [Assessed July 01, 2018].

Link T. (2014). Impact of E-waste recycling on Water and Soil, pp. 1e87. New Delhi, India. Available at: http://toxicslink.org/docs/Impact-of-E-waste-recycling-on-Soil-and-Water.pdf [Assessed July 03, 2018].

Lett BK, Bruce KR, Beach LO. (1992). Mechanistic steps in the production of PCDD and PCDF during waste combustion. Chemosphere. 25:1387. DOI: 10.1016/0045-6535(92)90158-N

Liu X, Tanaka M, Matsui Y. (2006). Generation amount prediction and material flow analysis of electronic waste: a case study in Beijing, China. Waste Management and Research. 24:434–45. DOI: 10.1177/0734242X06067449

Manomaivibool P. (2009). Extended producer responsibility in a non-OECD context: the management of waste electrical and electronic equipment in India. Resources, Conservation and Recycling. 53:136–44. DOI: 10. 1016/j.resconrec.2008.10.003

Marconi M, Gregori F, Germani M, Papetti A, Favi C. (2018). An approach to favor industrial symbiosis: the case of waste electrical and electronic equipment. Procedia Manufacturing. 21:502–9. DOI: 10. 1016/j.promfg.2018.02.150

Ministry of electronics and information technology (MeitY), Government of India. (2016), Electronics and information technology annual report 2015–2016. Available at: http://meity.gov.in/writereaddata/files/annual-report-2015%E2%80%9316.pdf [Assessed June 20 2018].

Ministry of electronics and information technology (MeitY), Government of India. (2017). Electronics and information technology annual report 2016–2017. Available at: http://meity.gov.in/writereaddata/files/AR 2016–17_English.pdf [Assessed June 20 2018].

Ministry of electronics and information technology (MeitY), Government of India. (2018). Electronics and information technology annual report 2017–2018. Available at: http://meity.gov.in/writereaddata/files/Annual_Report_2017%E2%80%9318.pdf. [Assessed June 20 2018].

MoEF. (2008). Guidelines for environmental friendly sound management of E-waste. As approved vide MoEF letter No. 23–23/2007-HSMD dt. March 12, 2008. Available at: http://cpcb.nic.in/displaypdf.php?id= UHJvamVjdHMvRS1XYXN0ZS9mdWxsLXRleHQucGRm [Assessed July 05, 2018].

Needhidasan S, Samuel M, Chidambaram R. (2014). Electronic waste–an emerging threat to the environment of urban India. Journal of Environmental Health Science and Engineering. 12:36. DOI: 10.1186/2052-336X-12-36

Network BA. (2005). The digital dump: exporting high-tech re-use and abuse to Africa. BAN Report. Available at: http://archive.ban.org/library/TheDigitalDump.pdf [Assessed July 15, 2018].

Ongondo FO, Williams ID, Cherrett TJ. (2011). How are WEEE doing? A global review of the management of electrical and electronic wastes. Waste Management. 31:714–30. DOI: 10.1016/j.wasman.2010.10.023

Osibanjo O, Nnorom, IC. (2007). The challenge of electronic waste (E-waste) management in developing countries. Waste Management and Research. 25:489–501. DOI: 10.1177/0734242X07082028

Pathak P, Srivastava, RR. (2017). Assessment of legislation and practices for the sustainable management of waste electrical and electronic equipment in India. Renewable and Sustainable Energy Reviews. 78:220–32.

Perkins DN, Drisse MNB, Nxele T, Sly PD. (2014). E-waste: a global hazard. Annals of Global Health. 80:286–95. DOI: 10.1016/j.aogh.2014.10.001

Pérez-Belis V, Bovea MD, Ibáñez-Forés V. (2015). An in-depth literature review of the waste electrical and electronic equipment context: trends and evolution. Waste Management and Research. 33:3–29. DOI: 10.1177/0734242X14557382

Pinto VN. (2008). E-waste hazard: the impending challenge. Indian Journal of Occuptional and Environmental Medicine. 12:65. DOI: 10.4103/0019-5278.43263

Pradhan JK, Kumar S. (2014). Informal E-waste recycling: environmental risk assessment of heavy metal contamination in Mandoli industrial area, Delhi, India. Environmental Science and Pollution Research. 21:7913–28. DOI: 10.1007/s11356-014-2713-2

Premalatha M, Abbasi T, Abbasi SA. (2014). The generation, impact, and management of e-waste: State of the art. Critical Reviews in Environmental Science and Technology. 44:1577-678. DOI: 10.1080/10643389.2013.782171

Puckett J, Byster L, Westervelt S, Gutierrez R, Davis S, Hussain A, et al. (2002). Exporting harm: the high-tech trashing of Asia. The Basel Action Network (BAN) and Silicon Valley Toxics Coalition (SVTC). SVTC report. Available at: http://svtc.org/wp-content/uploads/technotrash.pdf [Assessed June 12, 2018].

Rahul M. (2016). Digital India: Barriers and Remedies. In: International Conference on Recent Innovations in Sciences, Management, Education and Technology. Available at: http://data.Conference world. in/ICISMET/P256–261.Pdf [Assessed May 15, 2018].

Reddy RN. (2015). Producing abjection: E-waste improvement schemes and informal recyclers of Bangalore. Geoforum. 62:166–74. DOI: 10.1016/j.geoforum.2015.04.003

Robinson BH. (2009). E-waste: an assessment of global production and environmental impacts. Science of the Total Environment. 408:183–91. DOI: 10.1016/j.scitotenv.2009.09.044

Sarath P, Biswal M, Mohanty S, Nayak SK. (2018). Effect of silicone rubber based impact modifier on mechanical and flammability properties of plastics recovered from waste mobile phones. Journal of Cleaner Production. 171:209–19. DOI: 10.1016/j.jclepro.2017.10.024

Schecter A, Päpke O, Robert Harris T, Tung KC. (2006). Partitioning of polybrominated diphenyl ether (PBDE) congeners in human blood and milk. Toxicology and Environmental Chemistry. 88:319–24. DOI: 10.1080/02772240600605087

Schluep M, Hagelüken C, Kuehr R, Magalini F, Maurer C, Meskers CE, et al. (2009). Recycling–from E-waste to Resources. United Nations Environment Programme and United Nations University, Germany. Available at: http://www.unep.fr/shared/publications/pdf/DTIx1192xPA-Recycling%20from%20ewaste%20to%20Resources.pdf [Assessed May 04, 2018].

Secretariat RS. (2011). E-waste in India. India Research Unit (Larrdis), Rajya Sabha Secretariat, New Delhi. Available at: https://rajyasabha.nic.in/rs new/publication_electronic/E-Waste_in_india.pdf [Assessed July 06, 2018].

Senophiyah-Mary J, Loganath R, Meenambal T. (2018). A novel method for the removal of epoxy coating from waste printed circuit board. Waste Management and Research. 36:645–52. DOI: 10.1177/0734242X18782392

Sepúlveda A, Schluep M, Renaud FG, Streicher M, Kuehr R, Hagelüken C, et al. (2010). A review of the environmental fate and effects of hazardous substances released from electrical and electronic equipments during recycling: examples from China and India. Environmental Impact Assessment Review. 30:28–41. DOI: 10.1016/j.eiar.2009.04.001

Sjödin A, Patterson Jr DG, Bergman Å. (2003). A review on human exposure to brominated flame retardants—particularly polybrominated diphenyl ethers. Environment International. 29:829–39. DOI: 10.1016/S0160-4120(03)00108-9

Skinner A, Dinter Y, Lloyd A, Strothmann P. (2010). The challenges of E-waste management in India: can India draw lessons from the EU and the USA. Asien. 117:26. Available at: http://asien.asienforschung.de/wp-content/uploads/sites/6/2014/04/ASIEN_117_Skinner-Dinter-Llyod-Strothmann.pdf [Assessed July 08, 2018].

Socolof, ML, Overly, JG, Geibig JR. (2005). Environmental life-cycle impacts of CRT and LCD desktop computer displays. Journal of Cleaner Production. 13:1281–94. DOI: 10.1016/j.jclepro.2005.05.014

Song Q, Li J, Liu L, Dong Q, Yang J, Liang Y, et al. (2016). Measuring the generation and management status of waste office equipment in China:

a case study of waste printers. Journal of Cleaner Production. 112:4461–8. DOI: 10.1016/j.jclepro.2015.07.106

Song Q, Wang Z, Li J, Duan H, Yu D, Zeng X. (2017). Characterizing the transboundary movements of UEEE/WEEE: is Macau a regional transfer center?. Journal of Cleaner Production. 157:243–53. DOI: 10.1016/j.jclepro.2017.04.149

StEP Initiative. (2014). Solving the E-Waste Problem (Step) White Paper: One Global Definition of E-waste. Bonn, Germany. Step, 3576.

Suresh SS, Bonda S, Mohanty S, Nayak SK. (2018a). A review on computer waste with its special insight to toxic elements, segregation and recycling techniques. Process Safety and Environmental Protection. 116:477–93. DOI: 10.1016/j.psep.2018.03.003

Suresh SS, Mohanty S, Nayak SK. (2017a). Bio-based epoxidised oil for compatibilization and value addition of poly (vinyl chloride)(PVC) and poly (methyl methacrylate)(PMMA) in recycled blend. Journal of Polymer Research. 24;120. DOI: 10.1007/s10965-017-1282-8

Suresh SS, Mohanty S, Nayak SK. (2017b). Composition analysis and characterization of waste polyvinyl chloride (PVC) recovered from data cables. Waste Management. 60:100–11. DOI: 10.1016/j.wasman.2016.08.033

Suresh SS, Mohanty S, Nayak SK. (2017c). Investigation into the mechanical and thermal properties of poly (methyl methacrylate) recovered from light guidance panels with a focus on future remanufacturing and sustainable waste management. Journal of Remanufacturing. 7:217–33. DOI: 10.1007/s13243-017-0041-7

Suresh SS, Mohanty S, Nayak SK. (2017d). Preparation and characterization of recycled blends using poly (vinyl chloride) and poly (methyl methacrylate) recovered from waste electrical and electronic equipments. Journal of Cleaner Production. 149:863–73. DOI: 10.1016/j.jclepro.2017.02.057

Suresh SS, Mohanty S, Nayak, SK. (2018b). Influence of acrylonitrile butadiene rubber on recyclability of blends prepared from poly (vinyl chloride) and poly (methyl methacrylate). Waste Management and Research. 36: 495–504. DOI: 10.1177/0734242X18771164

Suresh SS, Mohanty S, Nayak SK. (2018). Synthesis and application of functionalised acrylonitrile-butadiene rubber for enhancing recyclability of poly (vinylchloride)(PVC) and poly (methylmethacrylate)(PMMA) in recycled blends. Clean Technologies and Environmental Policies. 20:969–79. DOI: 10.1007/s10098-018-1514-6

Tansel B. (2017). From electronic consumer products to E-wastes: global outlook, waste quantities, recycling challenges. Environment International. 98:35–45.DOI: 10.1016/j.envint.2016.10.002

Tiwary CS, Kishore S, Vasireddi R, Mahapatra DR, Ajayan PM, Chattopadhyay K. (2017). Electronic waste recycling via cryo-milling and nanoparticle beneficiation. Materials Today. 20:67–73. DOI: 10.1016/j.mattod.2017.01.015

TRAI. (2009). Telecome Regulatory Authority of India Annual report 2008–2009. Available at: https://trai.gov.in/sites/default/files/ar_08_09.pdf [Assessed 21 June 2018].

Verma HR, Singh KK, Mankhand TR. (2016). Dissolution and separation of brominated epoxy resin of waste printed circuit boards by using di-methyl formamide. Journal of Cleaner Production. 139:586–96. DOI: 10.1016/j.jclepro.2016.08.084

Wath SB, Dutt PS, Chakrabarti T. (2011). E-waste scenario in India, its management and implications. Environment Monitoring and Assessment. 172:249–62. DOI: 10.1007/s10661-010-1331-9

Watkins J, Kitner KR, Mehta D. (2012). Mobile and smartphone use in urban and rural India. Continuum. 26:685–97. DOI: 10.1080/10304312.2012.706458

Widmer R, Oswald-Krapf H, Sinha-Khetriwal D, Schnellmann M, Böni H. (2005). Global perspectives on E-waste. Environmental Impact Assessment Review. 25:436–58. DOI: 10.1016/j.eiar.2005.04.001

Yedla S. (2016). Development of a methodology for electronic waste estimation: a material flow analysis-based SYE-Waste Model. Waste Management and Research. 34:81–6. DOI: 10.1177/0734242X15610421

Yoshida A, Terazono A. (2010). Reuse of second hand TVs exported from Japan to the Philippines. Waste Management. 30:1063–72. DOI: 10.1016/j.wasman.2010.02.011

MODULE 2

Water-Recycle and Reuse

6

A Critical Review on Wastewater Treatment Techniques for Reuse of Water in Industries

Shanmugasundaram O. Lakshmanan[1]* and Sujatha Karuppiah[2]

[1]Department of Textile Technology, K.S. Rangasamy College of Technology, Tiruchengode, India
[2]Department of Physics, Vellalar College for Women, Erode, India
E-mail: mailols@yahoo.com
*Corresponding Author

Industries like textile, chemical, poultry, paper, leather, etc., are consuming large volume of fresh water and discharging wastewater into the environment which cause not only water pollution but also scarcity for fresh water. Therefore, research on wastewater treatment methods for reuse of treated water in industries is gaining importance worldwide. This chapter communicating an overview of extensive research has been carried out by various scientists on advanced wastewater treatment methods for recycle and reuse of water in industries. Different treatment techniques such as reverse osmosis, nanofiltration, ultrafiltration, activated carbon adsorption, electrochemical, membrane bioreactor, and their combinations were experimented on at pilot plant to test their efficiency in removing pollutants and verified the potential reuse of treated water for various processes in industries. Moreover, cause and remedies on the treatment methods were discussed in detail.

6.1 Introduction

Water is one of the basic needs for survival of living things. Currently, 55% of water is being used by industries and agriculture purpose. The pressure on industries is to recycle and reuse some quantity of wastewater for process due to high growth of population in India (Tang and Chen, 2002). A huge

125

volume of wastewater is generating by textile industries and is highly colored with inorganic salts. Textile effluents contain a high amount of color, contaminants, and pollutants. In order to control industrial effluents, several stringent environmental regulations are being enforced in our country.

Textile industries especially dyeing, printing, and finishing are generating a huge volume of wastewater that contains toxic substances, surfactants, salts, organic matter, and pH (Ranganathan et al., 2006; Chakraborty et al., 2013). Textile industries are generating effluent approximately 150–200 L per kilogram of fabric being processed and it contains color, H_2O_2, surfactants, dyes, soap, alkali, detergents, acids, high BOD, COD, and pH (Rajagopalan, 1989; Panchiao, 1994; Wang et al., 2011; Kant, 2012; Paul et al., 2012).

More than 1 lakh synthetic dyes with 70,000 tons of dyestuff produced annually (Papic et al., 2004; Lee and Choi, 2006). Ali (2010) reported that the textile industries are discharging more than 280,000 tons of dyes as effluent to the environment. Kant (2012) stated that textile industries are contributing 17–20% of total industrial water pollution. Many researchers pointed out the causes and effects of wastewater on irrigation, landscape, pipelines, fisheries, human being, and environment.

The major issues of discharging wastewater into water stream are disturbing photosynthesis in plant and impact on aquatic life and marine life, and hence treatment is necessary for textile effluent before their discharge (Holkar et al., 2016). Gaston (1979) and Hutson and Roberts (1990) discussed the major pollutants present in industrial effluents and their causes to the environment, aquatic life, marine life, plant, animals, and human beings. Very recently, researchers used fly ash and red mud as an adsorbent for removal of pollutants from water.

Vrcek et al. (2001), Dorange et al. (2004) and Rajamani et al. (2004) reported that tannery effluents contain chromium, sulfate, chloride, ammonia, emulsifiers, etc. Among the pollutants, chromium and sulfate are highly toxic substances which cause environmental pollution. The industries are facing shortage of water due to water scarcity and limited usage of ground water. Van der Bruggen et al. (2004) reported that the textile industries have to treat wastewater to meet the standards for reuse.

The industries, agricultural activities, and domestic are the sources of water pollution. Nowadays, ground water and surface water are being contaminated by the above sources and hence by 2020, the world may face severe water scarcity (Franklin, 1991; Droste, 1997). Gupta et al. (2012) explained the importance of wastewater recycling and guidelines for selection of appropriate treatment technologies to treat wastewater.

From the review, the authors have found that the industries are consuming a huge volume of freshwater and generating high volume of effluents, as a result lack of fresh water, environmental pollution, etc. By introducing new wastewater treatment techniques for recycle and reuse of water in industries thus reduce the consumption of fresh water and lower the volume of effluent discharge into the environment.

6.2 Importance of Wastewater Treatment

Water is the basis of all life. Life of the planet, Earth, hinges on the availability of clean water. But its faulty usage and the ensuing scarcity necessitate the treatment of waste water. In times of drought and in acute water shortage areas, waste water treatment may not only offer universal remedy for the existing water crisis but also prevent further complications in this front. This venture holds the twin benefits of restoring the water supply and cleansing the planet of water-based impurities.

Restoring the water supply through waste water treatment implies measures to establish water treatment plants wherever needed. This ensures the possibility of reusing or recycling waste water to boost up the water cycle mechanism. Protecting the planet is crucial in this regard because much of the domestic, commercial, and Industrial waste water is composed of pollutants, hazardous chemicals, pathogens, human waste, food scraps, oils, soaps, and storm runoff which may endanger the health of the fragile ecosystem.

Ninety seven percentage of the planet's clean water is either in the form of ice or as saline in ocean bodies. The major portion of the remaining 3% of fresh water is in glaciers, leaving only a meagre 0.01% as available fresh water for humans and his eco-partners' consumption. Sadly though, this scanty source is also polluted leading to water-borne diseases among humans and cattle. Besides such adverse factors, the unforeseen climate change floods or drowns certain places while leaving the other half of the globe parched dry.

In this context, judicious use of existing water resources supplemented with waste water treatment measures may be promising for the agricultural and food production sectors, human consumption, and sanitation purposes. This increasing demand for clean water reiterates that the supply of clean water is the need of the hour. So, Governments divert huge funds for research and development in waste water treatment-related projects. Strengthening the Agricultural and Food sectors, preserving the environment, and reducing, recycling, and reusing of water may help conserve energy in its treatment processes, reduce pollution, and conserve fuel resources.

6.3 Conventional Wastewater Treatment Techniques and its Drawbacks

Pagga and Taeger (1994) used activated sludge treatment to treat textile effluents and found to be ineffective in removing colors from wastewater. Ozone technique and power-activated carbon as an adsorbent are used for the removal of color in textile effluents (Wu et al., 1998). Rosell and Huertas (1995) mentioned that the activated sludge treatment is an effective method for removal of COD but ineffective in color removal. Few researchers suggested to use combined treatment techniques (UV-radiation and ozone) to treat textile wastewater. Hart et al. (1983) used Jar tests to evaluate the efficiency of physico-chemical treatment. Traditional treat methods are inadequate to treat textile effluents due to large variability of composition in effluents (Hao et al., 2000; Sakalis et al., 2005).

Textile industries are generating a huge volume of effluents and treatment methods are physico-chemical (Kim et al., 2002, 2004; Golob et al., 2005) and biological methods (Chang et al., 2002; Libra et al., 2004) and advanced treatment techniques. A high COD (95%) removal was achieved by treating denim wastewater using activated sludge process (Orhon et al., 2001; Pala and Tokat et al., 2003).

Ganesh et al. (1994) and Bruggen et al. (2004) attempted to decolorize textile effluents by combining biological and physico-chemical treatment methods, but a huge quantity of sludge was generated during the process. Several authors have adopted new technologies like coagulation, flotation (Suksaroj et al., 2005; Banerjee et al., 2007), advanced oxidation (Flous and Cabassud, 1999; Balcioglu et al., 2001), and sorption (Walker and Weatherly, 2000) for removal of color in textile effluents.

Micro filtration is used to remove contaminants having a particle size of 100–1000 nm. Bacteria and virus can be removed by ultrafiltration techniques, whereas the dissolved particles with a size of 1–5 nm are removed using nanofiltration. Vourch et al. (2008) stated that the dairy industry is generating a huge volume of effluents due to large water consumption and adopted reverse osmosis technique to treat wastewater and reused for cleaning, heating, and cooling process in dairy factories.

A high COD (90%) and color (80%) removal was achieved by treating denim wastewater using activated sludge process (as a pre-treatment) and nanofiltration technique. A high conductivity (8 mS/cm) was obtained by Sahinkaya et al. (2008) by implementing microfiltration techniques between activated sludge and nanofiltration processes to remove coarse particles.

Kumfer et al. (2010) and Samudro et al. (2010) stated that the ratio between BOD and COD value is <0.5 and hence the wastewater has to be treated either physically or chemically before the biological treatment.

Riera-Torres et al. (2010) attempted to treat textile wastewater using coagulation-flocculation and nanofiltration and their combination for removal of dye and reuse of treated water in textile industries and the results implied 98% color removal and a drastic reduction in sludge generation and COD value. The reactive dyes those contain azo-groups are having poor biodegradation property and hence new treatment techniques are needed for removal of color from effluents (Messina and Shculz, 2006; Yue et al., 2007).

Conventional treatment technologies such as physico-chemical and biological treatment methods are not suitable to recover and reuse of chemical and water from tannery effluents. In this context, many researchers adopted membrane technologies such as microfiltration, ultrafiltration, nanofiltration, and reverse osmosis and their combination processes to recover chemicals and water for reuse.

6.4 Advanced Wastewater Treatment Methods for Water Reuse

Several researchers/industrialists implemented membrane bioreactor technology (MBR) to treat industrial effluents (Sutton, 1996), food industries wastewater, and reuse of treated water for various applications. Sapari (1996) adopted combined treatment techniques (physico-chemical, biological, and land treatment) to treat textile wastewater and the results indicated that the treated water has a potential for reuse in fisheries and agriculture purpose and also meet the standards of regulatory bodies for discharge into the environment. Erswell et al. (1988) used charged membrane to treat reactive dye effluents and achieved high-quality water for reuse. Rozzi et al. (1999) achieved good quality permeate for reuse in textile industries (rinsing purpose) by treating wastewater using nanofiltration techniques.

Researchers are using advanced oxidation processes and electrochemical treatment techniques to treat industrial wastewater due to its unique features like cost effectiveness, versatility, and energy efficient (Gutierrez and Crespi, 1999; Lorimer et al., 2001). The electrochemical technique degrades organic substances present in textile effluents without generating byproduct or sludge. Fernandez et al. (2004) have successfully achieved 90% COD removal while treating the effluent using boron-doped diamond electrode and 94% dye removal was demonstrated by Anastasios et al. (2005) when treating the efflu-ents by electrochemical reactors. Scientists have successfully used graphite and noble metal anodes for oxidation of organic formulates (Naumczyk et al., 1996).

Mavrov et al. (2000) produced drinking water quality standards from wastewater of beverage industries by treating with nanofiltration and reverse osmosis. Few researchers produced water of reuse or recycle quality from wastewater of fruit juice industries (Noronha et al., 2002), fish meal industries (Afonso et al., 2002a,b), soybean soaking (Guu et al., 1997), chiller tank of poultry industries (Zhang et al., 1997), and wastewater of sausage products. Mavrov et al. (2001) using membrane bioreactor, nanofiltration, and reverse osmosis technology.

Tang and Chen (2002) obtained high-quality water for reuse (an average dye rejection of 98% and <14% of sodium chloride rejection) by treating the wastewater using thin-film composite polysulfone membranes. Bes-Pia et al. (2003a) attempted to treat wastewater by combining physico-chemical treatment and nanofiltration for reuse of water in textile industries and the results showed that the COD and conductivity level are within the allowable limits as per standards.

Bes-Pifi et al. (2002) obtained quality water for reuse in textile industry by treating the wastewater (socks, stockings, and panties industries) using physico-chemical and membrane technologies. Marcucci et al. (2001), Tang et al. (2002), Fersi et al. (2005), and Kim and Lee (2006) suggested to adopt membrane technologies for effective treatment of wastewater to control effluents and to reuse of treated water for industrial applications. High solid concentrations in textile effluents need very exhaustive pretreatments to avoid membrane deterioration and membrane fouling in membrane technologies (Petrov et al., 2003; Sostar-turk et al., 2005). Bes-Pia et al. (2003b) reported that membrane separation is the best wastewater treatment method to achieve high-quality water for reuse.

MBR system gave high efficiency compared to that of conventional systems and it produces high-quality water (Chang et al., 2006), better in removing contaminants, and generates less amount of sludge (Le-Clech et al., 2006). The treated water is reused for heat generation in boiler (Radjenovic et al., 2008). Fersi et al. (2005) stated that ultrafiltration process is an effective technique as a pre-treatment for nanofiltration process to treat textile effluents for reuse.

The coagulation-flocculation techniques are widely used for dye removal (Zhuang et al., 2006; Kim et al., 2007). The use of combined treatment techniques is widely reported to recover salts, water, and other compounds for reuse. Many scientists preferred to use coupling treatment techniques biological-membrane filtration (Zyłła et al., 2006), biological-adsorption (Chakraborty et al., 2005), and biological-ozonation (Bes-Pia et al., 2004; Barredo-Damas et al., 2005) to improve wastewater quality for reuse.

Mohan et al. (2007) and their team members found that the electro-chemical oxidation technique is more suitable to treat textile wastewater for the production of quality water for reuse in dyeing industries and critically examined the dye uptake and water characteristics during reuse of water in dyeing process. Sostar-Turk et al. (2008) stated that the effective removal of particles and macromolecules in textile effluents can be done by implementing ultrafiltration techniques.

The dead-end and cross-flow operation are the two types of operations in membrane technology. Mhurchu (2008) explained the usage of cross-flow operation, which reduces the formation of cake layer on the surface of the membrane. The textile and food industries wastewater can be successfully treated by MBR. Ceramic membranes show good performance in filtration of industrial effluents compared to that of polymer membranes. Ceramic membranes are inert, high resistance to chemical, and easy to clean, which was demonstrated by Ciora and Liu (2003), Jin et al. (2010), and Hofs et al. (2011).

Rodrigues et al. (2008) applied integrated photoelectrochemical oxidation-electrodialysis process to treat tannery effluents to obtain high-quality water for reuse in leather process [cleaning and washing processes] and found that removal of chromium, major pollutants, smell, and color in tannery effluents. Lin et al. (2012) recorded that MBR is superior to other conventional technologies.

Mutamim et al. (2012) published a review report on application of MBR for treating industrial wastewater and stated that the MBR is an efficient method in removing soluble organic waste while adding activated carbon as a fouling reducer. The solid retention time, hydraulic retention time, mix liquor suspended solids, membrane fouling, etc., are the factors to be considered during the treatment/operation. MBR is reliable, economical, and effective in removing pollutants and treated wastewater/effluents contain very low contaminants and free from pathogens, bacteria, and viruses.

6.5 Causes and Remedies of Advanced Methods

The membrane processes have some limitations like membrane fouling (due to pore blocking) and cake formation is responsible for rapid flux decline. Feris et al. (2008) tried a combination of ultrafiltration and nanofiltration membrane processes to reduce the effect of membrane fouling and found that removal of color, conductivity, and TDS were good and also permeate quality has been improved. The limitations of dual scheme process are the decline of flux and membrane fouling (Rodrigues et al., 2008).

Jegatheessan et al. (2016) and their team members critically reviewed the textile wastewater treatment techniques using membrane bioreactor and found that the major limitations are membrane fouling and flux decline and found that the lifespan of membrane has been decreased and concluded that combination of microfiltration/nanofiltration with membrane bioreactor recovered dyes and water efficiently. Limitation of the application of filtration technologies is the disposal of retentate/concentrate stream. Currently, the retentate is disposed either by evaporation or discharging into ocean and causes environmental pollution.

To overcome the membrane fouling and flux decline problem in membrane-based processes, an integrated separation technique has been introduced to achieve high-quality water for reuse (Dhale and Mahajami, 2000; Adbessemed and Nezzal, 2002; Meier et al., 2002; DasGupta et al., 2005). Apart from many advantages, MBR system has few disadvantages such as membrane fouling (Choo and Lee, 1996; Bouhabila et al., 2001; Hai et al., 2005; Sombatsompop, 2007), maintenance and operational cost, limitations in pressure, temperature, and pH (Kurian and Nakhla, 2006; Le-clech et al., 2006; Radjenovic et al., 2008). Yejian et al. (2008), Yuniarto et al. (2008), and Damayanti et al. (2011) attempted to modify the MBR system to minimize the above problems.

6.6 Conclusion

Due to large composition of textile effluents, no any specific treatment techniques are suitable for all kinds of textile wastewater and hence an integrated/combined treatment technique is essential to treat wastewater to meet pollution control board norms for discharge or reuse. The research in the area of wastewater treatment methods/techniques is being actively conducted to establish new technologies/techniques to treat wastewater from different process houses.

The stringent pollution control board regulations, highly polluted fresh water by industries, and increasing scarcity of drinking water are the major factors for the researchers to think further to elevate the quality of treated water to meet the standards for reuse.

In future, extensive research works have to be carried out to select suitable wastewater treatment technologies to treat various compositions of textile effluents at low cost and reuse of treated water for agriculture and drinking purposes.

References

Abdessemed D, Nezzal G. (2002). Desalination. 152:367–73.

Advances in Treating Textile Effluent. InTech, (2011) pp. 91e116.

Afonso MD, Borquez R. (2002a). Desalination. 142:29–45.

Afonso MD, Borquez R. (2002b). Desalination. 151:131–8.

Ali H. (2010). Biodegradation of synthetic dyes-a review. Water, Air, and Soil Pollution. 213(1–4):251–73.

Anastasios S, Konstantinos M, Ulrich N, Konstantinos F, Anastasios V. (2005). Evaluation of a novel electrochemical pilot plant process for azodyes removal from textile wastewater. Chemical Engineering Journal. 111:63–70.

Balcioglu FA, Arslan I. (2001). Partial oxidation of reactive dye bath by the O_3 and O_3/H_2O_2 processes. Water Science and Technology. 43:221–8.

Banerjee P, Dasgupta S, De S. (2007). Removal of dye from aqueous solution using a combination of advanced oxidation process and nanofiltration. Journal of Hazardous Materials. 140:95–103.

Barredo-Damas S, Iborra-Clar MI, Bes-Pia A, Alcaina-Miranda MI, Mendoza-Roca JA, Iborra-Clar A. (2005). Study of preozonation influence on the physical-chemical treatment of textile wastewater. Desalination. 182:267–74.

Bes-Pia A, Iborra-Clar A, Mendoza-Roca JA, Iborra-Clar MI, Alcaina-Miranda MI. (2004). Nanofiltration of biologically treated textile effluents using ozone as a pre-treatment. Desalination. 167:387–92.

Bes-Pia A, Mendoza-Roca JA, Alcaina-Miranda A, et al. (2003a). Combination of physico-chemical treatment and nanofiltration to reuse wastewater of a printing, dyeing, and finishing textile industry. Desalination. 157:73–80.

Bes-Pia A, Mendoza-Roca JA, Roig-Alcover L, Iborra-Clar A, Iborra-Clar MI, Alcaina-Miranda MI. (2003b). Comparison between nanofiltration and ozonation of biologically treated textile wastewater for its reuse in the industry. Desalination. 157:81–6.

Bes-Pifi A, Mendoza-Roca JA, Alcaina-Miranda MI, Iborra-Clar A, Iborra-Clal MI. (2002). Reuse of wastewater of the textile industry after its treatment with a combination of physico-chemical treatment and membrane technologies. Desalination. 149:169–74.

Bouhabila EH, Aim RB, Buisson H. (2001). Fouling characterisation in membrane biore-actors. Separation and Purification Technology. 22–23:123–32.

Bruggen BV, Curcio E, Driolli E. (2004). Process intensification in the textile industry: the role of membrane technology. Journal of Environmental Management. 73:267–74.

Chakraborty S, De S, Basu JK, DasGupta S. (2005). Treatment of a textile effluent: application of a combination method involving adsorption and nanofiltration. Desalination. 174:73–85.

Chakraborty S, Purkait MK, Dasgûpta S, De S, Basu JK. (2003). Nanofiltration of textile plant effluent for color removal and reduction in COD. Separation and Purification Technology. 31:141–51.

Chang J-S, Chang C-Y, Chen A-C, Erdei L, Vigneswaran S. (2006). Long-term operation of submerged membrane bioreactor for the treatment of high strength acrylonitrile-butadiene-styrene (ABS) wastewater: effect of hydraulic retention time. Desalination. 191:45–51.

Chang WS, Hong SW, Park J. (2002). Effect of zeolite media for the treatment of textile wastewater in a biological aerated filter. Process Biochemistry. 37:693–8.

Choo KH, Lee CH. (1996). Membrane fouling mechanisms in the membrane-coupled anaerobic bioreactor. Water Research. 30:1771–80.

Ciora RJ, Liu PKT. (2003). Ceramic membranes for environmental related applications. Fluid/Particle Separation Journal. 15:51–60.

Damayanti A, Ujang Z, Salim MR. (2011). The Influence of PAC, Zeolite, and Moringa oleifera as biofouling reducer (BFR) on hybrid membrane bioreactor of palm oil mill effluent (POME). Bioresource Technology. 102:4341–6.

DasGupta S, Chakraborty S, De S, Basu JK. (2005). Desalination. 174:73–85.

Dhahbi CFM. (2008). Treatment of textile plant effluent by ultrafiltration and/or nanofiltration for water reuse. Desalination. 222: 263–71.

Dhale AD, Mahajami VV. (2000). Waste Management. 20:85–92.

Dorange G, Maachi R, Cabon J, Chaabane T, Taha S, Taleb Ahmed M. (2004). Treatment of the tannery effluents from a plant near Algiers by nanofiltration (NF): experimental results and modeling. Desalination. 165:155e60.

Droste RL. Theory and Practice of Water and Wastewater Treatment, John Wiley & Sons, Inc., New York, 1997.

Erswell A, Brouchaert CJ, Buckley CA. (1988). The reuse of reactive dye liquors using charged ultra-filtration membrane technology. Desalination. 70:157–67.

Fernandes A, Morao A, Magrinho M, Lopes A, Goncalves I. (2004). Electrochemical degradation of C.I. Acid Orange 7. Dyes and Pigments. 61:287–96.

Fersi C, Gzara L, Dhahbi M. (2005). Treatment of textile effluents by membrane technologies. Desalination. 185:1825–35.

Flous V, Cabassud C. (1999). A hybrid membrane process for Cu (II) removal from industrial waste-water – comparison with a conventional process system. Desalination. 126:101–8.

Franklin LB. Wastewater Engineering: Treatment. Disposal and Reuse, McGraw Hill, Inc., New York, 1991.

Ganesh R, Boardman GD, Michelsen D. (1994). Fate of azo dyes in sludges. Water Research. 28:1367–76.

Gaston V. International Regulatory Aspects for Chemicals, Vol. I, CRC Press, Inc., New York, 1979.

Golob V, Vinder A, Simonic M. (2005). Efficiency of the coagulation/flocculation method for the treatment of dyebath effluents. Dyes and Pigments. 67:93–7.

Gupta VK, Ali I, Saleh TA, Nayak A, Agarwal S. (2012). Chemical treatment technologies for waste-water recycling-an overview. RSC Advances. 2:6380–8.

Gutierrez MC, Crespi M. (1999). A review of electrochemical treatments for colour elimination. Journal of the Society of Dyers and Colourists. 115:342–5.

Guu YK, Chiu CH, Young JK. (1997). Journal of Agricultural and Food Chemistry. 45:4096–100.

Hai FI, Yamamoto K, Fukushi K. (2005). Different fouling modes of submerged hollow-fiber and flat-sheet membranes induced by high strength wastewater with concurrent biofouling. Desalination. 180:89–97.

Hao OJ, Kim H, Chiang PC. (2000). Decolorization of wastewater. Critical Reviews in Environmental Science and Technology. 30(4):449–505.

Hart OO, Groves GR, Buckley CA, South-worth B. A guide for the planning, design and implementation of wastewater treatment plants in the textile industry. Part I: Closed loop treatment/recycle system for textile sizing/desizing effluents. Pretoria, 1983.

Hofs B, Ogier J, Vries D, Beerendonk EF, Cornelissen ER. (2011). Comparison of ceramic and polymeric membrane permeability and fouling using surface water. Separation and Purification Technology. 2011:365–74.

Holkar CR, Jadhav AJ, Pinjari DV, Mahamuni NM. (2016). A critical review on textile wastewater treatments: possible approaches. Journal of Environmental Management. 182:351–66.

Hutson DH, Roberts TR. Environmental Fate of Pesticides, Vol. 7, John Wiley & Sons, New York, 1990.

Jegatheesan V, Pramanik BK, Chen J, Navaratna D, Chang C, Shu L. (2016). Treatment of textile wastewater with membrane bioreactor: a critical review. Bioresource Technology. Available from: http://dx.doi.org/10.1016/j.biortech.2016.01.006.

Jin L, Ong SL, Ng HY. (2010). Comparison of fouling characteristics in different pore-sized submerged ceramic membrane bioreactor. Water Research. 44:5907–18.

Kant R. (2012). Textile dyeing industry an environmental hazard. Natural Science. 4:22e26. Available from: http://dx.doi.org/10.4236/ns.2012.41004.

Kim I, Lee K. (2006). Dyeing process wastewater treatment using fouling resistant nanofiltration and reverse osmosis membranes. Desalination. 192:246–51.

Kim T, Park C, Lee J, Shin E, Kim S. (2002). Pilot scale treatment of textile wastewater by combined process (fluidized biofilm process – chemical coagulation – electrochemical oxidation). Water Research. 36:3979–88.

Kim T, Park C, Yang J, Kim S. (2004). Comparison of disperse and reactive dye removals by chemical coagulation and Fenton oxidation. Journal of Hazardous Materials. 112:95–103.

Kim Y, Joo DJ, Shin WS, Choi J, Choi S, Kim MH, Ha TW. (2007). Dyes and Pigments. 73:59–64.

Kumfer B, Felch C, Maugans C. In: N.P.R.s. Association (Ed.). Wet air oxidation treatment of spent caustic in petroleum refineries, National Petroleum Refiner's Association Conference, Phoenix, AZ, 2010.

Kurian R, Nakhla G. (2006). Performance of aerobic MBR treating high strength oily wastewater at mesophilic—thermophilic transitional temperatures. Proceedings of the Water Environment Federation. 3249–55.

Le-Clech P, Chen V, Fane TAG. (2006). Fouling in membrane bioreactors used in waste-water treatment. Journal of Membrane Science. 284:17–53.

Lee J-W, Choi S-P, Thiruvenkatachari R, Moon H. (2006). Dyes and Pigments. 69:196–203.

Libra JA, Borchert M, Vigelahn L, Storm T. (2004). Two stage biological treatment of a diazo reactive textile dye and the fate of the dye metabolites, Chemosphere. 56:167–80.

Lin H, Gao W, Meng F, Liao BQ, Leung KT, Zhao L, Chen J, Hong H, Lorimer JP, Mason TJ, Plattes M, Phull SS, Walton DJ. (2001). Degradation of dye effluent. Pure and Applied Chemistry. 73(12):1957–68.

Lin H, Gao W, Meng F, et al. (2012). Membrane bioreactors for industrial wastewater treatment: a critical review. Critical Reviews in Environmental Science and Technology. 42(7):677–740.

Marcucci M, et al. (2001). Treatment and reuse of textile effluents based on new ultrafiltration and other membrane technologies. Desalination. 138:75–82.

Mavrov V, Bélières E. (2000). Desalination. 131:75–86.

Mavrov V, Fahnrich A, Chmiel H. (1997). Desalination. 113:197–203.

Meier J, Melin J, Eilers LH. (2002). Desalination. 146:361–6.

Messina PV, Shculz PC. (2006). Journal of Colloid and Interface Science. 299:305–20.

Mhurchu JN. In: Dead-End and Crossflow Microfiltration of Yeast and Bentonite Suspensions: Experimental and Modelling Studies Incorporating the Use of Artificial Neural Networks, School of Biotechnology, Dublin City University, 2008.

Mohan N, Balasubramanian N, Ahmed Basha C. (2007). Electrochemical oxidation of textile wastewater and its reuse. Journal of Hazardous Materials. 147:644–51.

Mutamim NSA, Noor ZZ, Hassan MAA, Olsson G. (2012). Application of membrane bioreactor technology in treating high strength industrial wastewater: a performance review. Desalination. 305:1–11.

Naumczyk J, Szpyrkowicz L, De Faveri M, Zilio Grandi F. (1996). Electrochemical treatment of tannery wastewater containing high strength pollutants. Transactions on I ChemE 74:58–9.

Noronha M, Britz T, Mavrov V, Janke HD, Chmiel H. (2002). Desalination. 143:183–96.

Orhon D, Babuna FG, Insel G. (2001). Characterization and modeling of denim-processing wastewaters for activated sludge. Journal of Chemical Technology and Biotechnology. 76:919–31.

Pagga U, Taeger K. (1994). Development of a method for adsorption of dyestuff on activated sludge. Water Research. 28:1051–7.

Pala A, Tokat E. (2003). Activated carbon addition to an activated sludge model reactor for color removal from a cotton textile processing wastewater. Journal of Environmental Engineering-ASCE. 129:1064–8.

Panchiao. Internal Technical Report for Eastern Textile Company. Taiwan, 1994, p. 15.

Papic S, Koprivanac N, Bozic AL, Metes A. (2004). Dyes and Pigments. 62:293–300.

Paul SA, Chavan SK, Khambe SD. (2012). Studies on characterization of textile industrial waste water in solapur city. International Journal of Chemical Sciences. 10:635e642.

Petrov S, et al. (2003). Ultrafiltration purification of waters contaminated with bifunctional reactive dyes. Desalination. 154:247–52.

Radjenovic J, Matosic M, Mijatovic I, Petrovic M, Barceló D. (2008). Membrane bioreactor (MBR) as an advanced wastewater treatment technology. The Handbook of Environmental Chemistry. 5:37–101.

Rajagopalan S. Pollution Management in Industries, vol. 7, Environmental Publication, Karad, India, 1989, pp. 31–3.

Rajamani S, Ravindranath E, Chita T, Umamaheswari B, Ramesh T, Suthantharajan R. (2004). Membrane application for recovery and reuse of water from treated tannery wastewater. Desalination. 164:151e6.

Ranganathan K, Karunagaran K, Sharma DC. Recycling of wastewaters of textile dyeing industries using advanced treatment technology and cost analysis – case studies. Resource Conservation and Recycling, in press (doi: 10.1016/j.resconrec. 2006.06.004).

Riera-Torres M, Gutierrez-Bouzan C, Crespi M. (2010). Combination of coagulation-flocculation and nanofiltration techniques for dye removal and water reuse in textile effluents. Desalination. 252:53–9.

Rodrigues MAS, Amado FDR, Xavier JLN, Streit KF, Bernardes AM, Ferreira JZ. (2008). Application of photoelectrochemical-electrodiaysis treatment for the recovery and reuse of water from tannery effluents. Journal of Cleaner Production. 16:605–11.

Rosell CM, Huertas JA. (1995). Ldpez, Indutria textil: depuracidn biol6gica o fisicoquimica? Revista de la lndustria Textil. 233:42–61.

Rozzi A, Antonelli M, Arcari M. (1999). Membrane treatment of secondary textile effluents for direct reuse. Water Science and Technology. 40(4–5):409–16.

Sahinkaya E, Uzal N, Yetis U, Dilek FB. (2008). Biological treatment and nanofiltration of denim textile wastewater for reuse. Journal of Hazardous Materials. 153:1142–8.

Sakalis A, Mpoulmpasakos K, Nickel U, Fytianos K, Voulgar-opoulos V. (2005). Evaluation of novel electrochemical pilot plant process for azodyes removal from textile wastewater. Chemical Engineering Journal. 111:63–70.

Samudro G, Mangkoedihardjo S. (2010). Review on BOD, COD and BOD/COD ratio: a tri-angle zone for toxic, biodegradable and stable levels. International Journal of Academic Research. 2:235–9.

Sapari N. (1996). Treatment and reuse of textile wastewater by overland flow. Desalination. 106:179–82.

Sombatsompop KM. In: Membrane fouling studies in suspended and attached growth membrane bioreactor systems, Asian Institute of Technology School of Environment, Resources & Development Environmental Engineering & Management, Thailand, 2007.

Sostar-Turk S, Simonic M, Petrinic I. (2005). Waste-water treatment after reactive printing. Dyes and Pigments. 64:147–52.

Suksaroj C, Heran M, Allegre C, Persin F. (2005). Treatment of textile plant effluent by nanofiltration and/or reverse osmosis for water reuse. Desalination. 178:333–41.

Sutton PM. (2006). Membrane bioreactors for industrial wastewater treatment: applicability and selection of optimal system configuration. Water Environment Federation. 3233–48.

Tang C, Chen V. (2002). Nanofiltration of textile waste-water for water reuse. Desalination. 143:11–20.

Van der Bruggen B, Kim JH, DiGiano FA, Geens J, Vandecasteele C. (2004). Influence of MF pretreatment on NF performance for aqueous solutions containing particles and an organic foulant. Separation and Purification Technology. 36:203–13.

Vourch M, Balannec B, Chaufer B, Dorange G. (2008). Treatment of diary industry wastewater by reverse osmosis for water reuse. Desalination. 219:190–202.

Vrcek IV, Bajza Z. (2001). Water quality analysis of mixtures obtained from tannery waste effluents. Ecotoxicology and Environmental Safety. 50:15e8.

Walker GM, Weatherly LR. (2000). COD removal from textile industry effluents pilot plant studies. Environmental Pollution. 108:201–11.

Wang Z, Xue M, Huang K, Liu Z. Textile dyeing wastewater treatment. In:

Wu MA, Eiteman, Law SE. (1998). Evaluation of membrane filtration and ozonation processes for treatment of reactive dye wastewater. Journal of. Environmental Engineering-ASCE. 124(3):272–7.

Yejian Z, Li Y, Xiangli Q, Lina C, Xiangjun N, Zhijian M, Zhenjia Z. (2008). Integration of biological method and membrane technology in treating palm oil mill effluent. Environmental Science. 20:558–64.

Yue Q-Y, Li Q, Su Y, Gao B-Y, Fu L. (2007). Journal of Hazardous Materials. 147:370–80.

Yuniarto A, Ujang Z, Noor ZZ. (2008). Performance of bio-fouling reducers in aerobic submerged membrane bioreactor for palm oil mill effluent treatment. Jurnal Teknologi. UTM 49:555–66.

Zhang SQ, Kutowy O, Kumar A, Malcolm I. (1997). Canadian Agricultural Engineering. 39(2):99–105.

Zhuang Y, Ren L, Shen J. (2006). Journal of Hazardous Materials. B136:809–15.

Zyłła R, Sojka-Ledakowicz J, Stelmach E, Ledakowicz S. (2006). Coupling of membrane filtration with biological methods for textile wastewater treatment. Desalination. 198:316–25.

7

Effect of Local Industrial Waste Additives on the Arsenic (V) Removal and Strength of Clay Ceramics for Use in Water Filtration

Amrita Kaurwar, Lovelesh Dave, Sandeep Gupta, and Anand Plappally*

Indian Institute of Technology Jodhpur, Jodhpur
E-mail: anandk@iitj.ac.in
*Corresponding Author

The study discusses investigations on the ability of modified clay ceramic water filtration units in removing arsenic (V) from water. The work involved manufacturing ceramic membrane plates of square shape with 100 cm^2 area and 1.5 cm thickness. The membranes were manufactured by sintering composites of local clay as matrix, carpentry residue as pore former, and additives such as marble processing slurry and ferrous mill waste added in order to affect functional properties. Two distinct sets of ceramics were separately prepared with additives of 10% marble slurry by volume and 10% ferrous mill waste by volume respectively to the control composition. The ceramic filtration plate manufactured from equal volumetric fraction of clay–carpentry residue mixture was the control. The influence of contact time and pH on the arsenic (V) removal was investigated. Arsenic removal from the water by using the modified ceramics was quantified using results from atomic absorption spectroscopy. It was found that 150 cm^3 of modified ceramics effectively removed up to 99% of arsenic in a stipulated time of 90 min. This provides an opportunity of scaling up these modified ceramics for use in arsenic removal on a large scale. The surface morphologies of these ceramic membranes were enumerated using imagery retrieved using scanning electron microscopy. The arsenic adsorption followed Freundlich isotherm in ceramics manufactured with marble slurry additive while those with ferrous mill waste followed Langmuir adsorption isotherm. The structural integrity of the modified filtration ceramics showed improvement with the addition of industrial additives.

7.1 Introduction

Solid waste accumulation does have preternatural adverse effects on ecology, environment, and health. India alone generates more than 100,000 metric tons of solid waste every day (Smith, 1972). Solid waste in toto includes animal and human waste, mining and industrial waste, household and construction waste, and agricultural waste. Marble from Rajasthan, India, contributes to 80% of the Earth's marble mining waste. Iron industries and manufacturing are a commonplace in every town in India. Tons of ferrous mill waste also contribute to the solid waste generation from industrial towns in Western Rajasthan. Rajasthan is well known for timber-based handicrafts. The heaps of timber waste generated from these handicraft enterprises usually remain unutilized (Pappu et al., 2007). Previous research on soils with high calcium and Fe found them to be used for construction of flexural stable cantilever structures and arsenic filtration, respectively (Zhang and Itoh, 2005; Escudero et al., 2009; Kaurwar et al., 2017; Satankar et al., 2017). Similarly, sawdust is a major raw material appended to clay to manufacture sterile water filtration structures (Plappally et al., 2011). The specific utility and character of elements contained in the solid wastes from Rajasthan discussed above provide goals within the ambit reflecting their probable use to resolve water filtration issues.

Occurrence of arsenic in distinct oxidation states in groundwater and subsurface aquifers has enforced many treatment challenges (Mohan and Pittman, 2007). Low-cost systems for efficient arsenic removal from water, especially for rural parts of developing nations (like India) are still being explored (Babel and Kurniawan, 2003). Clay-organic residue-based ceramics have been prepared for providing sustainable water solutions across the globe (Gupta et al., 2016). Low structural integrity of these ceramics is an impediment to this sustenance (Plappally, 2010). In order to jointly address the marble waste recycling, arsenic removal device, and structural integrity of clay ceramic water filtration materials, a novel idea is ushered through the following exegesis. Local industrial wastes (marble slurry, ferrous mill waste) from Western Rajasthan have been used as additives to the clay-organic residue-based mixture to develop modified clay ceramic water filtration membranes. Two distinct sets of membranes prepared with marble slurry and ferrous mill waste as additive have been evaluated and compared for their arsenic removal efficiency and structural integrity with the clay-organic residue-based ceramic membranes which act as a control.

7.2 Experimental

7.2.1 Apparatus

Atomic absorption spectroscopy (AA500 spectrophotometer, MRC, MNIT Jaipur, Jaipur, India) has been used to retrodict the influent concentration of As (V) and effluent As (V) presence. Standard stock solution of As (V) (5 g/L) was prepared using Na_2HAsO_4. $7H_2O$ (A6756, Merck) in distilled water. Other distinct concentrations were derived by diluting the basic solution to the desired concentration of As (V). The pH studies were performed using pocket-sized pH meter (Oakton pHTestr 30 Waterproof Pocket Tester). Scanning electron microscopy [Nova Nano FE-SEM 450 (FEI)] was used to determine the microstructure of the ceramics prior and post filtration process. A 50-ton Universal Material Testing Machine (Model EZ-50, Lloyd Instrument, Germany) was utilized to perform compression and flexural strength tests.

7.2.2 Materials and Fabrication

Local clay formed the matrix and the carpentry residue was the pore former (Plappally et al., 2011). The marble slurry and ferrous mill waste obtained from local mining and milling industry were separately (10%) added to the equal volume fraction of the clay–carpentry residue mixture to form the ceramics (Satankar et al., 2018). The three distinct sets of porous ceramics were fabricated, first using marble slurry-based-clay–sawdust mix (M-CC), second using iron powder based-clay–saw dust mix (F-CC), and the control with clay–saw dust mixture (CC). The materials were homogeneously mixed and blended together using water. The green blend was compacted in the form of square plates ($10 \times 10 \times 10$ cm^3) using 50-ton compaction machine (Model HBPO10, KEN-985-5000K, UK). The samples were kept for drying at ambient temperature conditions for a period of 1 week, and were then sintered at 850°C in an electric furnace [TEXCARE^{TM} Muffle Furnace 220 V, $250 \times 340 \times 180$ mm^3 (w $\times$ d $\times$ h)].

7.2.3 Adsorption Experiment

The effect of contact time was studied using 10 mg/L arsenic solution. The solution was allowed to percolate the membranes under the action of gravity (Plappally, 2010). To study adsorption kinetics, individual percolation experiments were conducted of the newly designed membranes by varying initial As (V) concentrations with arsenic concentrations of 50, 100, 150, and 200 mg L^{-1}, respectively (Liu et al., 2005; Mohan and Pittman, 2007; Mlilo et al., 2009; Plappally et al., 2011). The residual concentrations were noted after a contact time of 30 min. The quantity of adsorbed As (V) ions (mg/g) was calculated using Equation (7.1). The equilibrium sorption capacity q_e was calculated using the equation (Liu et al., 2005; Mlilo et al., 2009):

$$q_e = (C_o - C_e)V/M \qquad (7.1)$$

Here, C_o and C_e represent the influent As (V) concentration and membrane filtrate As (V) concentration after possible adsorption (mg/L). The parameter V represents the volume of the As (V) solution and M provides the mass (g) of the sorbent used in the reaction mixture. The removal efficiency was determined by computing the percentage adsorption using the formulae in Equation (7.2) (Liu et al., 2005).

$$\% \ As \ (V) = \frac{(C_o - C_e)}{C_o} \times 100 \qquad (7.2)$$

Further set of sorption studies were performed to exemplify percentage adsorption as a function of contact time. Here the shortest span of this sorption study was assumed to be 5-min contact time.

7.3 Result and Discussion

7.3.1 Effect of Contact Time

Arsenic ion removal efficiency increased with contact time. It was sufficient for each of the M-CC, F-CC, and control (CC) ceramics to reach an asymptote removal as depicted in Figure 7.1 after a contact time of 1.5 h. F-CC displayed highest removal efficiency ($\sim$99%) after 90 min of contact time during the batch experiment. Further $\sim$95% As (V) removal was observed while utilizing M-CC. It is posited that formation of complexes by additives (marble and ferrous additives) with hydrogen arsenate accentuated As removal (Goldberg, 2002; Jeong, 2005; Alexandratos et al., 2007).

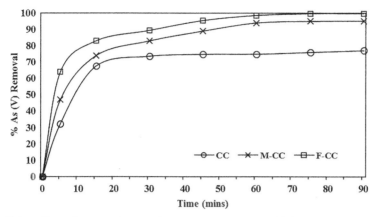

Figure 7.1 Effect of contact time on the arsenic (V) removal efficiency of distinct ceramic membranes.

Figure 7.2 SEM images from the surface of the ceramic membranes at 2000× before filtration (a) CC, (b) M-CC, and (c) F-CC.

7.3.2 Surface Morphology of Ceramics

7.3.2.1 Before filtration

The SEM images in Figure 7.2 show images of sintered ceramic membranes taken at 2000× magnification before the filtration process. The imagery showcases rough surface structures with irregularity in their shape, size, and distribution over the clay matrix. The addition of marble has revealed a comparatively dense microstructure with an average porosity of 39.36%. Large voids in the F-CC ceramic (Figure 7.2c) indicate in-homogeneity with an average porosity of 42.24%. The average porosity of CC ceramics (46.52%) experienced porosity reduction with addition of marble waste and iron scrap.

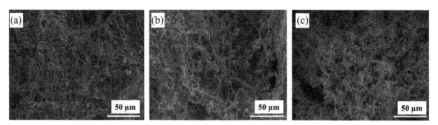

Figure 7.3 SEM images from the surface of the ceramic membranes at 2000× after filtration (a) CC, (b) M-CC, and (c) F-CC.

7.3.2.2 After gravity-based percolation

The SEM images of the ceramic surfaces after a contact time of 90 min are displayed in Figure 7.3. The visual difference in the surfaces before and post filtration can be observed from Figures 7.2 and 7.3. Adsorption of arsenic (V) onto the sides of the pores can be noticed along with the surface wetness possibly indicating new formations (Goldberg et al., 2007). The interior of the pore seems to be closed and the surface appears to be smoothly carpeted in comparison with their structures before filtration (Bothe and Brown, 1999; Goldberg et al., 2007).

The formation of bi-dentate As–Fe complexes is due to chelation and calcium arsenate formation, respectively (Bothe and Brown, 1999; Nriagu et al., 2007). These processes cover up the surface of the porous ceramic matrix supporting arsenic immobilization while using the marble slurry and milled iron scrap as additive materials for ceramic membrane manufacture (Bothe and Brown, 1999; Nriagu et al., 2007). This would mean that CC can be modified to MCC for arsenic and bacterial removal (Rizzotto, 2012).

7.3.3 Effect of pH

Figure 7.4 illustrates that addition of additives has accentuated arsenate removal efficiency over a large pH range in comparison to the control (CC). The addition of ferrous mill waste favored high arsenate adsorption in the acidic range, highest up to pH 7 (Goldberg, 2002).

The arsenate removal in case of M-CC increased with increasing pH up to a pH of 8. M-CC performed at its best in the range from pH of 6–8. The rise in arsenate removal efficiency in the extreme alkaline region in case of MCC may be attributed to co-precipitation of As (V) due to dissolved calcium as reported by other researchers (Nriagu et al., 2007).

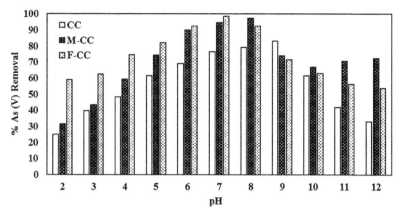

Figure 7.4 Effect of pH on the arsenic removal efficiency of distinct ceramic membranes.

In case of control (CC) sample, the decrease in the adsorption behavior at extreme acidic and alkaline pH values was likely due to the hydrolysis of clay, which resulted in dissolution of clay minerals (Bothe and Brown, 1999). The findings indicated that the change in pH influenced the surface charge of the adsorbent, which influenced sorption kinetics and sorption capacity of the heavy metal ions on to the solid clay ceramic surfaces (Bothe and Brown, 1999; Goldberg, 2002; Jeong, 2005; Alexandratos et al., 2007; Nriagu et al., 2007).

7.3.4 Adsorption Isotherm

The influence of As (V) uptake on its initial concentration on the sorption kinetics was deliberated using Freundlich and Langmuir adsorption isotherm analysis. The adsorption isotherm (Figure 7.5) in case of F-CC ceramic was better described by Freundlich isotherm. For M-CC and CC samples, Langmuir model displayed relatively better fit. From Table 7.1, high values of K_f and $1/n$ for M-CC and F-CC in comparison with CC indicated increased adsorption capacity and rise in intensity of adsorption. The rise in maximum adsorption capacity (q_e) as observed from Table 7.1 indicated that the addition of additives had a positive influence on arsenate adsorption.

7.3.5 Strength

The addition of industrial additive increased the compressive and flexural strength of ceramics as illustrated in Figure 7.6. The highest mechanical properties are obtained in case of M-CC samples. The presence of quartz

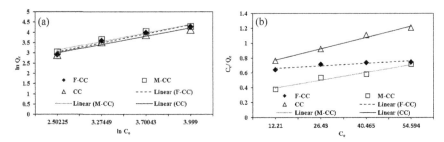

Figure 7.5 Freundlich adsorption isotherm (a) model for M-CC and Langmuir adsorption isotherm (b) model for F-CC.

Table 7.1 Constants for the corresponding adsorption isotherm models as depicted in Figure 7.5

Parameter	F-CC	M-CC	CC
Freundlich Isotherm Constants Corresponding to Figure 7.5a			
$1/n$	0.4417	0.4148	0.4048
K_f	13.486	15.36	13.4032
R^2	0.961	0.9564	0.9546
Langmuir Isotherm Constants Corresponding to Figure 7.5b			
B	0.053	0.3619	0.2434
q_e	29.85	9.4786	6.591
R^2	0.8487	0.9618	0.9872

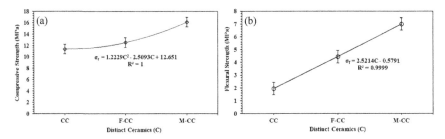

Figure 7.6 Compressive strength (a) and flexural strength behavior of distinct ceramics (b).

and calcium carbonates together in the M-CC ceramics is said to be one of the reasons for best mechanical properties (Saboya et al., 2007). The results resonated with the studies which involved incorporation of marble reject and iron tailings into the clay bricks for better structural properties (Chen et al., 2011). Addition of marble processing waste has been successfully incorporated to local clay ceramic water filters named G Filters in Rajasthan (Gupta et al., 2018; Kaurwar et al., 2018).

7.4 Conclusion

The investigation highlights the sustainable utilization of industrial wastes as additives in improving the arsenic (V) adsorption characteristics and mechanical properties of clay ceramics. The modified clay ceramics displayed up to 99% arsenic (V) removal for F-CC and $\sim$95% removal for M-CC in batch experiment study. The microstructure revealed adsorption and possible formation of complexes after batch studies. The Freundlich isotherm indicated increased As (V) intensity of adsorption and adsorption capacity. The modified ceramics displayed improved adsorption properties, with maximum adsorption capacity (for F-CC) almost approximately four times higher and maximum adsorption capacity for M-CC almost 1.4 times higher than the unmodified/controlled CC ceramics. The addition of additives enhanced the mechanical robustness of the clay ceramics. Marble processing waste addition to the control matrix may ensure development of new stronger materials capable of arsenic removal from contaminated drinking water.

References

Alexandratos VG, Elzinga EJ, Reeder RJ. (2007). Arsenate uptake by calcite: macroscopic and spectroscopic characterization of adsorption and incorporation mechanisms. Geochimica et Cosmochimica Acta. 71(17):4172–87.

Babel S, Kurniawan TA. (2003). Low-cost adsorbents for heavy metals uptake from contaminated water: a review. Journal of Hazardous Materials. 97(1–3):219–43.

Bothe JV, Brown PW. (1999). Arsenic immobilization by calcium arsenate formation. Environmental Science and Technology. 33(21):3806–11.

Chen Y, Zhang Y, Chen T, Zhao Y, Bao S. (2011). Preparation of eco-friendly construction bricks from hematite tailings. Construction and Building Materials. 25(4):2107–11.

Escudero C, Fiol N, Villaescusa I, Bollinger JC. (2009). Arsenic removal by a waste metal (hydr) oxide entrapped into calcium alginate beads. Journal of hazardous Materials. 164(2–3):533–41.

Goldberg S. (2002). Competitive adsorption of arsenate and arsenite on oxides and clay minerals. Soil Science Society of America Journal. 66(2):413–21.

Gupta S, Kaurwar A, Satankar RK, Usha K, Sharif MAR, Plappally AK. (2016). Flow, microbial filtration and petro-physical properties of ceramic plate ware gravity water filter during cyclic water loading events, in the Proceedings of From Pollution to Purification (ICW 2016), Dec. 12–15, 2016, Organized by IUIC, ASCEED & School of Environmental Sciences, Mahatma Gandhi University, Kottayam, Kerala.

Gupta S, Satankar R, Kaurwar A, Aravind U, Sharif M, Plappally A. (2018). Household production of ceramic water filters in Western Rajasthan, India. IJSLE. 13(1):53–66.

Jeong Y. (2005). The adsorption of arsenic (V) by iron (Fe2O3) and aluminum (Al2O3) oxides, PhD Thesis, Iowa State University.

Kaurwar A, Gupta S, Satankar R, Plappally A. (2017). Marble slurry as a potential ceramic water filtration material: Comparative analysis with machined Fe powder and clay ceramics for effectiveness in As removal from water at point of use, poster presented in: 3rd Int. Conf. Desalin. Using Membr. Technol., Spain, April, 2017.

Kaurwar A, Satankar RK, Dave L, Gupta S, Oomen J, Sharey M, Bodhankar S, Plappally AK. (2018). Use of clayey salty soils and its composite derivatives for construction and ceramics for household use in the Thar Desert in India. In Encyclopedia of Renewable and Sustainable Materials, Editors. Maleque et al., Elsevier.

Liu Y, Xu H, Tay JH. (2005). Derivation of a general adsorption isotherm model. Journal of Environmental Engineering. 131(10):1466–8.

Mohan D, Pittman Jr CU. (2007). Arsenic removal from water/wastewater using adsorbents—a critical review. Journal of Hazardous Materials. 142(1–2):1–53.

Mlilo TB, Brunson LR, Sabatini DA. (2009). Arsenic and fluoride removal using simple materials. Journal of Environmental Engineering. 136(4): 391–8.

Nriagu JO, Bhattacharya P, Mukherjee AB, Bundschuh J, Zevenhoven R, Loeppert RH. (2007). Arsenic in soil and groundwater: an overview. Trace Metals and Other Contaminants in the Environment. 9:3–60.

Pappu A, Saxena M, Asolekar SR. (2007). Solid wastes generation in India and their recycling potential in building materials. Building and Environment. 42(6):2311–20.

Plappally AK. (2010). Theoretical and Empirical Modeling of Flow, Strength, Leaching and Micro-Structural Characteristics of V Shaped Porous Ceramic Water Filters, PhD thesis, The Ohio State University.

Plappally AK, Yakub I, Brown LC, Soboyejo WO, Soboyejo ABO. (2011). Physical properties of porous clay ceramic-ware. Journal of Engineering Materials and Technology. 133(3):031004.

Rizzotto M. (2012). Metal complexes as antimicrobial agents. In *A Search for Antibacterial Agents*. Edited by V. Bobbarala, InTech, 355.

Saboya F, Xavier GC, Alexandre J. (2007). The use of the powder marble by-product to enhance the properties of brick ceramic. Construction and Building Materials. 21(10):1950–60.

Satankar RK, Kaurwar A, Gupta S, Plappally A. (2017). Horse dung and soil based composites for construction of aesthetic shelves in rural homes of Western Rajasthan. Journal of Environmental Nanotechnology. 6(2):43–7.

Satankar RK, Kaurwar A, Gupta S, Usha K, Azeko ST, Soboyejo WO, Soboyejo ABO, Plappally A. Role of Equine Ordure in Enhancing Physical and Mechanical Properties of Natural Bio-active Composites, in Advanced Polymeric Materials For Sustainability And Innovations, Editor(S): Rouxel D et al., Apple Academic Press, 370, 2018.

Smith VL. (1972). Dynamics of waste accumulation: disposal versus recycling. The Quarterly Journal of Economics. 86(4):600–16.

Zhang FS, Itoh H. (2005). Iron oxide-loaded slag for arsenic removal from aqueous system. Chemosphere. 60(3):319–25.

8

Treatment of Whey Water from Food Processing Units Using Hybrid Methods

Britika Mazumdar[1], Gargi Biswas[2], Rajnarayan Saha[3], and Susmita Dutta[2,*]

[1]Department of Earth and Environmental Studies, National Institute of Technology Durgapur, Durgapur, India
[2]Department of Chemical Engineering, National Institute of Technology Durgapur, Durgapur, India
[3]Department of Chemistry, National Institute of Technology Durgapur, Durgapur, India
E-mail: gargi129@gmail.com; susmita_che@yahoo.com
*Corresponding Author

Whey water discharging from various dairy processing units deteriorates water quality, as it has high chemical oxygen demand (COD), biological oxygen demand (BOD), and also promotes eutrophication, and harbors pathogens. Characterization of raw whey water gave significantly high values of COD (32,000 mg/L) and BOD_5 at 20°C (3120 mg/L) and BOD/COD ratio (0.098), unsuitable for direct biological treatment. In the present study, Fenton's oxidation process followed by biological treatments was done to treat the whey water using green algae and bacteria, individually. Fenton's oxidation showed a reduction in COD value (77.3%) and BOD value (16.08%) and increase in BOD/COD ratio (0.36), which made sample suitable for the biological treatment. Analysis of the whey water after Fenton's oxidation using ion chromatography showed the presence of Na^+, NH_4^+, K^+, Ca^{2+}, and PO_4^{3-} in an excess amount to that of the permissible limits. Algal growth was checked for different nitrate (NO_3^-) concentrations ranging from 0.002–1.0 g/L and at different dilutions of whey water with 5% inoculum at pH 7. Nitrate concentration of 0.01 g/L and 10 times diluted whey water showed maximum algal growth. Total chlorophyll, protein, carbohydrate, and lipid contents increased 15.35 times, 5.82 times, 10.16 times, and 13.68 times, respectively for 10 days cultivation. Fourier transform infrared (FTIR) analysis showed the involvement of functional groups such as $C-O$, $N=O$, $C=O$, $O-H$, $C=H$, and $N-H$, in the reduction of organic loading from whey water. Treatment with green algae showed 74.64% BOD and 65.98% COD removal. Bacteria showed maximum growth at 10% inoculum size at pH 7 when inoculum size was varied from 2.5 to 10%. Removal of BOD and COD was found as 83.02 and 75.82%, respectively, for bacterial degradation. Thus, it can be inferred that such hybrid treatment comprising Fenton's oxidation followed by either algal treatment or bacteriological treatment can be used to reduce pollution load in terms of COD and BOD. Dry biomass of micro-algae can further be used for biofuel production.

8.1 Introduction

Whey is the liquid portion obtained after curdling of milk. These effluents may create serious problems of organic burden on the local municipal sewage treatment systems due to the presence of a very high concentration of organic matter in them (Janczukowicz et al., 2008). The organic burden in terms of COD and BOD on the water bodies subsequently initiates the growth of pathogens, causes eutrophication, and creates a negative impact on aquatic environments. Whey wastewater causes pollution, mainly in the water bodies as whey contains sugar. Bacteria would use the whey sugars as food and reproduce in the water bodies. As the number of bacteria increased, it would use up the oxygen, so oxygen levels would decrease in the water. Fish and other living organisms would start to die as the oxygen level decreased. Green algae offer a promising step toward wastewater treatment. They provide a tertiary biotreatment and produce potentially valuable biomass, which can further be used for several purposes (de la Noüe et al., 1992). Bacterial metabolism determines the effectiveness of the biological treatment of wastewater. In the present study, Fenton's oxidation followed by biological treatments was done to treat the whey water using green algae and bacteria, individually. Fenton's oxidation is one kind of AOP. The main advantage of AOPs compared to other classical methods of wastewater treatment is their extremely destructive nature. The process results in total mineralization of organic waste with no or exceptionally low additional waste generation (Gogate and Pandit, 2004). COD to hydrogen peroxide ratio and hydrogen peroxide to Fe (II) ratio are the controlling parameters affecting the decomposition activity of the Fenton's reagent. The mundane world of wastewater treatment thrives on the cyclical, green economy. Algae have the potential to efficiently remove nutrients from wastewater, consume the carbon dioxide released from the bacterial activity, and provide a source of biomass. Algal biomass can be used for several purposes, including biogas, biofuels, fertilizers, and biopolymer production. Bacteria play an important role in the treatment of wastewater. The bacteria will remove a good portion of the organic matter present in the wastewater, consuming oxygen and releasing carbon dioxide. Wastewater treatment plants try to optimize this ability of bacteria to "eat" the organics in wastewater by providing ideal conditions for their growth and metabolism. Farmlands, subsequently, use the treated sludge as a form of fertilizer in the same manner as that of using horse manure in the household garden (Big River Citizens Education Guide: The Water Use Cycle, Missouri Department of Natural Resources).

8.2 Materials and Methods

8.2.1 Whey Water Sample Collection and Characterization

Whey water sample (grab sample) required for the project work was collected from sweet processing units. pH of the sample was measured as 5.6. The sample was kept inside the refrigerator for further analysis. BOD_5 at 20°C and COD were measured following the standard APHA protocols (Clesceri et al., 1996). A seed tank of 2 L was maintained with continuous aeration and supply of nutrients for preparing the seed water. Sludge for the seed tank was collected from the nearby sewage treatment plant. COD was measured by the closed reflux method.

8.2.2 Treatment of Whey Water using Fenton's Oxidation Method and its Analysis using Ion Chromatography

Fenton's oxidation was carried out to increase the biodegradability ratio of whey water for subsequent biological treatment. At first, the strength of hydrogen peroxide was determined to assess the ability of the commercial hydrogen peroxide to reduce the interference during COD determination. Raw sample (200 mL) was taken and calculated amount of 50% commercial grade H_2O_2 and $FeSO_4.7H_2O$ were added. COD: $H_2O_2 = 1:1$(wt:wt) and H_2O_2: $Fe^{2+} = 50:1$ (molar ratio) were used for calculating the dosage of H_2O_2 and $FeSO_4.7H_2O$. The mixture was stirred throughout the process. At definite time intervals, samples were withdrawn into separate conical flasks, followed by the addition of sulfuric acid solution (1:20) and KI, and was kept in dark for 15 min. After taking this out, few drops of ammonium molybdate solution were added and were titrated against 0.1 N sodium thiosulfate solution until the sample color turned pale. Then few drops of starch indicator were added and titrated as before until dark blue color changed to colorless. Thus, residual peroxide was determined after the treatment at different time intervals. Similarly, samples were also withdrawn at the same time intervals for BOD and COD determination. Whey water after Fenton's oxidation was analyzed for cations and anions using ion chromatography system (Metrohm India Ltd., India). Columns used for cations and anions were Metrosep C4-150 mm and Metrosep A Supp 5–250 mm, respectively.

8.2.3 Treatment using Green Algae

8.2.3.1 Treatment of whey water using green algae at various nitrate concentrations after Fenton's oxidation

Nitrate and phosphate are the two important nutrients used in BG 11 media for the growth of algae. Compositions of BG 11 media used per liter of distilled water are—$NaNO_3$ (1.5 g), K_2HPO_4 (0.04 g), $MgSO_4.7H_2O$ (0.075 g), $CaCl_2.2H_2O$ (0.036 g), citric acid (0.006 g), ferric ammonium citrate (0.006 g), EDTA (0.001 g), Na_2CO_3 (0.02 g), and Trace metal solution @ 1 mL/L [H_3BO_3 (2.86 g), $MnCl_2.4H_2O$ (1.81 g), $ZnSO_4.7H_2O$ (0.22 g), $NaMoO_4.2H_2O$ (0.39 g), $CuSO_4.5H_2O$ (0.079 g), $Co(NO_3)_2.6H_2O$ (49.4 g)] (Anjana et al., 2007). In the present study, ionic concentrations of whey water after Fenton's oxidation showed the presence of PO_4^{3-} along with Na^+, NH_4^+, K^+, and Ca^{2+}, in excess to that of permissible limits. Thus, an attempt was made to curtail the usage of PO_4^{3-} completely and also Na^+, NH_4^+, K^+, and Ca^{2+} during the preparation of BG 11 media to be used as a dilution medium for whey water. This results in the reduction of chemical cost and the process becomes economical.

To determine optimum nitrate concentration, algal growth was observed in pure BG 11 media at different NO_3^- concentrations ranging from 0.002 to 1 g/L with 5% inoculum size at pH 7, separately. After assessing the optimum NO_3^- concentration, treated whey water was diluted 0, 5, and 10 times using BG 11 media prepared by omitting the use of ions Na^+, NH_4^+, K^+, Ca^{2+}, and PO_4^{3-} which were present in excess in the whey water, at optimum NO_3^- concentration and at pH 7. To determine the optimum dilution factor, the green algal strain was incubated in diluted whey water with 5% inoculum size in an Algal incubator (Lab-X, Model No. BIL10LX) at 25 ± 1°C, 2400 lux, L/D 16/8 for 14 days. Samples were collected at 2 days interval and algal biomass was separated by centrifugation (ELTEK centrifuge TC 8100 F) at 5000 r/min for 15 min. Supernatants were collected for analysis of BOD and COD in the whey water after the treatment with green algae. The pellet was dried overnight at 60°C in a hot air oven (Universal Hot Air Oven) and dry biomass was weighed using a weighing balance (Precisa XR 205SM-DR).

8.2.3.2 Estimation of bio-molecules content of spent green algal biomass and its FTIR spectroscopy analysis

Samples of green algae grown at optimized NO_3^- concentration were collected for 14 days at a time interval of 2 days. The concentration of chlorophyll, protein, carbohydrate, and lipid present in the samples was then determined. Protein content in the green algae sample was estimated by Folin-Lowry method (Lowry et al., 1951). The procedures were given by Clesceri et al. (1996) and Mishra et al. (2013) and were followed for determination of carbohydrate and total chlorophyll content in green algae, respectively. Lipid content in green algae was measured as per Bligh and Dyer method (Bligh and Dyer, 1959).

The functional groups present on the cell wall of the green algae samples, before and after the addition of whey water, were studied using an FTIR spectrometer (Spectrum 100, Perkin-Elmer, USA). Green algae samples were centrifuged at 5000 r/min for 15 min. The pellets were dried at 60°C overnight in a hot air oven. The dry pellet was then used for FTIR analysis.

8.2.4 Treatment of Whey Water using *Bacillus Subtilis* at Various pH After Fenton's Oxidation

As a parallel step of green algal treatment, bacterial treatment of whey water was done after Fenton's oxidation. Whey water was treated with *Bacillus subtilis* to see its efficiency in removing organic loading in terms of BOD and COD. Bacterial sub-cultures were grown and maintained in LB broth (Miller), as well as agar plates. Compositions of LB media (per liter of distilled water) were peptone (5 g), yeast extract (3 g), dextrose as a carbon source (5 g), sodium chloride (10 g), and agar (20 g). Agar plates were done with 2% agar in distilled water and the bacterial strain was maintained on the plates after solidifying under aseptic conditions. OD at inoculum sizes ranging from 2.5 to 10% was taken for *B. subtilis* when grown in 10 times diluted whey water to check the optimum inoculum size. Readings were taken for 22 h at 2 h time interval using visible spectrophotometer (Genesys 20, Thermo Scientific) at a wavelength of 600 nm. All dilutions were done using

LB media. Bacterial broths were incubated in BOD Incubator & Shaker (Modern Instrument, Kolkata, India) at 30°C and shaken at 900 rps.

The effect of pH on the growth of *B. subtilis* was also tested by growing them in 10 times diluted whey water at pH 4, 7, and 10 and at optimum inoculum size for 22 h. Biomass of 50 mL sample was taken at 2 h time interval. The samples were centrifuged and supernatants were collected for BOD and COD analysis of the sample.

8.3 Results and Discussion

8.3.1 Whey Water Sample Collection and Characterization

The value for BOD_5 at 20°C for raw whey water was obtained as 3120 mg/L, COD as 32,000 mg/L, and thus, the ratio of BOD and COD becomes 0.098. As per IS:2490(Part I)–1981, tolerance limits for industrial effluents (http://www.worldenviro.com/effstd.htm), the value for BOD_5 at 20°C and COD for inland discharge must be within 30 and 250 mg/L, respectively. Compared to these values, both the values found in the present investigation for whey water were quite high for subsequent biological treatment. As a reason of this, Fenton's oxidation treatment was carried out to adjust the biodegradability ratio or BOD/COD ratio within the range of 0.3–0.5 and make the sample suitable for further biological treatment using algae or bacteria.

8.3.2 Treatment of Whey Water using Fenton's Oxidation Method and its Analysis using Ion Chromatography

The strength of raw commercial hydrogen peroxide was obtained as 782 mg/mL. For a 200 mL sample, the requirements of H_2O_2 and $FeSO_4.7H_2O$ were 8.2 mL and 0.051 g, respectively. Figures 8.1(a) and (b) show the variation of BOD and COD with time during Fenton's oxidation treatment for 3 h, respectively. BOD and COD were reduced to 2618 and 7274.1 mg/L, respectively, after Fenton's oxidation treatment for 3 h. Therefore, the final percentage reductions of BOD and COD after 3 h were 16.08 and 77.30%, respectively.

The ratio of BOD and COD ratio increased from 0.098 to 0.36, after Fenton's oxidation. Wastewater having a BOD/COD ratio within the range of 0.3–0.5 is considered to be easily treatable by biological means. If the ratio is below 0.3, either the wastewater has some toxic components or acclimated microorganisms may be required in its stabilization (Janna, 2016). Fenton's

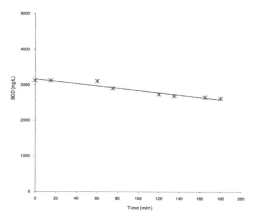

Figure 8.1(a) Concentration of BOD (mg/L) in whey water with time after Fenton's oxidation.

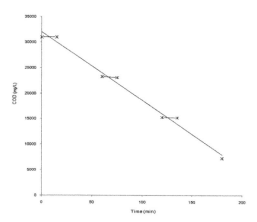

Figure 8.1(b) Concentration of COD (mg/L) in whey water with time after Fenton's oxidation.

oxidation was found to be a suitable treatment method for reducing the organic load present in the whey water in terms of BOD and COD. Moreover, it helped in achieving the required biodegradability ratio so that the same can then be used for further biological treatment.

Ion chromatography gave the cation concentrations for sodium as 474.077 ppm (permissible limit: 200 ppm as per WHO) (http://www.who.int/water_sanitation_health/dwq/chemicals/sodium.pdf), potassium as 1501.193 ppm (permissible limit: 10 ppm as per WHO) (http://www.who.int/water_sanitation_health/water-quality/guidelines/chemicals/potassium-background.

pdf), ammonium as 30.513 ppm (desirable limit: 0.5 ppm as per WHO) (http://www.who.int/water_sanitation_health/dwq/ammonia.pdf?ua=1), calcium as 775.922 ppm (permissible limit: 200 ppm as per IS 10500-1991), and magnesium as 83.251 ppm (permissible limit: 100 ppm as per IS 10500-1991) (http://www.indiawaterportal.org/sites/indiawaterportal.org/files/drinking_water_standards_is_10500_1991_bis.pdf). Similarly, it gave the anion concentrations for chloride as 929.343 ppm (permissible limit: 1000 ppm as per IS 10500-1991), nitrate as 6.815 ppm (permissible limit: 100 ppm as per IS 10500-1991), phosphate as 330.041 ppm (permissible limit: 5 ppm as per Environmental Protection Rules, EPR, 1986) (http://cpcb.nic.in/GeneralStandards.pdf), and sulfate as 225.790 ppm (permissible limit: 400 ppm as per IS 10500-1991).

Therefore, it is seen that ions Na^+, NH^{4+}, K^+, Ca^{2+}, and PO_4^{3-} were in excess to that of their permissible limits in the whey water and, thus, were not added while preparing BG 11 media for algal growth in subsequent steps, resulting in the reduction of chemical costs.

8.3.3 Treatment Using Green Algae

8.3.3.1 Treatment of whey water using green algae at various nitrate concentrations after Fenton's oxidation

Figure 8.2 shows the variation of biomass concentration of green algae with time at various values of NO_3^- concentration. The green algae showed maximum growth at 0.01 g/L NO_3^- concentration, when nitrate concentration was varied from 0.002–1 g/L at pH 7 and 5% inoculum size. The curve for biomass production showed a constant increment with time. Biomass concentration increased from 0.0031 g/L at the zeroth day to 0.0124 g/L on the 14th day, at optimum conditions. Thus, it can be concluded that green algae can be grown on BG 11 medium at 0.01 g/L NO_3^- concentration and pH 7, resulting in lesser usage of nitrate in the form of $NaNO_3$.

Figure 8.3 shows the variation of biomass concentration of green algae with time at different dilutions of whey water, at 0.01 g/L NO_3^- concentration, pH 7, and 5% inoculum size. BG 11 media, prepared by omitting ions Na^+, NH^{4+}, K^+, Ca^{2+} jmn and PO_4^3, was used as diluting medium. The green algae exhibited maximum growth at 1;0 times diluted whey water. The curve for biomass production showed a constant increment; with time. Biomass concentration increased from 0.078 g/L at the zeroth day to 0.572 g/L on the 14th day, compared to its growth on five times diluted whey water (0.501 g/L at 14th day) and in the said BG 11 media (0.529g/L at 14th day). Thus, it can

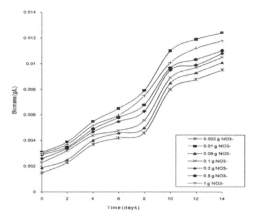

Figure 8.2 Biomass (g/L) versus time (days) plot for determining optimum NO_3^- concentration required for the growth of green algae in pure BG 11 media, at a constant pH of 7, and an inoculum size of 5%.

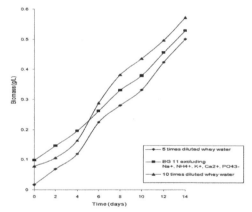

Figure 8.3 Biomass (g/L) versus time (days) plot for determining optimum dilution for whey water required for the growth of green algae when BG 11 media was prepared by not using ions Na^+, NH^{4+}, K^+, Ca^{2+}, and PO_4^{3-} at 0.01 g/L NO_3^- concentration, pH 7, and inoculum size of 5%.

be concluded that green algae can grow best on whey water at 0.01 g/L NO_3^- concentration and pH 7, when 10 times dilution is done using BG 11 media prepared by not using Na^+, NH^{4+}, K^+, Ca^{2+}, and PO_4^{3-}.

The values of BOD and COD in whey water reduced from 2618 to 663.93 mg/L and from 7274.1 to 2547.38 mg/L, respectively. Figure 8.4 shows that the percentage reductions of BOD and COD in whey water are 74.64 and

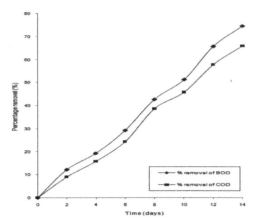

Figure 8.4   Percentage removal of BOD and COD from whey water with time (days) after treatment with green algae when dilution was made with BG 11 media prepared by not using ions Na^+, NH^{4+}, K^+, Ca^{2+}, and PO_4^{3-} at 0.01 g/L NO_3^- concentration, pH 7, and inoculum size of 5%.

64.98%, respectively, after algal treatment at 10 times dilution, pH 7, and 0.01 g/L NO_3^- concentration. Therefore, it can be concluded that green algae were found suitable for reducing BOD and COD from whey water.

8.3.3.2 Estimation of bio-molecules content of spent green algal biomass and its FTIR spectroscopy analysis

Variations of bio-molecules were checked for algae in BG 11 media, prepared by not using the ions Na^+, NH_4^+, K^+, Ca^{2+}, and PO_4^{3-}, and in 10 times diluted whey water, separately at optimum nitrate concentration of 0.01 g/L and pH 7. Figures 8.5(a)–(d) show the variation in total chlorophyll, protein, carbohydrate, and lipid contents with time, respectively, for 14 days. Total chlorophyll content increased 7.55 times (0.544–4.102 mg/L) for algae grown in BG 11 media prepared by not using Na^+, NH_4^+, K^+, Ca^{2+}, and PO_4^{3-} and 15.35 times (0.324–4.971 mg/L) for algae grown in diluted whey water. Protein content increased three times (38.598–115.125 mg/L) for algae grown on BG 11 media prepared by not using Na^+, NH_4^+, K^+, Ca^{2+}, and PO_4^{3-} and 5.82 times (23.256–135.125 mg/L) for algae grown in diluted whey water. Carbohydrate content increased four times (86.8–334.8 mg/L) for algae grown in BG 11 media prepared by not using Na^+, NH_4^+, K^+, Ca^{2+}, and PO_4^{3-} and 10.16 times (36–365.8 mg/L) for algae grown in diluted whey water. Lipid content increased 6.93 times (59.857–414.286 mg/L)

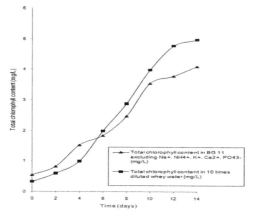

Figure 8.5(a) Variation of total chlorophyll content (mg/L) with time (days) for algae in BG 11 media, prepared by not using Na^+, NH_4^+, K^+, Ca^{2+}, and PO_4^{3-} and in 10 times diluted whey water, at 0.01 g/L NO_3^- concentration, pH 7, and inoculum size of 5%.

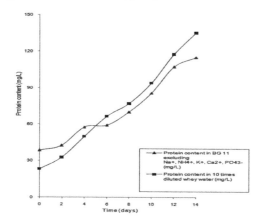

Figure 8.5(b) Variation of protein content (mg/L) with time (days) for algae in BG 11 media, prepared by not using Na^+, NH_4^+, K^+, Ca^{2+}, and PO_4^{3-} and in 10 times diluted whey water, at 0.01 g/L NO_3^- concentration, pH 7, and inoculum size of 5%.

for algae grown in BG 11 media prepared by not using Na^+, NH_4^+, K^+, Ca^{2+}, and PO_4^{3-} and 13.68 times (35.257–441.429 mg/L) for algae grown in diluted whey water. The concentration of the respective bio-molecules showed constant increments with time. It was observed that green algae were able to perform more photosynthetic activity when grown in 10 times diluted whey water after Fenton's oxidation, due to the presence of ions like Na^+, NH_4^+, K^+, Ca^{2+}, and PO^{3-} in excess to their permissible limits in the whey water itself. Since the whey water already contained the said ions in

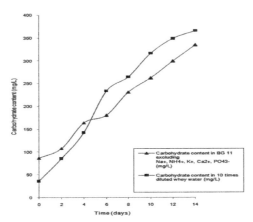

Figure 8.5(c)   Variation of carbohydrate content (mg/L) with time (days) for algae in BG 11 media, prepared by not using Na^+, NH_4^+, K^+, Ca^{2+}, and PO_4^{3-} and in 10 times diluted whey water, at 0.01 g/L NO_3^- concentration, pH 7, and inoculum size of 5%.

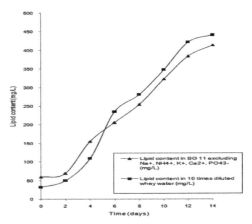

Figure 8.5(d)   Variation of lipid content (mg/L) with time (days) for algae in BG 11 media, prepared by not using Na^+, NH_4^+, K^+, Ca^{2+}, and PO_4^{3-} and in 10 times diluted whey water, at 0.01 g/L NO_3^- concentration, pH 7, and inoculum size of 5%.

greater amounts than their respective permissible limits, it itself acted as a supplement for those ions. Therefore, the present study shows an avenue to treat whey water in a more economical way.

Figures 8.6(a) and (b) show the FTIR spectra of dry biomass of green algae grown in BG 11 media, prepared by not using the ions Na^+, NH_4^+, K^+,

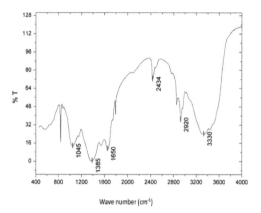

Figure 8.6(a) FTIR spectra of dry biomass of green algae grown in BG 11 media, prepared by not using Na^+, NH_4^+, K^+, Ca^{2+}, and PO_4^{3-}, at zeroth day, at 0.01 g/L NO_3^- concentration, pH 7, and inoculum size of 5%.

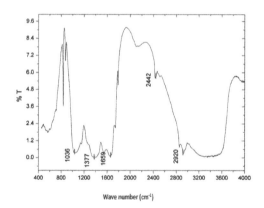

Figure 8.6(b) FTIR spectra of dry biomass of green algae grown in BG 11 media, prepared by not using Na^+, NH_4^+, K^+, Ca^{2+}, and PO_4^{3-}, at 10th day, at 0.01 g/L NO_3^- concentration, pH 7, and inoculum size of 5%.

Ca^{2+}, and PO_4^{3-}, at zeroth day and 10th day at varied wavenumber 400–4000 cm^{-1}, respectively. The characteristic bonds present on the cell wall of green algae culture showed shifting in wavenumbers when grown for 10 days. $C-O$ stretching vibration shifted from wave numbers of 1045 to 1036 cm^{-1}, $N=O$ shifted from 1385 to 1377 cm^{-1}, and $C=O$ shifted from 1650 to 1659 cm^{-1}. Three $O-H$ stretching vibrations were seen at wavenumbers 2434, 2920, and 3330 cm^{-1} at zeroth day, whereas two numbers $O-H$ bonds remained after the 10th day at wavenumbers 2442 and 2920 cm^{-1},

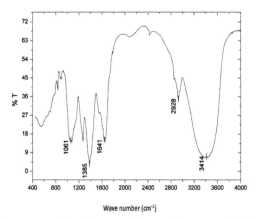

Figure 8.6(c)   FTIR spectra of dry biomass of green algae grown in 10 times diluted whey water at zeroth day, at 0.01 g/L NO$_3^-$ concentration, pH 7, and inoculum size of 5%.

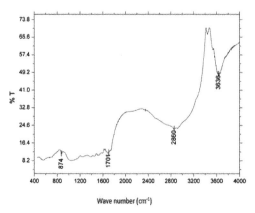

Figure 8.6(d)   FTIR spectra of dry biomass of green algae grown in 10 times diluted whey water at 10th day, at 0.01 g/L NO$_3^-$ concentration, pH 7, and inoculum size of 5%.

respectively. Shifting of wavenumbers for C−O, N=O, and O−H bonds showed their involvement in the treatment method by getting consumed for the said purpose.

In the same manner, the study was done for green algal biomass grown in 10 times diluted whey water. Figures 8.6(c) and (d) show the FTIR spectra for algae grown in 10 times diluted whey water at zeroth and 10th day of incubation. C=O stretching vibration shifted from wavenumbers of 1641 to 1701 cm^{-1}, O−H stretching vibration changed from a single number at 2928 cm^{-1} to double the number at 2860 and 3636 cm^{-1} wavenumbers, respectively. Moreover, C−O, N=O, and N−H stretching vibrations were seen to be present in the zeroth day sample but absent on 10th day. Bond C−H was present in the 10th day sample. C−O, N=O, and N−H stretching vibrations were completely used for treatment purpose and thus were seen to be absent in the 10th day sample.

8.3.4 Treatment of Whey Water using *Bacillus Subtilis* After Fenton's Oxidation

Figure 8.7 shows the variation of OD with time for determining the optimum inoculum size of *B. subtilis* in 10 times diluted whey water. *B. subtilis* gave maximum growth at 10% inoculum size in 10 times diluted whey water, when inoculum size was varied from 2.5 to 10%. The curve for OD showed constant increment with time, indicating the growth of the bacteria in whey water. Biomass concentration increased from 0.035 g/L at the zeroth hour to 0.901 g/L on the 22nd hour, at optimum conditions. It can be concluded that *B. subtilis* showed the best growth on 10 times diluted whey water when inoculum size is kept at 10%.

Figure 8.8 shows the variation in biomass concentration with time, for the growth of *B. subtilis* in whey water, at different pH values and 10% inoculum sizes. *B. subtilis* exhibited maximum growth at pH 7 in 10 times diluted whey water, when pH was varied from 4 to 10. The curve for biomass production shows constant increment with time. Biomass concentration increased from

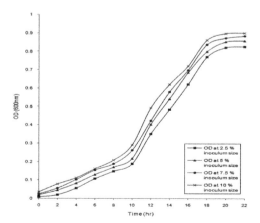

Figure 8.7 OD (nm) versus time (h) plot for determining the optimum inoculum size of *Bacillus subtilis* required for its growth in 10 times diluted whey water.

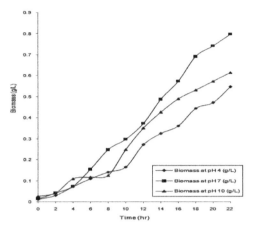

Figure 8.8   Biomass (g/L) versus time (h) plot for determining the optimum pH required for the growth of *Bacillus subtilis* in 10 times diluted whey water, at 10% inoculum size.

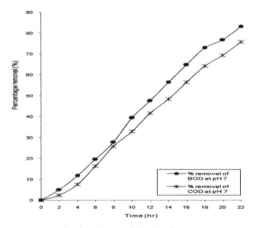

Figure 8.9   Percentage removal of BOD and COD from whey water with time (h) after treatment with *Bacillus subtilis* at pH 7 and inoculum size of 10%.

0.0141 g/L at the zeroth hour to 0.7976 g/L on the 22nd hour, at optimum conditions. The growth curve showed the extension of log phase roughly up to the 18th hour and stationary phase extended from the 18th to the 22nd hour.

The values for BOD and COD in whey water reduced from 2618 to 444.54 mg/L and from 7274.1 to 1758.87 mg/L, respectively, after treatment with *B. subtilis*. Figure 8.9 shows the percentage reduction of BOD and COD in whey water after treatment with *B. subtilis* at 10 times dilution, pH 7, and 10% inoculum size, as 83.02 and 75.82%, respectively. Therefore, it can be concluded that *B. subtilis* are found suitable for reducing BOD and COD from whey water.

8.4 Conclusion

In the present study, it was found that the organic loading of whey water in terms of BOD and COD was much higher than the permissible limits. Initially, Fenton's oxidation method was implemented to treat whey water to increase the ratio of its BOD and COD. Green algae and *B. subtilis* were then used separately to decrease BOD and COD. Fenton's oxidation showed a decrease in BOD and COD, from which it can be stated that Fenton's oxidation can be suitably used for reduction of organic loading in whey water. The biodegradability index after Fenton's oxidation made the sample fit for biological treatment.

Since treated whey water contained Na^+, NH_4^+, K^+, Ca^{2+}, and PO_4^{3-} more than permissible limit, while treating using green algae, salt of these ions was not used during the preparation of BG 11. This shows a new avenue to treat whey water in a more economical way. The bio-molecules' production from green algae showed an increase in their concentrations when grown in whey water. FTIR study on green algae confirmed the interaction between the bonds $C-O$, $N=O$, $O-H$, and $N=O$ during the treatment of whey water. *B. subtilis* showed maximum growth at 10% inoculum size and at pH 7. The use of *B. subtilis* for reducing organic loading from wastewater can also be made into use in several dairy industries along with the use of green algae, with the purpose of reusing the activated sludge for bacteria isolation and treating the effluent.

Acknowledgment

The authors greatly acknowledge the support of Department of Earth and Environmental Studies, Department of Chemical Engineering and Department of Chemistry of National Institute of Technology, Durgapur, to carry out the research work.

References

Ammonia in Drinking-water, Background document for development of WHO Guidelines for Drinking-water Quality. Available at:

http://www.who.int/water_sanitation_health/dwq/ammonia.pdf?ua=1 [Accessed on 29 July, 2018].

Anjana K, Kaushik A, Kiran B, Nisha R. (2007). Biosorption of Cr(VI) by immobilized biomass of two indigenous strains of cyanobacteria isolated from metal contaminated soil. Journal of Hazardous Materials. 148: 383–6. DOI:10.1016/j.jhazmat.2007.02.051.

Big River Citizens Education Guide, The Water Use Cycle, Missouri Department of Natural Resources. Available at: file:///C:/Users/User/ Downloads/how-bacteria-eat- waste-generic.pdf. [Accessed on 29 July, 2018].

Bligh EG, Dyer WJ. (1959). A rapid method of total lipid extraction and purification. Canadian Journal of Biochemistry and Physiology. 37(8):911–7. DOI: 10.1139/o59–099.

Clesceri LS, Greenberg AE, Trussell RR. (1996). Standard methods for the examination of water and wastewater, APHA, AWWA and WPCF, Washington, DC.

de la Noüe J, Laliberté G, Proulx D. (1992). Algae and wastewater. Journal of Applied Phycology. 4(3):247–54. DOI: 10.1007/BF02161210.

Gogate PR, Pandit AB. A review of imperative technologies for wastewater treatment I: oxidation technologies at ambient conditions, 2004. Advances in Environmental Research. 8:501–51. DOI: 10.1016/ S1093–0191(03)00032–7.

Indian Standard, Drinking Water-Specification (First Revision), IS 10500: 1991 Edition 2.1 (1993–01). Available at: http://www.indiawaterportal. org/sites/indiawaterportal.org/files/drinking_water_stand ards_is_10500_ 1991_bis.pdf [Accessed on 29 July, 2018].

Janczukowicz W, Zielin ski M, D bowski M. (2008). Biodegradability evaluation of dairy effluents originated in selected sections of dairy production. Bioresource Technology. 99(10):4199–205. DOI:10.1016/j. biortech.2007.08.077.

Janna H. (2016). Characterisation of raw sewage and performance evaluation of Al-Diwaniyah Sewage Treatment Work, Iraq. World Journal of Engineering and Technology. 4:296–304. DOI:10.4236/wjet.2016.42030.

Lowry OH, Rosebrough NJ, Farr AL, Randall RJ. (1951). Protein measurement with the Folin phenol reagent. Journal of Biological Chemistry. 193(1):265–75.

Mishra SS, Mishra KN, Mahananda MR. (2013). Chlorophyll content studies from inception of leaf buds to leaf-fall stages of Teak (Tectona grandis) of Kapilash forest division, Dhenkanal, Odisha. Journal of Global Biosciences. 2(1):26–30.

Potassium in Drinking-water, Background document for development of WHO Guidelines for Drinking-water Quality. Available at: http://www.who.int/water_sanitation_health/water-quality/guidelines/chemicals/potassium-background.pdf [Accessed on 29 July, 2018].

Sodium in Drinking-water, Background document for development of WHO Guidelines for Drinking-water Quality. Available at: http://www.who.int/water_sanitation_health/dwq/chemicals/sodium.pdf [Accessed on 29 July, 2018].

The Environment (Protection) Rules, 1986, Schedule VI, General Standards for discharge of environmental pollutants. Available at: http://cpcb.nic.in/GeneralStandards.pdf [Accessed on 29 July, 2018].

Tolerance Limits for Industrial Effluents, IS: 2490(Part I) – 1981. Available at: http://www.worldenviro.com/effstd.htm [Accessed on 29 July, 2018].

9

Bioremediation of High-strength Post-methanated Distillery Wastewater at Lab Scale by Using Constructed Wetland Technology

Aparna Bhardwaj[1], Mona Sharma[1,*], Chander Prakash Kaushik[1,2], and Anubha Kaushik[1,3]

[1]Department of Environmental Science and Engineering, Guru Jambheshwar University of Science and Technology, Hisar, India
[2]Amity School of Earth and Environmental Sciences, Amity University Haryana, Gurgaon, India
[3]University School of Environment Management, G.G.S. Indraprastha University, New Delhi, India
E-mail: drmonasharma1@gmail.com
*Corresponding Author

Constructed wetlands (CWs) are specifically designed systems with an aim to treat wastewater efficiently. They consist of a basin having substrate (soil, gravel) and macrophytes in which wastewater after being released is treated by natural attenuation in the presence of soil microorganisms and enzymes. In the present study, bioremediation of high-strength post-methanated distillery wastewater (PMDW) is done with the help of CW technology in microcosms (lab scale CW models). On the basis of preliminary tolerance studies, three macrophytes *Eichhornia crassipes*, *Typha latifolia* and *Canna indica* were selected and planted in mono-, bi- and poly-culture combinations so as to find out the best suitable plant/plant combination for the treatment of PMDW [half strength (S1) – 5000 mg/L chemical oxygen demand (COD); full strength (S2) – 10,000 mg/L COD]. The constructed wetland microcosms (CWMs), with a working volume of approximately 1.5 L, were planted in triplicate and experiments were run for 3 cycles of 12 days each using "-'*fill and drain method*'-". The wastewater treatment potential of CWMs (batch mode) was analyzed for the removal of COD, total Kjeldahl nitrogen (TKN), total phosphate (TP) from PMDW and changes in plant parameters like fresh plant biomass and leaf chlorophyll content following the treatment. It was found that plant combinations played a significant role in removal of COD, TKN and TP from wastewater up to 8 days at S1, and up to 4 days at S2. The overall results established that, CWM planted with *Canna-Typha* combination was most efficient in treating full strength PMDW.

9.1 Introduction

In India, most of the distilleries use sugar-based feed stocks (molasses from sugarcane) for ethanol production, as they contain readily available fermentable sugar. Such molasses-based units, however, produce extremely large quantity of high strength wastewater (100,000–150,000 mg/L COD), because of which they are ranked as one of the highly polluting "Red Category" industries by the MoEF, Government of India (Tewari et al. 2007). Therefore, since past few decades, many distilleries started switching over to grain-based distillery processes which generate wastewater of comparatively less strength (40,000–50,000 mg/L COD), yet it requires proper treatment process. Anaerobic biological degradation is most typically employed as a primary treatment for distillery effluents (both molasses based and grain based), but the spentwash generated after anaerobic treatment does not meet the stringent effluent standards laid down by CPCB, India (Asthana et al., 2001). Biomethanated wastewater from grain-based units generally have BOD and COD in the range of 3,000–3,600 mg/L and 10,000–13,000 mg/L respectively (CPCB, 2011) so, further treatment of post biomethanated wastewater is necessary. Many conventional methods have since been used for further treatment of PMDW, but they require a large quantity of water and are less efficient and quite expensive. CW Technology has emerged as one of the eco-technological, cost effective and efficient approaches for treatment of different types of wastewaters. CW typically consists of a basin which has wastewater, a substrate, and, most commonly, vascular plants. These components can however be manipulated in constructing a wetland. CWs have traditionally been used for domestic or municipal wastewater or storm water runoff treatment (Vymazal, 2005), but with advancement in designs and their low cost and energy intensive wastewater treatment potential, CWs have now been used to treat other types of wastewaters also.

In the present work, the potential of CW technology at lab scale was studied for the treatment of PMDW of full strength, S2 (approximately- COD – 10,000 mg/L) and half strength, S1 (approximately- COD – 5000 mg/L) from a grain-based distillery, in microcosm units using combination of some selected macrophytes. The experiments were conducted in batch mode and were aimed to find the best suitable combination of plants for effective wastewater treatment.

9.2 Materials and Methods

The CW microcosms used are indigenously fabricated PVC reactors as described in Figure 9.1. The microcosms were planted with three plant species, viz., *Typha latifolia, Canna indica* and *Eichhornia crassipes* the species, which successfully tolerated half-strength (5,000 mg/L COD) PMD wastewater in the preliminary studies conducted to assess tolerance of various macrophyte species. The plants were transplanted such that their root zone remained in the soil layer of microcosm unit. The CWMs planted in mono-culture with only *Canna indica, Typha latifolia* and *Eichhornia crassipes* are depicted as 'C', 'T' and 'E' respectively. Likewise, microcosm units planted in bi- and poly-culture combinations are depicted as CT, CE, EC and ECT. The total fresh biomass of macrophytes in each microcosm was 500 g at the time of plantation. The water level was maintained below the surface so as

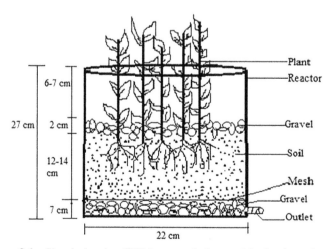

Figure 9.1 Sketch showing CWM reactor design used for batch mode study.

to imitate the subsurface CW like conditions. The microcosms each with a working volume of approximately 1.5 L were fed in the morning (0 d) and water samples (effluent) were taken from the outlet on 4, 8 and 12 days for analysis. Different parameters of water (viz. COD, TKN and TP) were analysed following the Standard Methods (APHA, 1998). The batch mode experiments were run for 3 cycles of 12 days each. The PMD wastewater collected for these batch experiments showed 10,275 mg/L COD which was taken as S2; whereas on diluting it with tap water (1:1), it was taken as S1 (5,200 mg/L COD).

9.3 Results and Discussion

9.3.1 Changes in Quality of PMDW after Treatment in Different CW Microcosms

Removal of COD:

All the CWMs exhibited excellent COD removal efficacy and more than 60% removal of COD, in both S1 and S2, was achieved in 4 days, more than 80% in 8 days and more than 95% in 12 days (Table 9.1).

Treatment potential of CW microcosms depend upon the initial COD concentration of wastewater, treatment duration (days) and the type of macro-phytes used. In order to test the significance of differences in COD removal, as influenced by the above three factors, three-way ANOVA was applied to the COD data as shown in Table 9.2. The ANOVA results show that each of the three factors and their interactions had a very important role to play in determining COD treatment efficacy. All the differences due to the factors and their interactions were found to be statistically significant ($P < 0.001$).

Table 9.1 COD removal from distillery wastewaters (S1 and S2) by CW microcosms with different plant combinations after 12 d batch treatment

Plant Combinations	COD Removal (%)	
	S1	S2
C	94.8	95.0
T	96.8	95.2
E	96.5	97.6
CT	93.1	94.6
EC	97.0	98.9
ET	97.1	97.9
ECT	98.2	98.3

Table 9.2 Three-way ANOVA showing significance of variations in COD of distillery wastewater due to initial COD strength, treatment time and plant combinations

Source	df	Anova SS	Mean Square	F Value	Pr>F
Influent Concentration	1	89546181	89546181	1078.12	0.0001
Treatment time (days)	3	1693351046	564450349	6795.88	0.0001
Plant combinations	6	1907670	317947	3.83	0.0016
Concentration × days	3	202662801	67554267	813.34	0.0001
Concentration × plant combinations	6	3124673	520779	6.27	0.0001
Days × plant combinations	18	41287821	2293768	27.62	0.0001
Concentration × days × plant combinations	18	22057222	1225401	14.75	0.0001
ERROR	112	9302467	83058		

Table 9.3 TKN removal from distillery wastewaters (S1 and S2) by CW microcosms with different plant combinations after 12 d batch treatment

Plant Combinations	TKN Removal (%)	
	S1	S2
C	91.4	95.1
T	94.0	92.0
E	90.2	97.3
CT	87.7	91.5
EC	92.6	97.5
ET	93.4	97.6
ECT	94.3	95.0

Removal of Total Kjeldahl Nitrogen (TKN):

TKN in PMDW in S1 was 1568 mg/L and that in S2 was 3248 mg/L. It was observed that microcosms planted with *Eichhornia* performed exceptionally well in removal of TKN from the wastewater at both S1 and S2 concentrations. At lower concentration, removal of TKN was comparatively less at 4 d for all microcosm units than that at higher concentration; however, it increased as the treatment progressed on 8 d and 12 d. By the end of the 12th day, it was observed that all the microcosm units removed more than 90% TKN from the effluent at both concentrations S1 and S2 (Table 9.3).

Removal of Total Phosphorous (TP):

The initial concentration of TP in the feed water at S1 and S2 were 28.3 and 51.26 mg/L respectively. The removal of TP from effluent in all microcosms was quite gradual at both the concentrations but eventually at 12 d more than 80% TP removal was exhibited by all microcosms except for

Table 9.4   TP removal from distillery wastewaters (S1, S2) by CW microcosms with different plant combinations after 12 d batch treatment

Plant Combinations	TP Removal (%)	
	S1	S2
C	80.1	88.6
T	93.3	85.6
E	75.1	96.6
CT	87.3	87.9
EC	79.3	94.8
ET	88.0	95.4
ECT	92.9	87.3

E which removed 75% TP. However, in mixed cultures, *Eichhornia* showed better results (Table 9.4).

When three-way ANOVA was applied to the data for TKN and TP removal by CWMs, similar observation was made as with COD. In both the cases (TKN and TP) all the three factors viz., plant combinations, initial TKN and TP concentration in wastewater and duration of treatment period are important as indicated by their high F-values for all the factors and their interactions and the differences due to these are statistically significant ($P < 0.001$). This indicates that all the factors were significantly important in removal of TKN and TP, when considered as individual factor or in combination. Moreover, the presence of different plant species in CW systems may result in efficient root partitioning of soil which enhances the bioremediation capability of mixed culture systems (Coleman et al., 2001). The results further show that *Eichhornia* and its combinations removed COD, TKN and TP from PMDW most efficiently. Even in monocultured microcosms E outperformed C and T microcosms. Results of 12 d batch treatment of full strength (S2) and 50% PMDW (S1) in CW microcosms with different plants showed that in general, it is more beneficial to grow polyculture plants as they show better treatment efficiency as indicated by high removal of COD, TKN and TP from both S1 and S2 concentrations of the feed water.

9.3.2 Changes in Leaf Chlorophyll Content of Plants After Treatment in Different CW Microcosms

Chlorophyll content of leaves is considered as one of the important parameters of plant growth and hence determined to assess the tolerance and general ability of the plants to grow well in the presence of the PMDW. In order to calculate the change in the total chlorophyll content of plants in each

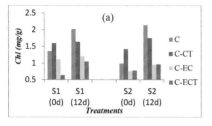
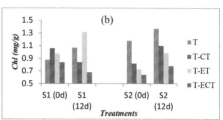
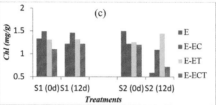

Figure 9.2 Changes in total chlorophyll content of *Canna* (a), *Typha* (b) and *Eichhornia* (c) in their different plant combinations in CW microcosm units.

microcosm during the experiment, plant leaf samples were collected at the beginning and end of the experiment from each microcosm. Total chlorophyll content in plants varied between 0.5 to 2.0 mg/g in different microcosms (Figure 9.2).

It was observed that total chlorophyll content in *Eichhornia* was more than in other plant species. With increase in concentration of wastewater and by the end of the experiment, the chlorophyll content increased in most of the combinations of *Canna* and *Typha* while it decreased in *Eichhornia* except for ET. It suggests that with increasing load of wastewater strength, C and T in all their combinations specifically CT, survived well. Thus, chlorophyll data indicates that both *Canna* and *Typha* plant species tolerated both the strengths of effluent however, *Eichhornia* failed miserably to withstand especially at higher concentration of effluent.

9.3.3 Changes in Fresh Plant Biomass After Treatment in Different CW Microcosms

Plant biomass in CWs is another very important factor which affects its treatment efficiency as whole. Total fresh biomass of all the plant species was calculated before and at the termination of the experiment after destructive sampling. At the start of the experiment, an attempt was made to equally distribute the weight of the plant in each microcosm such that each microcosm has a total of approximately 500 g of plant material in it. In order to

Table 9.5 Relative change in total fresh biomass of plants in CWMs after 12d batch treatment of S1 and S2 PMD wastewater (value at 0 d taken as 100)

Plant Combinations in Microcosms	Relative Change S1	Relative Change S2
C	117.6	81.4
T	58.0	60.4
E	24.5	10.8
CT	148.5	89.7
EC	54.0	43.4
ET	34.1	31.9
ECT	61.8	46.7

compare the relative change in total fresh plant biomass in microcosms, the 0 day value of total fresh plant biomass in each microcosm unit is taken as 100 (as depicted in Table 9.5).

It is clear from Table 9.5, that the total fresh biomass of plants in C and CT microcosms increased by the end of the treatment at lower concentration. At *S2*, although all microcosms observed a decrease in their plant biomass but decrease was least in CT followed by *C. Eichhornia* could barely survive in mono or in mixed cultures at S1 and S2. *Canna* and *Typha* however, survived better in both monoculture and bi-culture at both the treatments of PMDW (S1 and S2). Zhang et al. (2007) also reported an increase in biomass of Canna both in mono- and mixed culture at different nutrient concentration. CT CW microcosm was thus found to be best for plant survival, growth and COD removal efficiency.

9.4 Conclusion

Based on the present work, it is concluded that it is possible to treat the post methanation distillery wastewater with the help of CW technology, which is highly efficient and cost-effective bioremediation technique. The constructed wetland microcosms with CT combination was most efficient for treatment of high-strength PMDW (full strength – 10,000 mg/L COD; half strength – 5000 mg/L COD) in batch mode. The results showed that *Canna* and *Typha* managed to survive even at full-strength PMDW, up to three wastewater treatment cycles in batch mode, while *Eichhornia* tended to die off. Moreover, the CT combination showed superior biomass and chlorophyll content, indicating a synergistic interaction between the two plant species.

References

American Public Health Association (APHA). (1998). American Water Works Association. Water Pollution Control Federation. Standard Methods for the Examination of Water and Wastewater. Washington. DC. USA.

Asthana A. et al. (2001). Treatment of colour and biochemical oxygen demand of anaerobically treated distillery effluent by aerobic bacterial strains. Indian Journal of Environment Protection 21:1070–1072

Central Pollution Control Board (CPCB report). (2011). Report on: Assessment of grain based fermentation technology, waste treatment options, disposal of treated effluents. Delhi.

Coleman, J. et al. (2001). Treatment of domestic wastewater by three wetland plant species in constructed wetlands. Water, Air and Soil Pollution. 128:283.

Tewari, P. K. et al. (2007). Water management initiatives in sug-arcane molasses based distilleries in India. Resources, Conservation and Recycling, 52:351.

Vymazal, J. (2005). Horizontal sub-surface flow and hybrid constructed wetlands systems for wastewater treatment. Ecological Engineering, 25:478.

Zhang, Z. Rengel, Z., and Meney, K. (2007). Growth and resource allocation of *Canna indica* and *Schoenoplectusvalidus* as affected by interspecific competition and nutrient availability. Hydrobiology, 589:235.

10

Reuse of Magnetite (Fe_3O_4) Nanoparticles in De-Emulsification of Emulsion Effluents of Steel-rolling Mills

Parsanta Verma and Ashok N. Bhaskarwar

Department of Chemical Engineering, Indian Institute of Technology Delhi, Hauz Khas, New Delhi, India
E-mail: parsantaiitd@gmail.com; ashoknbhaskarwar@yahoo.co.in

The magnetite (Fe_3O_4) nanoparticles are superparamagnetic in nature. These can be synthesized by several methods such as co-precipitation, sol–gel, and thermal decomposition. Among these, co-precipitation method is the best method if Fe_3O_4 nanoparticles are required in bulk amount. The present work is based on co-precipitation method for the synthesis of magnetite nanoparticles at room temperature. These nanoparticles are characterized by X-ray diffraction (XRD), scanning electron microscopy (SEM), and transmission electron microscopy (TEM) techniques, and used for the de-emulsification of emulsion effluents of steel-rolling mills. In steel industries, relatively stable oil/water emulsions, generally spontaneously formed, help in cooling and lubrication in hot-rolling mills. Such emulsions, containing mineral oils, can withstand adverse environmental conditions, and thus can cause hazardous effects in the environment if disposed-off untreated. According to the Hazardous Waste and Management Society, the maximum waste generated is related to oil (oil spills or emulsions), i.e., 1000 kg/year. According to the Ministry of Environment and Forests' notification, the permissible (concentration) limit for oily waste discharge is 10 mg/L. Such emulsified oils are, however, often difficult to treat. The conventional processes used for de-emulsification are either costly, non-eco-friendly, time-consuming, or energy intensive. Our method of de-emulsification by Fe_3O_4 nanoparticles is quick, eco-friendly, and cost-effective. These nanoparticles can easily be retrieved from the processed stream by simple application of an external magnetic field. The recovered nanoparticles can be reused several times depending upon the concentration of oil in the emulsion effluent. In this work, we treated 10% (v/v) emulsion effluents and also checked the mass balance of Fe$_3$O$_4$ nanoparticles.

10.1 Introduction

In steel industries, large amounts of relatively stable oil-in-water emulsions are used for cooling and lubrication in cold- and hot-rolling mills. As these types of emulsions can withstand high temperature and pressure conditions, the same properties also make them tough for further treatment for oil recovery. Although oil spills can be controlled by many surface-skimming devices, the emulsified oil are often quite difficult to treat. In the present work, we have treated such relatively stable o/w emulsion effluents using Fe$_3$O$_4$ nanoparticles. The advantages of this novel process over the conventional processes are that it is quick, cheap, eco-friendly, and much less energy intensive. Fe$_3$O$_4$ nanoparticles, also known as MNPs, are black in color, and have a cubic inverse spin structure (Ahn et al., 2012; Sundar et al., 2014). These are superparamagnetic and eco-friendly in nature, qualities which make them an excellent candidate even as a catalyst. These nanoparticles can be synthesized by several methods such as co-precipitation, sol–gel, thermal decomposition, hydrothermal, electron-beam lithography, and gas-phase deposition (Koutzarova, 2005). If Fe$_3$O$_4$ nanoparticles are required in large amounts, the co-precipitation method is the best because it is safe and quick.

10.2 Experimental

10.2.1 Synthesis of Uncoated Fe$_3$O$_4$ Nanoparticles by Co-precipitation Method at Room Temperature

We prepared separately the solutions of ferric chloride (2 M) and ferrous chloride (1 M) by taking 32.2 g of black-colored ferric chloride in 200 mL of deionized water and 16.2 g of light-green colored ferrous chloride in 50 mL of deionized water and then mixed them thoroughly using a magnetic stirrer. Ammonium hydroxide solution was then added to it dropwise while continuously stirring it by a mechanical stirrer until the color of the solution turned black. The black-colored precipitate is the indication of the synthesis of Fe_3O_4 nanoparticles (magnetite, iron oxide) (Talekar et al., 2012). MNPs were then air dried and stored at room temperature.

Co-precipitation method:

$$2FeCl_3(s) + 6H_2O \rightarrow 2Fe(OH)_3 + 6HCl \tag{10.1}$$

$$FeCl_2 + 2H_2O \rightarrow Fe(OH)_2 + 2HCl \tag{10.2}$$

$$2FeCl_3(s) + FeCl_2 + 8NH_3 + 4H_2O \rightarrow Fe_3O_4 + 8NH_4Cl \tag{10.3}$$

These nanoparticles were analyzed by XRD for their confirmation as Fe_3O_4 nanoparticles, by the SEM to check the morphology of the Fe_3O_4 nanoparticles, and by the TEM to determine the size of the Fe_3O_4 nanoparticles.

10.2.2 Treatment of Emulsion Effluents

A synthetic, stable o/w emulsion (10%, v/v) was prepared in the lab using cutting oil and RO water by mixing in a homogenizer. A slurry (8%, w/v) of Fe_3O_4 nanoparticles in RO water was prepared using a mechanical stirrer. The emulsion and the slurry were then mixed in a beaker for 10 min using a mechanical stirrer. A sticky paste (Fe_3O_4 nanoparticles with oil) was formed (Figure 10.1(a)). It was separated with the help of a wire mesh from the rest of the emulsion and washed with acetone.

The mass balance of recovered Fe_3O_4 nanoparticles was finally done. The total amount of all recovered Fe_3O_4 nanoparticles was reused with the rest of the emulsion. These steps were repeated seven times (Figure 10.1(b)), and after that, no more sticky paste was formed.

(a) (b)

Figure 10.1 (a) Sticky paste of Fe_3O_4 nanoparticles with oil. (b) Seventh-time recovered Fe_3O_4 nanoparticles with oil without a sticky paste left behind.

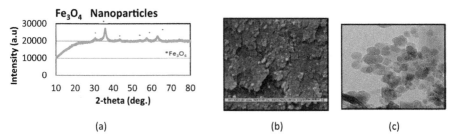

(a)   (b)   (c)

Figure 10.2   (a) XRD of uncoated Fe_3O_4 nanoparticles. (b) SEM image of uncoated MNPs. (c) TEM image of uncoated MNPs.

10.3 Results and Discussion

10.3.1 Characterization of Fe₃O₄ Nanoparticles

XRD: The precipitated fine particles were characterized by XRD for the determination of structure and for the estimation of the crystallite sizes (Figure 10.2(a)). The six characteristic peaks at $2\theta \approx 30°, 35°, 43°, 53°, 57°$, and $62°$ were the ones corresponding to the (220), (311), (400), (422), (511), and (440) crystal planes of a pure Fe_3O_4 with a spinal structure (JCPDS file PDF no. 65–3107). No characteristic peaks of impurities were detected in the XRD pattern (Hariani et al., 2013).

SEM: The surface morphology of the Fe_3O_4 nanoparticles was determined by SEM analysis. Figure 10.2(b) shows that the particles are agglomerated due the lack of a capping agent (Hariani et al., 2013).

TEM: The microscopic analysis was done to investigate the morphology of the synthesized particles by TEM (Figure 10.2(c)), which shows that the diameter of Fe_3O_4 nanoparticles is in the range of 5–20 nm (Andradeet al., 2014).

10.3.2 Total Mass Balance of Oil and Fe₃O₄ Nanoparticles

It was found that the percentage oil recovery successively increased up to the sixth time of reuse of MNPs and after that, there was no significant recovery observed (Figure 10.3). The MNPs' oil-adsorption/de-emulsifying capacity decreased with each reuse and no further de-emulsification was observable with the MNPs beyond the seven times of reuse. No sticky paste was formed after seven times of MNPs' reuse. The percentage of Fe_3O_4 –

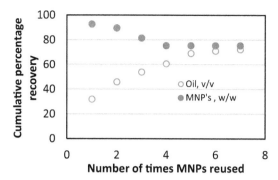

Figure 10.3 Effect of reuse of Fe_3O_4 nanoparticles on the de-emulsification of o/w emulsion and recovery of MNPs.

Table 10.1 Results of treatment of 10% (v/v) emulsion using Fe_3O_4 nanoparticles

S. No.	Contents	Feed	Amount Recovered	Treatment of 10% Oil Emulsion
1	Oil	250 mL	181 mL	72.4%
2	MNPs	148 g	112 g	75.5%

nanoparticles' recovery decreased after initial few reuses, but after the fourth time, it becomes approximately constant. We finally got a total of 72.4% (v/v) and 75.5% (w/w) of oil and MNPs' recovery, respectively.

10.4 Conclusion

The cumulative percent age of oil recovery successively increased up to a seven times of MNP's reuse after which there was no significant recovery, and the MNP's recovered amount in each successive reuse decreased. MNP's efficiency to adsorb oil or to de-emulsify the emulsion by an as-yet unestablished mechanism decreased with each reuse. We could reuse the MNPs for a maximum of seven times. The final recovery of oil we got 72.4% (v/v) and of MNPs 75.5% (w/w) (Table 10.1).

Acknowledgments

The authors would like to thank the Department of Chemical Engineering, Indian Institute of Technology Delhi (IITD), for providing us with labs and support to carryout the work. Ms. Parsanta would also like to thank Petrotech for providing her with their research fellowship, and Professor

A. Sethuramaiah of Center for Industrial Tribology, Machine Dynamics and Maintenance engineering Centre (ITMEC) of IITD for suggesting the difficult and challenging practical problem to one of us (Professor Ashok N. Bhaskarwar).

References

Sundar S, Mariappan R, Piraman S. (2014). Synthesis and characterization of amine modified magnetite nanoparticles as carriers of curcumin-anticancer drug. Powder Technology. 266:321–8.

AhnT, Kim JH, Yang H, Lee JW, Kim J. (2012). Formation pathways of magnetite nanoparticles by coprecipitation method. The Journal of Physical Chemistry. 116(10):6069–76.

Koutzarova T, et al. (2005). Microstructural study and size control of iron oxide nanoparticles produced by microemulsion technique. Current Topics in Solid State Physics. 3(3):1302–7.

Talekar S, et al. (2012). Novel magnetic cross-linked enzyme aggregates (magnetic CLEAs) of alpha amylase. Bioresource Technology. 123:542–7.

Hariani PL, Faizal M, Setiabudidaya D. (2013). Synthesis and properties of Fe$_3$O$_4$ nanoparticles by co-precipitation method to removal procion dye. International Journal of Environmental Science and Development. 4(3):336–40.

AndradeVAJSPL, Bustamante A, Valladares LDLS, Cavalcanti MMIASKPS, Aguiar MPCSJA. (2014). Magnetic and Mössbauer studies of fucan-coated magnetite nanoparticles for application on antitumoral activity. 224:227–38.

11

Application of Agro-Residues-Based Activated Carbon as Adsorbents for Phenol Sequestration from Aqueous Streams: A Review

Pushpa Jha

Sant longowal Institute of Engineering and Technology, India

Phenol is one of the common constituents from the effluents of various chemical industries. Due to its good solubility in water, it is one of the most hazardous constituents to aquatic lives. Among the various methods available in the literature, adsorption is the most popular one. However, usually, adsorbents cost high. For the last three decades, much research has been reported toward investigating cheap materials as adsorbents. Agro-residues are abundantly available and cause a disposal problem after the harvest, all over the world. They being the high source of carbon can be considered as the low-cost material for the purpose.

This review compiles the work reported in the literature related to various agro-residues (with their characterization), used for batch-wise phenol removal from aqueous streams. To make the agro-residues-based adsorbents more effective, thermo-chemical treatments are given to them. These treatments make the adsorbent comparable to the commercial grade carbon concerning their BET surface area, ash and fixed carbon contents, phenol numbers, methylene blue test values, and iodine numbers of various agro-residues. Various parameters that affect the process of sequestration of phenol from aqueous streams are presented. This review also summaries the equilibrium and kinetic studies. Thermodynamic studies indicate the feasibility of the adsorption process. The need for regeneration has also been stressed for the economic reasons. Finally, the paper is concluded.

Future research needs are discussed for the application of agro-residues for phenol sequestration from aqueous streams.

11.1 Introduction

Phenol and its compounds are present in water bodies and effluents in the form of hydroxybenzene and substituted hydroxyl-benzenes. It is known as one of the most toxic pollutants. Animal and plant lives are affected by it. Phenols and its derivatives can be found abundantly in the effluents of petroleum refineries, gas, coke, plastic, paints, dyes, pharmaceutical, and textile industries (Anne et al., 2017).

These types of phenolic wastes lead to severe consequences. Phenol irritates skin and causes its necrosis; it damages kidneys, liver, muscle, and eyes. Damage to the skin is caused by its coagulation related to reaction to phenol with amino acids contained in the keratin of epidermis and collagen in inner skin (Clayton and Clayton, 1994). Clinical data have shown that people exposed to chlorophenols influence fall ill with tumors, sarcoma, and lung cancer.

As reported in various documents (Mahajan, 1985; Jha, 1996), concentration of phenolic wastes in effluents varies from 3 to 10,000 mg/L. It adversely affects plant and aquatic lives. Even at a very low concentration of value 0.1 mg/L, chlorophenols cause undesirable effects in potable water. The acceptable limit of phenol in drinking water is 0.002 mg/L (Shawabkeh and Abu-Namesh, 2007).

A variety of methods which fall under chemical, physical, and biological methods are available for sequestration of all levels of initial phenol concentrations. All these methods are exploited commercially (Karabacakoglu et al., 2008). The methods include steam stripping, solvent extraction, oxidation (O_3, H_2O_2, and ClO_2), ion exchange, biodegradation, and adsorption methods. Among the methods mentioned, adsorption methods are most popular in use all over the world (Bhatnagar et al., 2105; Orchando-Pulido et al., 2017). However, generally, adsorbents cost high. For the last three decades, much research has been reported toward investigating cheaper materials such as fly ash, peat, soil, agro-residues, and tendu leaf for their use as adsorbents (Dass and Jha, 2015a).

Agro-residues are abundantly available all over the world after the harvest of the crop. They are rice husk, bagasse, corn-cob, coconut-coir, coconut shell, wheat straw, cotton stalk, date stones, date pit, olive pomace, rice-straw, *Acacia nilotica* branches, saw-dust, orange peel, and pistacia mutica shells (Ganesh, 1990). These wastes as biomass render disposal problems. There is a need to have various decentralized technologies to make use of them (Iyer et al., 2002). In the absence of any economic technology, farmers of developing countries burn the agro-residues off as a solution to the disposal problem, creating air pollutions (Stubble burning in North India, 2014).

All agro-residues are sources of carbon which makes them fit to be the precursor for activated carbon. These agro-residues can be thermo-chemically treated by combinations of various methods (Miretzky and Cirelli, 2010; Santos et al., 2015; Palma et al., 2016; Singh and Balomajumder, 2016; Franco et al., 2017; Kumar et al., 2017) to tailor it to a particular need. To check its absorbability, they have to be characterized, before proceeding for their study as an adsorbent. After satisfactory characterization, preliminary batch adsorption is done at a standard temperature of 25°C (Sáenz-Alanís, 2017).

There is a need to study the effects of various parameters: pH, adsorbent dosage, contact time, initial phenol concentration, and temperature on adsorption of phenol. This study gives the optimized condition at which adsorption needs to be studied for an adsorbent (Sahu et al., 2017). For further analysis, there is need to fit the experimental data in various mathematical models, e.g., Langmuir, Freundlich, Temkin, and Dubinin–Radushkevich isotherm model (Inam et al., 2017). This enables us to know various possible mechanism through which adsorption can take place. Effect of contact time enables us to conduct kinetic studies. Mathematical models for kinetics most commonly used are pseudo-first-order model and pseudo-second-order models (Chiou and Li, 2003; Ho, 2004; Tseng et al., 2010). These studies enable one to know whether adsorption is physical or chemical. To study their thermodynamics analysis, one has to study its adsorption at various temperatures including the normal one (Kilic et al., 2011). This study will tell us the spontaneity of the process. It also helps us to know whether the process is exothermic or endothermic.

One can also study whether there is the possibility of regeneration of the adsorbent (Lata et al., 2015). For this desorption needs to be done at the same optimized conditions at which adsorption took place. The perfect regeneration process is the one which does not change the original properties of a particular adsorbent.

11.2 Methods Available for Removal of Phenol

Treatment technologies for phenolic wastes are widely reported in the literature. Many methods are in successful full-scale industrial use, and high efficiencies of treatments are reported. Industrial waste may contain an extensive range of phenolic materials (Zouboulis et al., 2015; Lofrano et al., 2016). Therefore, treatment technology is discussed under the headings: concentrated, intermediate, and dilute phenolic wastewaters.

Phenol recovery from concentrated wastes (500–10,000 mg/L) usually is made using extractive recovery into an immiscible organic solvent. Phenol recovery from intermediate wastes water (5–500 mg/L) is usually treated biologically. The biological process includes lagoons, oxidation ditches, trickling filters, and activated sludge (Jha, 1996).

Dilute phenolic wastes (0–5 mg/L) are given biological treatment. This treatment generally reduces phenol concentration down to 1–2 mg/L level. Other processes can achieve removal below 1–0.1 mg/L. Ordinarily chemical or physicochemical methods replace the biological processes for the treatment of these dilute phenolic wastes (Fernadez et al., 2003).

Activated charcoal is often used for adsorption of phenol (Bhatnagar et al., 2015) but the use of coal, calcium carbonate, coke, fly ash, iron, and other metallurgical ores to reduce the cost of operation has also been reported. The adsorption can be carried out in columns or vessels lined with plastics and ceramics.

Agro-residues, especially like bagasse fly ash, rice husk, coconut waste, jute fiber, olive pomace, date pits, EFBs of oil palm, corn cob, tamarind nutshell, and sawdust, are among reported adsorbents for phenol (Nor et al., 2013). These agro-residues are either grossly under-utilized or wholly unutilized and are disposed of as such, as many parts of the developing countries (Stubble burning in North India, 2014). This biomass is the source of carbon and considered as suitable material for sorption of phenol from aqueous solutions.

11.3 Adsorption as a Cost-Effective Method for Removal of Phenol

Adsorption is a surface phenomenon. It increases the concentration of a substance as adsorbate from solution onto the surfaces called adsorbent

(Treybal, 1985). Separation takes place due to unequal distribution of adsorbate on solid and liquid. This process can be applied to separate phenol from effluents. It has been reported widely in the literature of various kinds that this process is very efficient and economical methods for removal of phenols (Fernadez et al., 2003; Kilic et al., 2011; Zouboulis et al., 2015; Lofrano et al., 2016; Singh et al., 2016).

Adsorption processes are generally categorized as (1) physical adsorption and (2) chemical adsorption. Both types of adsorptions are explained in Table 11.1.

Adsorption being the surface phenomenon, commercial adsorbents are characterized by large surface areas, the majority of which is composed of internal surfaces bonding the large pores and capillaries of highly porous solids. The performance characteristics of adsorbents are related to their intraparticle properties. Surface area and the distribution of area concerning pore size generally are primary determinants of adsorption capacity (Wong, 1998; Tseng and Tseng, 2005).

Commercial carbon typically has total surface areas in the range of 400–1500 m^2/g as measured by nitrogen adsorption methods (Dass and Jha, 2015b; Gisi et al., 2016). Phenol being many organic types of the low-cost carbon-based adsorbent can be utilized for its sorption. Agricultural wastes fall under low-cost adsorbents which are rich in carbon content. By characterizing these agro-residues concerning proximate analysis, one can determine fixed carbon content (carbon in the solid form) which is mainly responsible for adsorption. Ash content determines any scope for its modification. Higher is the ash content in an agro-residue/agro-residue-based adsorbent, higher are the chances of getting modified activated carbon due to the creation of pores after their removal (Santos et al., 2015).

Table 11.1 Difference between physical and chemical adsorption

Physical Adsorption	Chemical Adsorption
Intermolecular force (Van der Waals forces) of attraction between molecules of adsorbate and adsorbent are present	There is a formation of chemical bonds between molecules of adsorbate and adsorbent
The process is rapid and reversible	The process is slow and irreversible
Amount of heat liberated as a result of the process is small	Amount of heat liberated as a consequence of this process is significant
Formation of multi-layer of adsorbate is possible	Single-layer formation of adsorbate is there

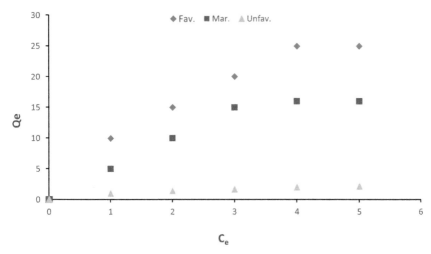

Figure 11.1   Adsorption trends of phenol on agro-residue-based adsorbent for (a) favorable adsorbent, (b) marginal adsorbent, and (c) unfavorable adsorbent.

11.3.1 Quantification of Phenol Adsorbed

For batch adsorption study, a fixed amount of solution is taken in a conical flask with dosage variation in it. Temperature, pH, and agitation speed are fixed. Generally, to study the equilibrium for an adsorbent–adsorbate system for the first time, 24 h is fixed for adsorption to take place. In all reported cases, the equilibrium of adsorption reached before 24 h (Das, 2016). The adsorption capacities of all the adsorbents at equilibrium are calculated using the equation (Treybal, 1985):

$$q_e(\mathrm{mg/g}) = [(C_0 - C_e)/W]V \qquad (11.1)$$

where q_e is the amount of phenol adsorbed (mg,) per unit mass of adsorbent (derived from agro-residues-based adsorbent) in grams, C_o and C_e are the initial and final concentration of phenol in the solution, respectively, V is the volume of solution, and W is the mass of adsorbent (in grams) used.

Figure 11.1 shows the general trend for equilibrium isotherm of adsorption of phenol on activated carbon (Das, 2016).

11.4 Agro-residues as Adsorbents for Phenol

11.4.1 Baggasse Fly Ash

Baggasse fly ash obtained from sugar industries could be utilized as an adsorbent for the removal of phenol. Karabacakoglu et al. (2008) investigated hazelnut bagasse carbon as an adsorbent for the removal of phenol from the water. The sequestration efficiency increases with increase in adsorbent dose and phenol concentration. The adsorption data follow both Langmuir and Freundlich models. The adsorption capacities of 97.36, 91.32, and 99.27 mg/g were obtained at 25, 35, and 45°C, respectively (Karabacakoglu et al., 2008). Srivastava et al. (2006) have studied the adsorption of phenol on carbon-rich bagasse fly ash at the optimum conditions of pH 6.5, adsorbent dosage 10 g/L of the solution, and equilibromophenolium for a time of 5 h. Gupta et al. (2006) also found bagasse fly ash generated from the sugar industry for the removal and recovery of phenol and *p*-nitrophenol from wastewater. Data fitted the pseudo-second-order kinetics.

11.4.2 Rice Husk

Rice is one of the primary crops grown throughout the world. Rice husk is the outermost layer of paddy grains. Many researchers have found that rice husk can be used as an adsorbent for the removal of phenol. Ahmaruzzaman and Sharma (2005) investigated rice husk and rice husk char as an adsorbent and found the adsorption capacities of 4.7 and 7.9 mg/g for rice husk and rice husk char, respectively. Langmuir model fits adsorption data better than Freundlich (Ahmaruzzaman and Sharma, 2005). Shiundu et al. (2002) found the adsorption capacity of 14.382 mg/L and 0.1532 mg/g for phenol and chlorophenol, respectively, on rice husk char. Jha (2011) studied that thermo-chemically treated rice husk can be used as a low-cost adsorbent in the removal of the initial concentration of phenol in no small extent. The adsorption capacity of activated rice husk char found was 150 mg/g at a residual phenol concentration of 8000 mg/L (Jha, 2011).

11.4.3 Coconut Waste

The harvest of coconut crop generates coconut waste. Coconut waste can be obtained in the form of a shell, coconut coir, and pith. All these wastes can be used as adsorbents for phenol. Many researchers had studied that by chemical activation of coconut shell, one can increase the surface area of adsorbent. Activated carbon possesses much higher adsorption capacity for phenol, 4-chlorophenol, and 4-nitrophenol from aqueous solutions. Mohd Din et al. (2009) investigated that coconut shell activated carbon has a maximum adsorption capacity of 205.8 mg/g. The adsorption isotherms were confirmed by both Langmuir and Freundlich, and adsorption system followed the pseudo-second-order (Mohd Din et al., 2009). Hameed and Rahman (2008) found that coconut husk-based activated carbon has a maximum adsorption capacity of 716.10 mg/g for the removal of 2, 4, 6 trichlorophenol from aqueous solution. Namasivayam and Kavitha (2003) have used coir pith carbon as an adsorbent which has adsorption capacities of 48.31, 19.12, and 3.66 mg/g for phenol, 2, 4-dichlorophenol, and p-chlorophenol, respectively.

11.4.4 Olive Pomace

Olive pomace is one of the agricultural byproducts. Stasinakis et al. (2008) studied the adsorption of phenol on olive pomace as an adsorbent. Different forms of physically and chemically treated olive pomace were used in the form of (OP-1), (OP-2), and (OP-3) as adsorbents. Removal efficiency was improved by increasing sorbent concentration and decreasing particle size of the sorbent material (Stasinakis et al., 2008).

11.4.5 Date Stones and Date Pits

Alhamed (2009) studied that conversion of date stones into a carbonaceous adsorbent and its use for the removal of phenol from wastewater using four different activated carbon particle sizes of 1.4, 0.8, 0.45, and 0.255 mm under packed bed studies. The initial rate of adsorption was predicted from the pseudo-second-order model with an adsorption capacity of 90.4 mg/g. The adsorption increases with decreasing particle diameter as a result of the higher interfacial area provided by particles with smaller diameter (Alhamed, 2009).

El-Naas et al. (2010) explored date pits to prepare effective adsorbent for the removal of phenol from the refinery and synthetically prepared aqueous wastewater. Pseudo-second-order kinetics fits the kinetic data and adsorption capacities of 16.64 and 12.6 mg/g were found for refinery and aqueous solutions, respectively. For regeneration of activated carbon, various chemical and thermal methods were tested, and ethanol was reported to be most effective with more than 86% regenration efficiency (El-Naas et al., 2010).

11.4.6 Oil Palm Empty Fruit Bunches

Activated carbon prepared from oil palm EFBs is environmentally friendly with a useful application for adsorption of phenol. Alam et al. (2009) investigated that adsorbent from oil palm EFB was effective for the removal of phenol from water when adsorbent was physically activated carbon. The adsorption capacity was found to be 4.68 mg/g (Alam et al., 2009).

11.4.7 Corncob

Corncob is waste obtained after removing corn from the maize crop. Tseng and Tseng (2005) explored that corncob carbon nutshell can be used as an adsorbent for the removal of phenolic compounds. Activation of corncob carbon was done with KOH/char ratios from 0.5 to 6. It was found that the adsorbent capacities at an impregnation ratio of 2–6 were found to be 232 and 340 mg/g for phenol, 362.8 and 485 mg/g for chlorophenol, and 333.7 and 451.2 mg/g, for 2, 4 dichlorophenol, respectively (Tseng and Tseng, 2005).

11.4.8 Tamarind Nutshell

Goud and Mohanty (2005) have studied mass transfer coefficients in a packed bed using tamarind nutshell activated carbon to remove phenol from wastewater. They reported work on the preparation and characterization of activated carbon from tamarind nutshell, an agricultural waste by-product, its use in a packed bed for the removal of phenol. Adsorption of phenol is dependent on both the flow rate and the particle size of the adsorbent. Phenol removal yield decreases with increasing flow rate of adsorbate and particle size of the adsorbents (Goud and Mohanty, 2005). Kulkarni et al. (2011) have also studied tamarind bean activated carbon as an adsorbent for phenol removal in a fluidized bed. Various parameters affecting adsorption on tamarind bean were concentration, fluid flow rate, and adsorbent's particle size. It is observed that as the concentration of adsorbate increases, the percent saturation of adsorbent increases (Kulkarni et al., 2011).

11.4.9 Sawdust

Dutta et al. (2001) have confirmed that the sawdust is a promising low-cost material for the removal of phenols from wastewater. They carried out studies on adsorption of *p*-nitrophenols on charred saw dust. Data fitted in Langmuir adsorption isotherm (Dutta et al., 2001). Jadhav and Vanjara (2004) investigated the adsorption of phenol on sawdust, polymerized sawdust, and sawdust carbon. The material proved to be very promising adsorbents.

11.4.10 Orange Peel Waste

Loriane et al. studied the adsorption of phenol on activated carbon obtained from orange peel. The adsorbent was characterized by X-ray diffraction,

Raman spectroscopy, SEM, and energy dispersive spectroscopy (EDS). Effects of sorbent dosage and pH on sorption were studied. The sorption process followed a pseudo-second-order model (Rebeiro et al., 2017).

11.4.11 Pistacia Mutica Shells

Ramin et al. (2018) studied the possibility of Pistacia mutica biomass as an alternative adsorbent for phenol removal from aqueous solution. The effects of various parameters, contact time, pH, adsorbent dosage, etc., were investigated in this experimental-lab study. Also, the isotherm and kinetic investigations were performed for phenol adsorption process. The adsorption equilibriums were analyzed by Langmuir, Temkin, Freundlich, and Harkins Jura isotherm models (Ramin et al., 2018).

11.4.12 Rice-straw

Sarker and Fakhruddin (2017) studied the adsorption of phenol on rice straw. For this rice straw, as such and physically and thermally activated rice straw was used for phenol removal. The highest phenol removal of 84% was obtained with the most treated rice-straw (in the form of ash). Particle size selected was <1 mm of particles. Freundlich isotherm was the best fit for the experimental data. It was also proved to be cost-effective adsorbent (Sarker and Fakhruddin, 2017).

11.4.13 *Acacia nilotica* Branches

Dass and Jha (2015b) studied activated *A. nilotica* branches for sorption of phenol from aqueous solutions. The powdered residue was activated thermochemically to tailor it to high adsorption capacity adsorbent. SEM analysis indicated its highly porous structure. Various sorption parameters of dosage, pH of the adsorbate-adsorbent system, contact time, initial phenol concentration, and the rotation speed of system were studied. Adsorption equilibrium models of Langmuir, Freundlich, Temkin, and Dubinin Radushkevich were applied on experimental data. Kinetic study on sorption indicated the presence of chemisorption. The thermodynamic study indicated the feasibility of sorption process. Regeneration of adsorbent with a maximum efficiency of 92% was also possible using acidic and basic solutions Dass and Jha (2015b).

11.5 Characterization of Agro-residues-based Adsorbents for Phenol

Before selecting any agro-residue-based adsorbents for phenol sorption, there is a requirement of characterizing it. Phenol being organic, any compound having carbon content can be considered as a suitable adsorbent. Agro-residues being a source of carbon can be considered an excellent source of the adsorbent for phenol sorption. There is good reporting of physical properties of adsorbents along with SEM and FTIR analysis (Namane et al., 2005; Namasivayam and Sangeetha, 2006; Kalderis et al., 2008; Srihari and Das, 2009; Sivakumar et al., 2012). Physical properties' considerations are concerning the proximate analysis of adsorbents on a dry basis (percentages of ash, volatile and fixed carbon), BET surface area analysis, average particle size, phenol number, and methylene blue test value. SEM analysis shows the porous structure of the adsorbent. FTIR analysis has been reported for ascertaining the presence of hydroxyl, amino, carbonyl, carboxyl, chlorides, aromatic, and cellulosic structures. Physical properties of agro-residues-based adsorbent are shown in Table 11.2.

11.6 Thermo-chemical Treatment to Agro-residues to Improve Its Adsorption Characteristics

It has been reported that the CGC has a much higher adsorption capacity of 317 mg/g compared to that of PPRH of 50 mg/g (Jha, 1996). The need was felt to compare the physico-chemical properties with agro-residues proposed to be used for adsorption of phenol. Based on the comparison, the properties of PPRH are far different from those of CGC.

To make the use of PPRH, as an effective adsorbent for phenol, it was considered appropriate to adopt a combination of thermal and chemical treatment (thermochemical) together. To bring the characteristics of rice husk, close to that of CGC the thermo-chemical treatment was given to rice husk. The rice husk was taken, washed thoroughly, and oven dried. Dried rice husk was subjected to pyrolysis at a temperature of 600°C for 90 min. After pyrolysis, steam at 0.5–0.7 kg/cm^2 (gage) was passed through the char for 2 h at 650°C. The char obtained after gasification with steam is leached with 3% hydrochloric acid at 60°C for 1 h. After leaching, the char is washed, filtered, and dried. Finally, the leached char in a batch of 70 g was digested with

Table 11.2　Physical properties of agro-residues-based activated carbon adsorbent

Agro-residues-based Activated Carbon	Ash (%)	Fixed Carbon (%)	BET Surface Area (m²/g)	Average Particle Size (μm)	Phenol Number	Methylene Blue Test Value (mg/g)	pH of Slurry	Iodine Number	Reference
Acacia niloticabranches	1.3	85.0	403	47.01	1.0	150	7.1	866	(Dass and Jha, 2015)
Martynia annua L.	13.9	–	1142	–	–	38	7.32	1024	(Sivakumar et al., 2012)
Hemidesmus indicus	2.08	97.92	627	–	5.3	82	6.3	204	(Srihari and Das, 2009)
Rice husk	27.8	–	811	41.21	–	–	5.3	–	(Kalderis et al., 2008)
Sugarcane bagasse	40.6	–	–	54.87	–	–	5.9	864	(Kalderis et al., 2008)
Coffee grounds	–	–	950	–	–	>90	5.7	590	(Namane et al., 2005)
Coir pith	3.2	81	–	–	30	137	3.3	203	(Namasivayam and Sangeetha, 2006)
Rubber seed coat	0.93	99.07	598	–	9.0	8.32	8.1	–	(Rengaraj et al., 2002)

Table 11.3 Comparison of physical properties of CGC with partially pyrolyzed char of rice husk (Jha, 1996)

Physical Properties	PPRH	Activated Rice Husk	CGC
Ash (% on dry basis)	24.84	6.22	1.46
Volatile matter (% on dry basis)	48.96	10.59	7.28
Fixed carbon (% on dry basis)	42.26	83.19	91.26
BET surface area (m^2/g)	1.20	290.3	1055.4
Methylene blue test value (mg/g)	0.0	52.5	210
Average particle size	0.729 mm	20.39 microns	27.2 microns
Phenol sorption capacity (mg/g)	50	150	317

10% sodium hydroxide at 70°C for 4 h. The solid residue was washed and filtered until neutral pH was obtained. The activated carbon was oven dried at 110°C. The physicochemical properties of the char are shown in Table 11.3. Here it is observed that activated rice husk has increased fixed carbon, BET surface area, methylene blue test values and decreased ash content, volatile matter and particle size values have. When phenol sorption was conducted, the sorption capacity increased to 150 from 50 mg/g. It is justifying the need for thermo-chemical treatment.

The following details show various literature regarding thermo-chemical treatment given to agro-residues for converting them into activated carbon having properties near to CGC with BET surface area in the range of 400–1500 m^2/g (various activating agents are H_3PO_4, H_2SO_4, KOH, NaOH, and $ZnCl_2$).

11.6.1 Date Stones

Dried date stones were ground to obtain a material with an average particle size of 1.71 mm. The powder was impregnated with 50% $ZnCl_2$ solution followed by carbonization at 700°C for 3 h. Obtained activated carbon was washed with diluted HCl (3wt%) and followed by washing with distilled water until wash is free of chloride ions. The activated carbon was dried at 110°C. BET surface area of the adsorbent obtained was 951 m^2/g (Alhamed, 2006).

11.6.2 Corncob

Corncobs were cut into small pieces, washed severally with distilled water and dried at 70°C. The dried samples were powdered and sieved to the particle size of 500–700 μm. Consequently, the sample soaked in orthophosphoric acid with impregnation ration of 1:1(w/w) for 24 h, dried in an oven at 105°C. The resultant sample was activated in an electric furnace at 500°C with a constant heating rate of 10°C/min under the purified nitrogen flow of 150 cm^3/min for 2 h. The activated carbon produced was cooled to room temperature and washed with 0.1 M HCl and successively with hot distilled water. The final product was oven dried. BET surface area of the adsorbent obtained was 1273.91 m^2/g (Njoku and Hameed, 2011).

11.6.3 Coconut Shell

The material was washed, sundried, and crushed to a particle size of 1–2 mm. The shell was carbonized at 700°C under the influence of nitrogen flow (1500 cm^3/min) for 1 h. The char produced was impregnated with KOH with an impregnation ration of 1:1. The mixture was dehydrated in the oven at 110°C. The dried solid mixture was employed in a stainless steel vertical tubular reactor and pyrolyzed in the atmosphere of high-purity nitrogen with a flow rate of 150 cm^3/min till it reached the desired temperature of 850°C. The sample was also activated with the help of CO^2 for 2 h. The activated carbon was washed with deionized water and 0.01 M hydrochloric acid to unclog the pores from tar and other chemicals. The washed activated carbon was finally dried and stored. BET surface area of the adsorbent obtained was 1026 m^2/g (Mohd Din et al., 2009).

11.6.4 Tobacco Residues

Tobacco residues with a particle size of 0.85–0.425 mm were activated using chemical activation technique. K_2CO_3 and KOH were used as chemical activation agents, and impregnation ratio of 75wt% was applied on this biomass sample. The sample was kept overnight at room temperature for 24 h and then dried at 85°C for 72 h. The impregnated sample was carbonized at 700°C under a nitrogen flow of 100 cm^3/min and heating of 10°C/min in a stainless steel fixed bed reactor. After being cooled, all the carbonized samples were washed several times with hot water until the pH became neutral and finally washed with cold water to remove residual chemicals. The resulting samples were dried at 105°C for 24 h to obtain the final activated carbons. BET surface area of the adsorbent obtained was 1634 m^2/g when impregnated with K_2CO_3 and 1474 m^2/g when impregnated with KOH (Kilic et al., 2011).

11.6.5 Rice Husk

Dried and ground rice husk ($\leq$0.5 mm) was impregnated with $ZnCl_2$ at ratios of 0.25–1. Distilled water was added at a quality amounting to 10 times the total weight of the solid mixture. The mixture was homogenized and heated to make homogeneous and impregnated at a temperature of 80–100°C. A thick uniform paste was obtained after 2 h. A sample of wet paste (around 75% moisture) was weighed before being fed to the reactor for a single step of carbonization/activation. The optimum $ZnCl_2$-to rice husk ratio was 1:1. Nitrogen gas was allowed to pass through the reactor (4 L/min) to remove air from the system. The furnace was heated to a temperature of 700°C, and once the target temperature was reached (in approximately 30 min), the gas supply was changed from nitrogen to CO_2 (5 L/min), and feeding of the paste commenced. The residence time required for activation was 30 min. The activated carbons were obtained in the lump form (0.2–1.5 cm) and were allowed to cool to room temperature. Then they were mechanically crushed and thoroughly washed with a solution of HCl (0.1 M) to remove excess $ZnCl_2$ and other impurities. Finally, the activated carbon was washed with distilled water and dried at 105°C for 6 h. BET surface area of the adsorbent obtained was 811 m^2/g (Kalderis et al., 2008).

11.6.6 Sugarcane Bagasse

The process was the same as that of rice husk except with a difference of optimum impregnation ratio of bagasse to $ZnCl_2$ as 0.75:1. BET surface area of the adsorbent obtained was 864 m^2/g (Kalderis et al., 2008).

11.6.7 Plum Kernels

The dried plum kernels were placed in a sealed ceramic oven and heated at a rate of 5 cm^3/min to a temperature of 900°C. The plum kernels were thermally decomposed to porous materials and some hydrocarbons under oxygen-deficient conditions. The steam was introduced at the same rate. The resulting activated carbon materials were ground in a mill followed by washing and drying. They were sieved in the range of 0.25–0.42 mm. BET surface area of the adsorbent obtained was 1162 m^2/g (Juang et al., 2000).

11.6.8 Coffee Husk

Coffee husk was dried at $110°C$ for 24 h and then treated with a $ZnCl_2$ solution at a 1:1 and pyrolyzed in a tubular oven under an N_2 flow of 100 mL/min at $550°C$ for 3 h. After the activation, the excess $ZnCl_2$ was removed with a 0.1 M solution of hydrochloric acid and the product was washed with hot distilled water until neutral pH was obtained. The same procedure was followed when $FeCl_3$ was used with a difference of activation temperature of $280°C$. BET surface area of the adsorbents obtained was 1522 m^2/g when treated with $ZnCl_2$ and 965 m^2/g when treated with $FeCl_3$ (Oliveira et al., 2009).

11.6.9 Root Residue of Hemidesmus Indicus

This agro-residue was collected, cleaned, and washed thoroughly to remove water-soluble substances. The material was treated with 1:1 weight ratio of concentrated H_2SO_4 and subsequently allowed to soak for 24 h, followed by cooling, washing (with distilled water), and soaking with 1% $NaHCO_3$ solution. The thoroughly washed sample was dried and subsequently subjected to pyrolysis at $850–900°C$ for 30 min. The sample thus obtained was ground in a ball mill, and the particles having an average diameter of 0.5 mm were collected and stored for further studies. BET surface area of the adsorbent obtained was 627 m^2/g (Srihari and Das, 2009).

11.6.10 *Acacia nilotica* Branches

Dried branches of *A. nilotica* branches were dried, cut into small pieces, put into grinding machine, and the powder was passed through a 250 mesh screen. This powder was taken to activate it. The powder was soaked in 30% H_3PO_4 for 4 h with agitation at a temperature of $35°C$. The resultant char obtained was filtered and washed thoroughly to bring its pH to 7. The sample was oven dried and kept for charring at $600°C$ in the muffle furnace for 3 h. The dried sample was digested with 10% NaOH solution for 4 h at $70°C$. The residue was filtered out, washed thoroughly, and dried. The dried sample was passed through a 250 mesh screen. The obtained sample was stored for further studies. BET surface area obtained was 403 m^2/g (Dass and Jha, 2015).

11.7 Effects of Various Parameters on Adsorption of Phenol

Adsorption process depends on various parameters, namely, adsorbent dosage, pH of the adsorbate–adsorbent system, the temperature of the system, the effect of initial phenol concentration, and contact time. For maximum sequestration of phenol from aqueous streams, there is a need to optimize the parameters affecting the process. They are discussed as follows.

11.7.1 Effect of Adsorbent Dosage

The adsorption dosage is an important parameter because it determines the capacity of the adsorbent for a given phenol concentration and also determines sorbent–sorbate equilibrium of the system. The effect of adsorbent dosage for agro-residues-based activated carbon has been studied for the adsorbent dosage range of 0.1–10 g/L depending upon the requirement of removal of phenol from the effluents. It also depends upon the convenience in which practically a particular adsorbent can be studied for the purpose. It is known that the increase in the amount of adsorbent concentration resulted in the higher availability of the exchangeable sites or surface area, but studies show that there is a decrease in the adsorption capacity due to the partial aggregation of particles of activated carbon, which results in a decrease in the active surface area for the adsorption (Kilic et al., 2011).

11.7.2 Effect of pH

The concentration of hydrogen ion in the adsorption is considered to be one of the most critical parameters that influence the adsorption behavior of phenol in aqueous solutions. It affects the solubility of phenol ions in the solution, replaces some of the positive ions found in the active sites, and affects the degree of ionization of the adsorbate during the reaction (Vimala and Das, 2009). The effect of pH on the adsorption of phenol is reported to be determined within the pH range of 2–12 and the results report that more than 90% of phenol is removed near the pH of 7. This may be due to the difference in the concentration of H+ and OH− ions in the solutions. Adsorbent particles have active sites with negative charges. The H+ ions within low pH environments can neutralize those negative sites and cannot hold phenol adsorbed on the adsorbent. On the other hand, high pH environment led to the high concentration of OH−, which can increase the hindrance to the diffusions of phenol ions and thus reduce the chances of their adsorption (Mukherjee et al., 2007).

11.7.3 Effect of Temperature

Effects of temperature variation on adsorption of phenol on agro-residues-based activated carbon have great importance. It has generally been reported in the literature that very low and very high temperatures do not favor the adsorption process. At very low temperatures, diffusion is very low which decreases the ability of phenol to get adsorbed on the adsorbent. At very high temperatures, adsorbents are not able to retain phenols, i.e., adsorption and desorption keep taking place simultaneously. Also at increased temperatures, there is a distortion of active sites. The experimental results have been reported for adsorption in the range of 20–50°C (Singh et al., 2016). The optimum temperatures for phenol sorption on agro-residues-based activated carbons have been reported in the range of 20–35°C (Demirbas, 2008; Kilic et al., 2011; Wang, 2017).

The effects of temperature on the adsorption study of phenol on agro-residue-based activated carbon enable us to conduct a thermodynamic study of adsorption. Details of this study are explained in Section 11.10.

11.7.4 Effect of Initial Phenol Concentration

The initial adsorbate concentration provides an essential driving force to over-come all mass transfer resistance of phenol between the aqueous and solid phase. The equilibrium adsorption efficiency of agro-residues-based activated carbon has been reported to increase with an increase in initial phenol concentrations. It is found that there is one optimum value (depending on the adsorbate–adsorbent system), beyond which adsorption of phenol decreases with increase in the concentration. This may be due to the availability of less vacant sites with increased phenol concentrations. This parameter also restricts the maximum adsorption of phenol (Kilic et al., 2011).

11.7.5 Effect of Contact Time

Contact time is a significant parameter to determine the equilibrium time of adsorption processes. The characteristics of activated carbon and its open sorption sights affect the time needed to reach the equilibrium.

Adsorption capacity increases with increasing contact time. It is generally reported that more than 90% of the maximum amount of phenol to be removed is in the first 60 min and equilibrium reaches in 120–360 min (Juang, 2000; Stasinakis et al., 2008). This parameter is a critical parameter which helps in kinetic studies of adsorption.

11.8 Mathematical Models for Adsorption Equilibrium Studies

Adsorption isotherm indicates the graphical representation of the relationship between the amount adsorbed by a unit weight of adsorbent and the amount of adsorbate remaining in a test medium at a constant temperature under an equilibrium condition. The mathematical model gives us the analysis of actual adsorption process which has taken place. Experimental data of adsorption can fit one or more than one adsorption model to explain the mechanism of the adsorption. The most common mathematical models reported in the literature are Langmuir, Freundlich, Temkin, and Dubinin–Radushkevich.

11.8.1 Langmuir Isotherm Model

This model describes the formation of a monolayer adsorbate on the outer surface of the adsorbent quantitatively, and after that, no further adsorption takes place. The Langmuir represents the equilibrium distribution of adsorbate between the solid and liquid phases (Vermeulan et al.,1996). The Langmuir adsorption isotherm is the most widely used isotherm for the biosorption of pollutants from a liquid solution based on the following hypotheses:

1. Monolayer adsorption.
2. Adsorption takes place at specific similar sites on the adsorbent.
3. Once the adsorbate occupies a site, no further adsorption can take place on that site.
4. Adsorption energy is fixed and does not depend on the degree of occupation of an adsorbent's active centers.
5. The forces of the intermolecular attractive forces are believed to fall off rapidly with distance.
6. The adsorbent has a fixed capacity for the pollutant.
7. All sides are identical and energetically equivalent.
8. The adsorbent is structurally homogeneous.
9. There is no interaction between molecules adsorbed on neighboring sites.

Based upon these assumptions, Langmuir represented the following equation:

$$C_e/q_e = (1/Q_0)b + C_e/Q_0 \qquad (11.2)$$

where C_e is the equilibrium concentration of adsorbate (mg/L), q_e denotes the adsorbate adsorbed per gram of the adsorbent at equilibrium (mg/g), Q_0 represents the monolayer coverage capacity (mg/g), and K_L is the Langmuir isotherm constant (L/mg).

The significant features of the Langmuir isotherm may be expressed concerning equilibrium parameter R_L, a dimensionless constant, referred to as separation factor or equilibrium parameter (Webber and Chakravarti, 1974):

$$R_L = 1/(1 + K_L C_0) \qquad (11.3)$$

where C_0 is the adsorbate initial concentration (mg/L), K_L is the constant related to the energy of adsorption (Langmuir constant). R_L value indicates the adsorption nature of the process.

The process is unfavorable if $R_L > 1$, linear if $R_L = 1$, favorable if $0 < R_L < 1$ and irreversible if $R_L = 0$.

11.8.2 Freundlich Isotherm Model

The Freundlich equilibrium isotherm equation is an empirical equation used for the description of multilayer adsorption with the interaction between adsorbed molecules:

$$\ln q_e = \ln k + 1/n \ln C_e \qquad (11.4)$$

where k [(mg/g). (L/g)] $1/n$ is an indication of sorption capacity, n is a measure of sorption intensity, C_e (mg/L) is the concentration of supernatant solution at equilibrium, and q_e (mg/g) is the amount of adsorbate collected per unit mass of adsorbent.

It is indicative of the presence of multilayer adsorption. This isotherm can be explained until saturation concentration is reached.

The model applies to the adsorption on heterogeneous surfaces by a uniform energy distribution and reversible adsorption. The Freundlich equation implies that adsorption energy exponentially decreases on the finishing point of adsorptional centers of an adsorbent (Freundlich, 1906). The Freundlich constants are empirical constants and depend on many environmental factors. The value of $1/n$ ranges between 0 and 1 and indicates the degree of non-linearity between solution concentration and adsorption (Treybal, 1985).

If the value of $1/n$ is equal to 1, the adsorption is linear (Al-Duri, 1996). Henry's isotherm or one-parameter isotherm is applicable for linear adsorption under the condition when $n = 1$ in Equation (11.4)

11.8.3 Temkin Isotherm Model

This is the early model proposed to depict adsorption of hydrogen on platinum electrodes within acidic solutions. The derivation of the Temkin isotherm is based on the assumption that the decline of the heat of sorption as a function of temperature is linear rather than logarithmic, as implied in the Freundlich equation (Basha et al., 2008; Aljboree et al., 2014). The following equation gives the model:

$$q_e = B \ln A + B \ln C_e \tag{11.5}$$

where A (L/g) and B (dimensionless) $= RT/b_1$, b_1 (J mol^{-1}) are constant indicating heat of sorption, C_e (mg/L) is the concentration of supernatant solution at equilibrium, and q_e (mg/g) is the amount of adsorbate adsorbed per unit mass of adsorbent.

This isotherm takes into account adsorbate–adsorbent interactions. This model assumes that heat of adsorption of all molecules would decrease linearly with coverage. The Temkin equation is often not suitable for representation of experimental data of the liquid phase adsorption in complex systems, since the derivation for the Temkin equation is based on simple assumption, and the complex phenomenon in liquid phase adsorption is not taken into account (Kiran and Kaushik, 2008).

11.8.4 Dubinin–Radushkevich Isotherm Model

The Dubinin–Radushkevich isotherm can be applied for the estimation of apparent free energy and the characteristics of adsorption (Dubinin, 1960; Rand, 1976). The following Radushkevich equation defines the Dubinin–Radushkevich equation:

$$\ln q_e = \ln q_m - \beta \varepsilon^2 \tag{11.6}$$

where β (mmol^{2}J^{-2}) is the D-R constant, ε (Jmmol^{-1}) is Polanyi potential, $\varepsilon = RT \ln(1 + 1/C_e)$, and q_e (mg/g) is the amount of adsorbate collected per unit mass of adsorbent.

This equation indicates that adsorption has taken place on microporous material especially of carbonaceous origin. It is mainly based on the assumption of the change in the potential energy between adsorbate and the adsorbent.

11.9 Mathematical Models for the Kinetics of Adsorption

The kinetics of adsorption describes the adsorbate uptake on activated carbon, and it controls the equilibrium time. The pseudo-first-order and pseudo-second-order models are most commonly used to describe the kinetics of the adsorption process (Suzuki, 1990).

11.9.1 Pseudo-first-order Kinetic Model

The pseudo-first-order kinetic model has been reported very frequently to predict sorption kinetics. Langergren, and Svenska is defined as:

$$Ln\,(q_e - q\,) = ln\,q_e - k \tag{11.7}$$

where q_e and q (mg/g) are the amount of adsorbate, adsorbed at equilibrium time and at any time, t (min^{-1}), respectively, and k_1 (min^{-1}) is the adsorption rate constant. The plot of ln $(q_e - q\,)$ versus t gives the slope of k_1 and the intercept value of q_e. The value of the coefficient of least square fit of $R^2 > 0.9$ confirms the fitting of data in pseudo-first-order kinetic model.

11.9.2 Pseudo-second-order Kinetic Model

The pseudo-second-order equation based on equilibrium adsorption is expressed as:

$$/q\, = 1/(k_2\,q_e^2) + \,/q_e \tag{11.8}$$

where k_2 (g/mg. min) is the rate constant of second-order adsorption. The linear plot of $/q$ versus t gives $1/q_e$ as the slope and $1/k\,q^2$ as the intercept. The slope enables us to find the value of q_e. Once the value of q_e is known, the value of k_2 can be found. The value of the coefficient of least square fit of $R^2 > 0.9$ confirms the fitting of data in pseudo-second-order kinetic model.

11.10 Thermodynamics of the Adsorption Process

Knowledge of thermodynamics of a process shows the feasibility of the process. It needs to be done to confirm the process of adsorption.

Heats of adsorption can also be calculated from this study. The data of effect of temperature on the adsorption process can be utilized to calculate the heat of adsorption (ΔH), the entropy of the process (ΔS), and adsorption constant K_d.

Thermodynamic parameters: free energy (ΔG°), enthalpy (ΔH°), and entropy (ΔS°) change in biosorption can be evaluated from the following equations (Kilic et al., 2011):

$$\Delta G^\circ = -RT \ \ln K_d \tag{11.9}$$

where R is the gas constant (8.314 J mol^{-1} K^{-1}), T is the temperature (K), and K_d is the equilibrium constant. The value of K_d is calculated using:

$$K_d = q_e/C_e \tag{11.10}$$

where q_e and C_e are the equilibrium concentrations of phenols in the solutions, respectively. Also, we know that

$$\Delta G^\circ = \Delta H^\circ - T\Delta S^\circ \tag{11.11}$$

From Equations (11.9), (11.10), and (11.11)

$$Ln \ K_d = (\Delta S^\circ/R) - (\Delta H^\circ/RT) \tag{11.12}$$

The values of (ΔH°) and (ΔS°) can be calculated from the slope and intercept, respectively, of Equation (11.12). The value of (ΔG°) can be calculated from Equation (11.11). A negative value of ΔG° confirms the spontaneity of the process. The values of (ΔH°) and (ΔS°) can be calculated by plotting ln K_d versus $1/T$ (Figure 11.2). The slope gives the value of ΔH° and intercepts ΔS°, respectively, of Equation (11.12). The value of (ΔG°) can be calculated from Equation (11.11). A negative value of ΔG° confirms the spontaneity of the process. A positive value of ΔH° shows that the process of adsorption is endothermic and the negative value of ΔH° tells that the process is exothermic. Value of ΔS° should be positive indicating the randomness of the process.

11.11 Regeneration of Adsorbents

One significant way to reduce the cost of the adsorption process is to efficiently desorb the retained substances and by regenerating the adsorbent for repeated use. Thus, the regeneration of porous carbons is a crucial issue

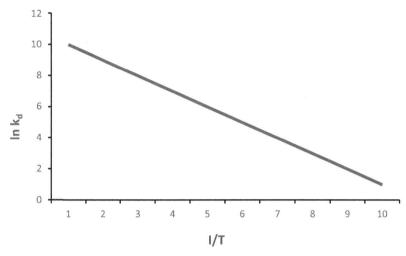

Figure 11.2   Trend of ln k_d versus 1/T.

and requires a detailed investigation to ascertain if such procedures are economically attractive. Generally spent activated carbon can be regenerated by several methods such as chemical regeneration, thermal regeneration, and electrochemical treatment. Among all the methods, thermal regeneration is by far the most commonly used because of its simplicity, high efficiency, and solvent-free property (Torrents et al., 1997; Chiang et al., 2007). If the adsorbent is regenerated chemically, one should confirm that after desorption process for regeneration of adsorbent, removal of all the bound sorbate from the adsorbent should be maximum. The regenerated adsorbent should be close to its original form, both morphologically and effectually. If this does not happen, a complete uptake cannot be expected in the next cycle. Therefore, non-destructive recovery by mild and cheap desorbing agents is desirable for regeneration of biomass for use in multiple cycles (Laurence et al., 2006; El-Naas et al., 2010). Generally, chemical desorption should be carried out precisely at the same conditions in which adsorption was carried out using the same amount of dosage. Mild solutions of chemicals used for the desorption process are NaOH, KOH, HCl, H_2SO_4, and HNO_3.

11.12 Conclusions

Application of agro-residues for phenol sequestration from industrial efflu-ents is very promising. Agro-residues being the source of carbon and lignocelluloses can prove to be low-cost adsorbents. Before suggesting any agro-residues-based adsorbents for any commercial applications, various parameters such as effect of adsorbent dosage, pH of the adsorbate, contact time, initial concentration of the adsorbate, and the effect of temperature are to be optimized for the maximum sequestration of phenol from aqueous streams. There is a great need to understand the mechanism of sorption by analyzing the process of sorption through mathematical modelling of Langmuir, Freundlich, Temkin, and D-R models. Also, kinetic studies give the mode of uptake of phenol with time. It also helps one to know whether the adsorption is physical or chemical. For complete research on the application of agro-residues for removal of phenol, one should also study continuous adsorption through column studies using agro-residues/activated agro-residues as adsorbents. Even regeneration is more real through this study. Exact optimization through various latest statistical techniques will enable us to get maximum sequestration of phenol using a specific agro-residue.

References

Afsharnia M, Saeidi M, Zarei A, Narooie MR, Biglari H. (2016). Phenol removal from aqueous environment by adsorption onto pomegranate peel carbon. Electronic Physician. 8(11):3248. DOI: 10.19082/3248

Ahmaruzzaman M, Sharma DK. (2005). Adsorption of phenol from wastewater. Journal of Colloid and Interface Science. 287:14–24. DOI: 10.1016/j.jcis.2005.01.075

Alam MZ, Ameem SE, Muyibi SA, Alam MZ, Ameem SE, Muyibi SA. (2009). The factors affecting the performance of activated carbon prepared from oil palm empty fruit bunches for adsorption of phenol. Journal of Chemical Engineering. 155:191–8. DOI: 10.1016/j.cej.2009.07.033

Alhamed YA. (2006). Activated carbon from dates' stone by $ZnCl_2$ activation. Journal of King Abdulaziz University Engineering Sciences. 17:75–100.

Alhamed YA. (2009). Adsorption kinetics and performance of packed bed adsorber for phenol removal using activated carbon from date's stones. Journal of Hazardous Materials. 170:763-70. DOI: 10.1016/j.jhazmat.2009.05.002

Ali A. (2017). Removal of Mn (II) from water using chemically modified banana peels as an efficient adsorbent. Environmental Nanotechnology, Monitoring and Management. 7:57–63. DOI: https://doi.org/10.1016/j.enmm.2016.12.004

Aljeboree AM, Radi N, Ahmed Z, Aklaim AF. (2014). The use of sawdust as a by-product adsorbent of an organic pollutant from wastewater: adsorption of maximum blue dye. International Journal of Chemical Science. 12(4):1239–52.

Anne P, Nadia M-C, Coline D, Sophie G, Corina B, Céline L, Giangiacomo T, Peter W, Grégorio C. (2017). Alkylphenol and alkylphenol polyethoxylates in water and wastewater: a review of options for their elimination. Arabian Journal of Chemistry. 10:S3749-73. DOI: 10. 1016/j.arabjc.2014.05.011

Araújo CST, Almeida ILS, Rezende HC, Marcionilio SMLO, Léon JJL, Matos TN. (2018). Elucidation of the mechanism involved in adsorption of Pb(II) onto lobeira fruit (Solanum lycocarpum) using Langmuir, Freundlich and Temkin isotherms. Microchemical Journal. 137:348–54. DOI: 10.1016/j.microc.2017.11.009

Basha S, Murthy ZVP, Jha BK. (2008). Biosorption of hexavalent chromium by chemically modified seaweed. crystoseira indica. The Chemical Engineering Journal. 137(3):480–8. DOI: 10.1016/i.cej.2007.04.038

Bhatnagar A, Sillanpää M, Witek-Krowiak A. (2015). Agricultural waste peels as versatile biomass for water purification – A review. Chemical Engineering Journal. 270:244–71. DOI: 10.1016/j.cej.2015.01.135

Bradl H, Xenidis A. "Remediation techniques" in Interface Science and Technology, ed. Bradl HB (Elsevier) ISBN 9780120883813, DOI: 10. 1016/S1573-4285(05)80022–5

Cheng YS, Yang SC, Hseih CT. (2006). Thermal regeneration of activated carbons exhausted with phenol compounds. Separation Science and Technology. 42(3):639–52. DOI: 10. 1080/014963906001070059

Chiou MS, Li HY. (2003). Adsorption behavior of reactive dye in aqueous solution on chemical cross-linked chitosan beads. Chemosphere. 50(8):1095–105. DOI: 10.1016/S0045-6535(02)00636–7

Clayton GD, Clayton FE. (1994). Patty's Industrial Hygiene and Toxicology. 4th ed. (John Wiley & Sons Inc: New York), vol. 2: 132. ISBN-13: 978-0471012801.

Das B. (2016). Characterization of Biomass/Agro Residues and Application of Selected Biomass for sorption of Phenol from Aqueous Solutions. PhD thesis, Department of Chemical Engineering, SLIET, Longowal, India.

Dass B and Jha P. (2015a). Adsorption of phenol by biomass (Acacia Nilotica Branches) based activated carbon for water purification. R. J. Pharmaceutical, Biological and Chemical Sciences. 6(4):1361–72.

Dass B and Jha P. (2015b). Batch adsorption of phenol by improved acacia nilotica branches char: Equilibrium, Kinetic and Thermodynamic studies. International Journal of ChemTech Research. 8(12):269–79.

Demirbas A. (2008). Heavy metal adsorption onto agro-based waste materials: A review. Journal of Hazardous Materials. 157(2-3):220-9. DOI: 10.1016/j.jhazmat.2008.01.024

Dubinin MM. (1960). The Potential theory of adsorption of gases and vapours for adsorbents with energetically nonuniform surfaces. Chemical Reviews. 60(2):235–41. DOI: 10.1021/acs.chemrev.8b00253

Dutta S, Basu JK, Ghar RN. (2001). Studies on adsorption of p-nitrophenol on charred sawdust. Separation and Purification Technology. 21(3):227–35.

El-Naas MH, Al-Zuhair S, Alhaija MA. (2010). Removal of phenol from petroleum refinery wastewater through adsorption on date-pit activated carbon. Journal of Chemical Engineering. 162:997–1005. DOI: 10.1016/j.cej.2010.07.007

Fernadez E, Hugi-Cleary D, López-Ramón M, Stoeckli F. (2003). Adsorption of phenol from dilute and concentrated aqueous solutions by activated carbons. 19(23):9719–29. DOI: 10.1021/la030137d

Franco DS, Dortzbacher GF, Dotto GL. (2017). Adsorption of Co (II) from aqueous solutions onto rice husk modified by ultrasound assisted and supercritical technologies. Process Safety and Environmental Protection. 109:55–62.

Freitas JV and Farinas CS. (2017). Sugarcane bagasse fly ash as a no-cost adsorbent for removal of phenolic inhibitors and improvement of biomass saccharification. ACS Sustainable Chemistry and Engineering. 5(12):11727–36. DOI: 10.1021/acssuschemeng.7b03214

Ganesh A. (1990). Studies on Characterization of Biomass for gasification. PhD thesis, Department of Chemical Engineering, IIT Delhi, India.

Gisi SB, Lofrano G, Grassi M, Notarnicola M. (2016). Characteristics and adsorption capacities of low-cost sorbents for wastewater treatment: A review. Sustainable Materials and Technologies. 9:10–40. DOI: 10.1016/j.susmat.2016.06.002

Goud VV, Mohanty K. (2005). Prediction of mass transfer coefficients in a packed bed, using tamarind nut shell activated carbon to remove phenol. Chemical Engineering and Technology. 28(9):991–7. DOI: 10.1002/ceat.200500099

Gupta VK, Ali I. (2013). "Water Treatment for Inorganic Pollutants by Adsorption Technology" in Environmental Water, ed. V.K. Gupta, I. Ali

(Elsevier) ISBN: 9780444593993, DOI: 10.1016/B978-0-444-59399-3.00002-7

Gupta VK, Sharma S, Yadav IS, Dinesh M. (1998). Utilization of bagasse fly ash generated in the sugar industry for the removal and recovery of phenol and *p*-nitrophenol from wastewater. Chemical Technology and Biotechnology. 71(2):180–6. DOI: 10.1002/(SIC)1097-4660(199802)71:2<180::AID-JCTB798>3.0.CO;2-1

Hameed BH, Rahman AA. (2008). Removal of phenol from aqueous solutions by adsorption onto activated carbon prepared from biomass material. Journal of Hazardous Materials. 160:576–81.

Ho YS. (2004). Citation review of Lagergren kinetic rate equation on adsorption reactions. Scientometrics. 59(1):171–7. DOI: 10.1023/B:SCIE.0000013305.99473.cf

Inam E, Etim UJ, Akpabio EG, Umoren SA. (2017). Process optimisation for the application of carbon from plantain peels in dye abstraction. Journal of Taibah University of Science. 11(1):173–85. DOI: 10.1016/j.jtusci.2016.01.003

Iyer PVR, Rao TR, Grover PD. (2002). Biomass Thermo-Chemical Characterisation. Published under MNES sponsored, Chemical Engineering Department, IIT Delhi, India.

Jadhav DN, Vanjara AK. (2004). Removal of phenol from wastewater using sawdust, polymerised sawdust and sawdust carbon. International Journal of Chemical Technology. 11:35–41.

Jha P. (1996). Biomass characterisation and the application of biomass char for sorption of phenol from aqueous solutions. PhD thesis, Department of Chemical Engineering, IIT, Delhi.

Jha P. (2011). Rice husk as an adsorbent for phenol removal. International Journal of Science and Nature. 2(3):593–6.

Juang R-S, Wu F-C, Tseng R-L. (2000). Mechanism of adsorption of dyes and phenols from water using activated carbons prepared from Plum kernels. Journal of Colloid and Interface Science. 227:437–44. DOI: 10.1006.jcis.2000.6912

Kalderis D, Koutoulakis D, Paraskeva P, Diamadopoulos E, Otal E, Olivaris V, Pereira CF. (2008). Adsorption of polluting substances on activated carbons prepared from rice husk and sugarcane bagasse. Journal of Chemical Engineering. 144:42–50. DOI: 10.1016/j.cej.2008.01.007

Karabacakoglu B, Tumsek F, Demiral H, Demiral I. (2008). Liquid phase adsorption of phenol by activated carbon derived from hazelnut bagasse.

Journal of International Environmental Applications and Science. 3(5):373–80.

Kilic M, Apaydin-Varol E, Putin AE. (2011). Adsorptive removal of phenol from aqueous solutions on activated carbon prepared from tobacco residues: equilibrium, kinetics and thermodynamics. Journal of Hazardous Materials. 189:397–403. DOI: 10.1016/j.jhazmat.2011.02.051

Kulkarni SJ, Patil SV, Shubhangi SK. (2011). "Studies on adsorption for phenol removal using activated carbon in batch and fluidised bed adsorption" in Proceedings of J. I. Conference on Current Trends in Technology, NIRMA University, Ahmedabad.

Kumar R, Sharna RK, Singh AP. (2017). Cellulose-based grafted biosorbents – Journey from lignocellulose biomass to toxic metal ions sorption applications – A review. Journal of Molecular Liquids. 232: 62–93. DOI: 10.1016/j.molliq.2017.02.050

Lata S, Singh PK, Samadder SR. (2015). Regeneration of adsorbents and recovery of heavy metals: a review. International Journal of Environmental Science and Technology. 12(4):1461–78. DOI: 10.1007/s13762-014-0714-9

Laurence LC, Ngoc HP, Sebastien R, Catherine F, Pierre LC, Thanh HN. (2006). Production of fibrous activated carbons from natural cellulose (jute, coconut) fibres for water treatment applications. Carbon. 44(12):2569–77. DOI: 10.1016/j.carbon.2006.05.048

Lofrano SDGG, Grassi M, Notarnicola M. (2016). Characteristics and adsorption capacities of low-cost sorbents for wastewater treatment: a review. Sustainable Materials and Technologies. 9:10–40. DOI: 10.1016/j.susmat.2016.06.002

Mahajan SP. Pollution Control in Processes Industries. (1985). Tata McGraw-Hill, New Delhi, 235–54.

Miretzky P, Cirelli F. (2010). Cr(VI) and Cr(III) removal from aqueous solution by raw and modified lignocellulosic materials: a review. Journal of Hazardous Materials. 180(1–3):1–19. DOI: 10.1016/j.jhazmat.2010.04.060

Mohd Din AT, Hameed BH, Ahmad AL. (2009). Batch adsorption of phenol onto physiochemical-activated coconut shell. Journal of Hazardous Materials. 161:1522–9.

Mukherjee S, Kumar S, Misra AK, Fan M. (2007). Removal of phenols from water environment by activated carbon, bagasse ash and wood charcoal. Journal of Chemical Engineering. 129:133–42. DOI: 10.1016/j.cej.2006.10.030

Namane A, Mekarzia A, Benrachedi K, Belhaneche N, Hellal A. (2005). Determination of the adsorption capacity of activated carbon made from the coffee ground by chemical activation with $ZnCl_2$ and H_3PO_4. Journal of Hazardous Materials. 119(1–3):189–94. DOI: 10. 1016/j.jhazmat.2004.12.006

Namasivayam C, Kavitha D. (2003). Adsorptive removal of 2-chlorophenol by low-cost coir pith carbon. Journal of Hazardous Materials. 98(1–3):257–74. DOI: 10.1016/S0304-3894(03)00006-2

Namasivayam C, Sangeetha D. (2006). Recycling of agricultural solid waste, coir pith: removal of anions, heavy metals, organics and dyes from water by adsorption onto $ZnCl_2$ activated coir pith carbon. Journal of Hazardous Materials. 135(1–3):449–52. DOI: 10.1016/ j.jhazmat.2005.11.066

Narwade VN, Khairnar RS, Kokol V. (2017). In-situ synthesised hydroxyapatite-loaded films based on cellulose nanofibrils for phenol removal from wastewater. Cellulose. 24(11):4911–25.

Neetu S, Chandrajit B. (2017). Biosorption of phenol and cyanide from synthetic/simulated wastewater by sugarcane bagasse—equilibrium isotherm and kinetic analyses. Water Conservation Science and Engineering. 2(1):1–14. DOI: 10.1007/s41101-017-0019-1

Nguyen, TAH, Ngo HH, Guo WS, Zhang J, Liang S, Yue QY, Li Q, Nguyen TV. (2013). Applicability of agricultural waste and by-products for adsorptive removal of heavy metals from wastewater. Bioresource Technology. 148: 574–585. DOI: 10.1016/j.biortech.2013.08.124

Njoku VO, Hameed BH. (2011). Preparation and characterisation of activated carbon from corncob by chemical activation with H_3PO_4 for 2,4-dichlorophenoxyacetic acid adsorption. Journal of Chemical Engineering. 173:391–9.

Nor NM, Chung LL, Teong LK, Mohamed AR. (2013). Synthesis of activated carbon from lignocellulosic biomass and its applications in air pollution control- review. Journal of Environmental Chemical Engineering. 1(4):658–66. DOI: 10.1016/j.jee.2013.09.017

Oliveira LCA, Vallone A, Sapag K. (2009). Preparation of activated carbons from coffee husks utilising $FeCl_3$ and $ZnCl_2$ as activating agents. Journal of Hazardous Materials. 165:87–94. DOI: 10. 1016/j.jhazmat.2008.09.064

Orchando-Pulido JM, Pimental-Moral S, Verardo V, Martinez-ferez A. (2017). A focus on advanced physicochemical processes for olive mill wastewater treatment. Separation and Purification Technology. 179: 161–74. DOI: 10.1016/j.seppur.2017.02.004

Palma C, Lioret L, Puen A, Tobar M, Contreras E. (2016). Production of carbonaceous material from avocado peel for its application as an alternative adsorbent for dyes removal. Chinese Journal of Chemical Engineering. 24(4):521–8. DOI: 10.1016/j.cjche.2015.11.029

Ramin S, Elhamh D, Shahin A. (2018). Adsorption isotherm and kinetics study: removal of phenol using adsorption onto modified pistacia mutica shells. Iranian Journal of Earth Sciences. 6(1):33–42.

Rand B. (1976). On the empirical nature of the Dubinin – Radushkevich equation of adsorption. Journal of Colloid and Interface Science. 56(2):337–46. DOI: 10.1016/0021-9707(76)90259-9

Rebeiro LAS, Rodriques LA, Thim GA. (2017). Preparation of activated carbon from orange peel and its application for phenol removal. International Journal of Engineering Research. 3(3):122–9.

Rengaraj S, Moon SH, Arabindoo B, Murugesan V. (2002). Removal of phenol from aqueous solution and resin manufacturing industry wastewater using an agricultural waste: rubber seed coat. Journal of Hazardous Materials. 89(2–3):185–96.

Sáenz-Alans CA, García-Reyes RB, Soto-Regalado E, García-González A. (2017). Phenol and methylene blue adsorption on heat-treated activated carbon: Characterization, kinetics, and equilibrium studies. Adsorption Science and Technology. 35(9–10):789–805. DOI: 10. 1177/0263617416684517

Sahu O, Rao DG, Engidayehu A, Teshale F. (2017). Sorption of phenol from synthetic aqueous solution by activated sawdust: optimizing parameters with response surface methodology. Biochemistry and Biophysics Reports. 12:46–53. DOI:10.1016/j.bbrep.2017.08.007

Santos CM, Dewek J, Viotto RS, Rosa AH, Morais LC. (2015). Application of orange peel waste in the production of solid biofuels and biosorbents. Bioresource Technology. 196:469–79. DOI: 10. 1016/j.biortech.2015.07.114

Sarker N, Fakhruddin ANM. (2017). Removal of phenol from aqueous solutions using rice-straw as an adsorbent. 7(3):1459–65. DOI:10.1007/s13201-015-0324-9

Shawabkeh RA, Abu-Namesh SM. (2007). Adsorption of phenol and methylene blue by activated carbon from Pecan shells. Colloid Journal.69(3):355–9.

Shin W. (2017). Adsorption characteristics of phenol and heavy metals on biochar from Hizikia fusiformis. Environmental Earth Sciences. 76(22):782. DOI: 10.1007/s12665-017-7125-4

Shiundu PM, Mbui DN, Ndonye RM, Kamau GN. (2002). Adsorption and detection of some phenolic compounds by rice husk ash of Kenyan origin. Journal of Environmental Monitoring. 4:978–84.

Singh N, Balomajumder C. (2016). Simultaneous biosorption and bioaccumulation of phenol and cyanide using coconut shell activated carbon immobilised Pseudomonas putida (MTCC 1194). Journal of Environmental Chemical Engineering. 4(2):1604–14. DOI: 10.1016/j.jece.2016.02.01

Singh N, Kumari A, Balomajumder C. (2016). Modeling studies on mono and binary component biosorption of phenol and cynamide from aqueous solution and activated carbon derived from sawdust. Saudi Journal of Biological Sciences (in Press). DOI: 10.1016/j.sjbs.2016.01.007

Sivakumar V, Asaithambi M, Sivakumar P. (2012). Physicochemical and adsorption studies of activated carbon from Agricultural waste. Journal: Advances in Applied Science Research. 3(1):219–26.

Srihari V, Das A. (2009). Adsorption of aqueous phenol media by an agro-waste (Hemidesmus Indicus) based activated carbon. Jounal of Applied Ecology and Environmental Research. 7(1):13–23.

Srivastava VC, Swamy MM, Mall ID, Prasad B, Mishra IM. (2006). Adsorptive removal of phenol by bagasse fly ash and activated carbon: equilibrium, kinetics and thermodynamics. Colloids and Surfaces: Physicochemical and Engineering Aspects. 272(1–2):89–104. DOI: 10.1016/j.colsurfa.2005.07.016

Stasinakis AS, Elia L, Halvadakis CP. (2008). Removal of total phenols from olive- mill wastewater using an agricultural by-product, olive pomace. Journal of Hazardous Materials. 160:408–13. DOI: 10.1016/j.jhazmat.2008.03.012

Stubble burning in north India: Home (NASA Earth Observatory. Images). Available at: https://earthobservatory.nasa.gov/images/84680/stubble-burning-in-northern-india [Accessed October 30, 2014].

Suzuki M. (1990). "Kinetics of adsorption in a vessel" in Adsorption Engineering, Kodansha Ltd, Tokyo and Elsevier Science Publishers, ISBN: 4-06-201485-8 (Japan).

Torrents A, Damera R, Hoa OJ. (1997). Low-temperature thermal desorption of aromatic compounds from activated carbon. Journal of Hazardous Materials. 54:141–53.

Treybal Robert E. (1985). "Adsorption and Ion Exchange" in Mass Transfer operations, International student edition, Mc. GrawHill Book Company, Mc. GrawHill, Chemical Engineering Series, 8th printing, Singapore, 565–612. ISBN: 0-07-066615-6.

Tseng RL, Tseng SK. (2005). Pore structure and adsorption performance of the KOH activated carbons prepared from corncob. Journal of Colloid and Interface Science. 287(2):428–37. DOI: 10.1016/j.jcis.2005.02.033

Tseng RL, Wu KT, Juang RS. (2010). Kinetic studies on the adsorption of phenol, 4-chlorophenol, and 2, 4- dichlorophenol from water using activated carbons. Journal of Environmental Management. 91(11):2208–14. DOI: 10.1016/j.jenavman.2010.05.018

Wang F. (2017). Novel high performance magnetic activated carbon for phenol removal: equilibrium, kinetics and thermodynamics. Journal of Porous Materials. 24(50):1309–17. DOI: 10.1007/s10934- 017-0372-7

Wong SH. (1998). Characterisation of activated carbon Adsorption processes for removal of 2- methylisoborneol and microcystin from model drinking water. PhD thesis, School of Chemical Technology, University of South Australia, Australia.

Zouboulis AI, Peleka EN, Samaras P. (2015). "Removal of Toxic Materials from Aqueous Streams" in Minerals Scales and Deposits, ed. Z. Ahmad, KD Demadis (Elsevier), ISBN: 9780444632289, DOI: 10.1016/B978-0-444-63228-9.00017-6.

MODULE 3

Solid Waste Management – New Breakthrough

12

Photocatalytic Degradation of Plastic Polymer: A Review

Tarun Parangi and Manish Kumar Mishra

Department of Chemistry, Sardar Patel University Vallabh Vidyanagar, Gujarat, India
E-mail: manishorgch@gmail.com

Environmental pollution and demolition as well as the lack of appropriate clean and natural energy resources are some of the most severe issues currently faced on a global scale. The uncontrolled use of plastic products produces a huge terrible thing called "white pollution," which has become a serious environmental problem in today's world. It should, therefore, be our aim to contribute to the development of environmentally harmonious, ecologically clean and safe, sustainable, and energy-efficient chemical technologies. The efficient and eco-friendly technologies are in great demand for safe disposal and removal of the plastic waste from the environment. Photocatalytic degradation of plastics, in which the inexhaustibly abundant, clean, and safe energy of the sun can be harnessed for sustainable, non-hazardous, and economically viable technologies, is a major advance in this direction. The main aim of the present paper is to review the use of different photocatalytic materials and processes for the degradation of various plastic wastes.

12.1 Introduction

Polymers are a broad class of materials which are made from repeating units of smaller molecules called monomers (IUPAC Recommendations, 1996). Polymers are useful because of their strength and durability in many applications. In recent years, polymers are receiving an increased attendance in the

225

chemical industry with their highly defined properties and utilities. Natural and synthetic polymers find a vital and pervasive use in day to day life due to their versatile traits (McCrum et al., 1997; Painter and Coleman, 1997). Polymers are widely used in petrochemical industries, oil and gas companies, steel and aluminum industries, and plastic industries which involve bags, boxes, routine consumer/home uses, etc. Also they find their adequacy in packaging of food items, agricultural products, electrical supplies, and so on.

Plastic, a polymer has become an essential ingredient of modern life with broad range of utilities. Central Pollution Control Board stated that in 2017 plastic usage in India itself is 8,500,000 tones (Kulkarni and Dasari, 2018). However, some of their most useful features, the chemical, physical, and biological inertness, durability, low density, and low cost, has made their production efficient and easy, resulting in their accumulation in the environment if not recycled. The management of wastes generated by polymers is difficult whereas recycling can be feasible, environmentally, economically, and technically.

Plastic is mainly segregated into two groups based on its back bone structure, plastics with a carbon–carbon backbone and plastics with heteroatoms in the main chain (Bandara et al., 2017). PE, PS, and PP PVC have a backbone which is solely built of carbon atoms. PET, PU, and PAM plastics have heteroatoms in the main chain.

Use of PE (LDPE and HDPE) is increasing day by day and now it can be seen that PE is being used in almost every activity of life like grocery bags, shampoo bottles, food wrapping material, power cable sheathing, laboratory containers, bullet proof vests, etc. (Kemp and McIntyre, 2006; Zhao et al., 2007; Piringer and Baner, 2008; Ali et al., 2016). PP is widely used in food packaging, carpets, rugs, and containers (Kamrannejad and Hasanzadeh, 2014). PS, one of the most important materials from the modern plastic industry, has been used all over the world, due to its excellent physical properties and low cost (Bandyopadhyay and Basak, 2007). PS is a versatile plastic used to make a wide variety of consumer products. As a hard, solid plastic, it is often used in products that require clarity, such as food packaging (Fa et al., 2013) and laboratory ware (Wunsch, 2000). PVC is widely used as heat shrink membranes, transparent slices, one-off medical products, emulsion gloves, cable insulation, and as a building material (Kemp and McIntyre, 2006; Chakrabarti et al., 2011; Fa et al., 2011; Gilani , 2017). A huge amount of PVC is now produced globally and is consumed with exceedingly low recovery value (Owen, 1984). PAM is used in genetics, genetic engineering, and molecular biology laboratories as a matrix for separating nucleic acid components during DNA sequence analysis and protein identification known as gel electrophoresis (Dearfield et al., 1995).

Unfortunately, the accumulation of plastics has become a severe day-to-day problem for all developing and the developed nations in the world. Plastic is versatile, lightweight, flexible, moisture resistant, strong, and relatively inexpensive. Those are the attractive qualities that lead us, around the world, to such a greedy appetite and over-consumption of plastic goods. However, durable and very slow to degrade, plastic materials that are used in the production of so many products all, ultimately, become waste with staying power.

In the last half-century, there have been many drastic changes on the surface of the planet, but one of the most instantly observable is the omnipresence and abundance of plastic fragments (Barnes et al., 2009). Plastics themselves contribute to approximately 10% of discarded waste. Plastic waste contributes to about 9% of the total 1.20 lakh tons per day of the municipal solid waste generated in India (Central Pollution Control Board, 2010). Plastic products bring convenience, but also produce a huge disastrous thing called "white pollution," which has become a serious environmental problem in today's world (GESAMP 2015; UNEP 2016 and 2018) (Zan et al., 2006; Yuan et al., 2013). White pollution refers to solid waste which comes from the usage of various types of life plastic products. Due to their chemical stability and non-biodegradability, waste plastic is being mainly disposed by incineration, which releases a lot of toxic byproducts in environment (Zan et al., 2006; Zhao et al., 2008). PE plastic wastes buried in soil cause negative effects to soil quality and may affect the drainage patterns leading to declined agricultural yield (Seymour, 1989). Excessive use of plastics in domestic, industrial, and agriculture sectors exerts pressure on capacities available for plastic waste disposal which causes an additional burden on the environment (Zan et al., 2006; Shah et al., 2008; Ali et al., 2016).

It is difficult to degrade and dispose of resulting in severe urban environmental consequences. Even the simplest form of plastics, i.e., PE is strong and highly durable and takes up to 1000 years for natural degradation in the environment (Phonsy et al., 2015). Now, most waste polymers are burned for disposal, but they produce toxic compounds like ketones, acrolein, dioxins, furans, polychlorinated biphenyls, and methane, and pollute the air which causes serious environmental hazards (Zan et al., 2006; Ali et al., 2016; Verma et al., 2016). The disposal of wasted plastic such as landfill, incineration, or recycle (where only a small percentage of the plastic waste is currently being recycled) is neither economically valuable nor acceptable due to causing more serious air pollution (Sun et al., 2003; Chakrabarti et al., 2008; Xiao-Jing et al., 2008; Fa et al., 2011).

Thus, there is considerable need to degrade plastic waste in a more economical and ecofriendly way. Therefore, attention has been focused on alternative means of degrading the plastic material, which includes thermal, catalytic, mechano-chemical, ozone induced, and photo-oxidative degradation. Thermal decomposition of plastic results in evolution of harmful gases during fires or waste burning like carbon monoxide, chlorine, furans, dioxins, CCl_4, etc., resulting into breathing problems (Ali et al., 2016).

In view of the environmental management and the sustainable development of human society, it is very important to find an eco-friendly disposal of plastic waste where they degrade to carbon dioxide and water under the sunlight irradiation without producing toxic by-products. Recently, Gewert et al. (2013) have developed laboratory protocol that simulates the exposure of plastic floating in the marine environment to UV light to identify degradation products of plastic polymers in water. In this photodegradation study, the plastic pellets of PE, PS, PP, and PET suspended in water were exposed to sunlight for effectively 5 days, which resulted to the release of compounds, those are potentially hazardous. The study directing toward photocatalytic degradation of plastic that stands out as the most promising and environment friendly pathway (Kim et al., 2006; Chakrabarti et al., 2008; Zhao et al., 2008; Asghar et al., 2011; Grover et al., 2015; Mehmood et al., 2015; Ali et al., 2016). Recently, the photocatalytic techniques have been successfully applied in the disposal of waste plastics. Therefore, it is worthwhile to study the solid phase photocatalytic degradation of waste plastic under the condition of the atmosphere and sunlight (sunlight is a very important issue, because sun is an unending resource).

Photocatalysis, as a research area, has witnessed a sea change over the past two decades with significant advancements being made in the synthesis of novel materials and nano-structures, and the design of efficient processes for the degradation of environmental pollutants and the generation of useful sources of fuel and energy (Hoffmann et al., 1995; Linsebigler et al., 1995; Fujishima et al., 1999; Fujishima et al., 2000; Shang et al., 2003a; Vinu and Madras, 2010; Yousif et al., 2013; Schneider et al., 2014; Zhu and Wang, 2017). There are many reports on the principles and mechanism of photocatalysis, with special importance on the electron transfer processes, lattice and electronic structure of various photocatalysts, surface chemistry of semiconductor metal/metal oxides, generation of reactive radicals, chemisorption of small and large molecules, surface modification by doping, photo oxidation of organic and inorganic substrates, green synthesis of organic compounds, and the generation of hydrocarbons and hydrogen (Hoffmann et al., 1995; Linsebigler et al., 1995; Fujishima et al., 1999; Fujishima et al., 2000; Shang et al., 2003a; Chen and Mao, 2007; Anpo and Kamat, 2010; Vinu and Madras, 2010; Yousif et al., 2013; Schneider et al., 2014; Zhu and Wang, 2017). Hence, photocatalysis can be regarded a well understood area where today also immense challenges and opportunities exist in realizing this technology for large-scale practical applications in the decontamination of the environment and the generation of clean energy (Schneider et al., 2014). The present review is humble attempts to report the status of the research on the degradation of plastic focusing on various photocatalysts used for degradation processes.

12.2 Degradation of Plastic Polymers Using Various Photocatalytic Materials

Among various types of photocatalysts, titanium dioxide (TiO_2; commonly known as titania) is one of the most widely used inorganic materials in the world (Fujishima et al., 2000; SCCNFP, 2000; Shang et al., 2003a; Kemp, 2006; Chen and Mao, 2007; Anpo and Kamat, 2010; Daghrir et al., 2013). The most common form is pigmentary TiO_2, but in recent years there has been a growing demand for ultrafine TiO_2. Ultrafine TiO_2 is known for its many versatile applications emanating from its very small (nano) particle size and semiconductor properties. Titania is a promising photocatalyst due to its high photo activity (UV absorbing capacity), high stability, low cost, and absence of toxicity (Herrmann, 1999; Zhao et al., 2005; Kudo and Miseki, 2009; Daghrir et al., 2013). In the past decades, TiO_2 was used for accelerating photodegradation of synthetic polymers, a promising method that resolved the problem of plastic waste disposal (Hagfeldt and Graetzel, 1995; Cho and Choi, 2001; Shang et al., 2003b; Kemp and McIntyre, 2006; Kim et al., 2006; Bandyopadhyay and Basak, 2007; Chakrabarti et al., 2008; Zhao et al., 2008; Asghar et al., 2011; Thomas and Sandhyarani, 2013; Kamrannejad and Hasanzadeh, 2014; Leong et al., 2014; Mehmood et al., 2015; Verma et al., 2016; Ning et al., 2017). In the solid-state degradation, a composite of the polymer and TiO_2 is used, while in the liquid state, the TiO_2 particles are suspended in the polymer solution. However, the mechanism of photodegradation is not much affected by the state in which the polymer is degraded.

Other than TiO_2, a number of other representative metal oxides (ZnO, CeO_2, MoO_3, ZrO_2, WO_3, α-Fe_2O_3, SnO_2, and $SrTiO_3$) and metal chalcogenides (ZnS, CdS, CdSe, WS_2, and MoS_2) are being used as photocatalysts due to their stability in aqueous solution, favorable excellent electrical, mechanical, and optical properties similar to TiO_2, and their low cost (Mills and Hunte, 1997; Bhatkhande et al., 2002; Wada et al., 2002; Carp et al., 2004; Kabra et al., 2004; Byrappa et al., 2006; Palmisano et al., 2007; Kudo et al., 2009; Vinu and Madras, 2010; Chakrabarti et al., 2011; Soltaninezhad and Aminifar, 2011; Wahab et al., 2011; Rajabi et al., 2013; Sathishkumar et al., 2013; Bandara et al., 2017). However, according to the thermodynamic requirement, the VB and CB of the semiconductor photocatalyst should be positioned in such a way that the oxidation potential of the hydroxyl radicals $(E^0 (H_2O/OH^\bullet) = 2.8$ V vs NHE) and the reduction potential of superoxide radicals $(E^0 (O_2/O_2^{\bullet-}) = -2.8$ V *vs* NHE) lie well within the bandgap. In other words, the redox potential of super oxide radicals the VB holes must be sufficiently positive to generate hydroxyl radicals and that of the CB electrons must be sufficiently negative to generate superoxide radicals (Vinu and Madras, 2010).

Figure 12.1 shows the band structure diagram of various inorganic materials, along with the potentials of the redox couples. It is clear that, TiO_2

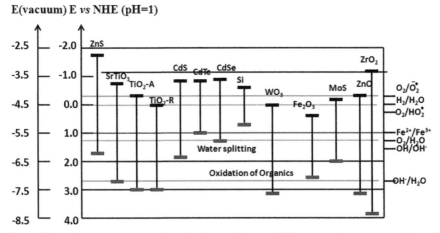

Figure 12.1 Bandgap energy and band edge positions of different semiconductor oxides and chalcogenides, along with selected redox potentials (Mills and Hunte, 1997; Bhatkhande et al., 2002; Wada et al., 2002; Carp et al., 2004; Kabra et al., 2004; Palmisano et al., 2007; Kudo and Miseki, 2009; Vinu and Madras, 2010).

and ZrO_2, ZnO and $SrTiO_3$ exhibit favorable bandgap positions compared to the other materials. The material selection is also based on the stability of the material toward photocatalytic degradation conditions (Vinu and Madras, 2010).

In a typical metal oxide, the VB and conduction band have O $2p$ and metal s character and therefore relatively ionic bonded materials have a large bandgap. ZnO is an excellent semiconductor having a broad direct bandgap width (3.37 eV), large excitation binding energy (60 meV), and deep violet/borderline UV absorption at room temperature (Vinu and Madras, 2010). A few researchers have reported better photocatalytic activity of ZnO compared to TiO_2 (Gouvea et al., 2000; Lizama et al., 2002; Sakthivel et al., 2003).

12.3 Solid-Phase Photocatalytic Degradation of Plastic Polymers–Photocatalyst Composites

12.3.1 Degradation of PE–Photocatalyst Composites

Zhao et al. (2007) and Zhao et al. (2008) have used TiO_2 and CuPc sensitized TiO_2 photocatalysts (TiO_2/CuPc) for degradation of PE in the ambient air under solar and UV light irradiation. In both cases, CO_2 and volatile organic compounds are formed as main products during the photocatalytic degradation through chain scission results in weight loss and average molecular weight decrease. The study gives a potential and promising way to solve "white pollution" for the plastic industry and environmental purification without producing any hazardous waste.

Fa et al. (2010) have prepared PE-OPW-TiO_2 nanocomposite film by embedding the organically modified TiO_2 nanoparticles into commercial PE in the presence of OPW. The photocatalytic degradation behavior was examined under UV light through the weight loss of the composite films. The PE-OPW-TiO_2 composite film demonstrates much higher photodegradation efficiency and much better mechanical property than either the PE-TiO_2 composite film or the pure PE film.

Thomas and Sandhyarani (2013) have prepared LDPE-TiO_2 nanocomposite film through a simple solution casting technique with 0.1 wt% TiO_2 nanoparticles. The film when exposed to solar radiation, the composite films shows a significant weight loss of 68% within a period of 200 h. The breakage of polymer chains, and hence the formation of small-size molecular segments, makes the nanocomposite favorable for degradation. The study aids in the development of degradable plastics which can limit the extent of plastic pollution.

Liang et al. (2013) have prepared LDPE/PAM-g-TiO$_2$ nanocomposite film using PAM-g-TiO$_2$ embedded into an LDPE plastic and it was treated under UV light irradiation in ambient conditions for 520 h. The properties of composite films were compared with those of the pure LDPE films by measuring the changes in weight loss and morphological analysis. The results show that PAM-g-TiO$_2$ embedded LDPE has enhanced photocatalytic degradation. This is due to the addition of PAM-g-TiO$_2$ brought about the good dispersion of TiO$_2$ in LDPE matrix and improved the hydrophilicity of composite films, which are able to facilitate the degradation of LDPE (Liang et al., 2013). The development of this new kind of composite polymer can lead to an eco-friendly disposal of polymer wastes.

Zapata et al. (2014) have prepared nanocomposites of PE and poly(ethylene-co-1-octadecene) copolymer with TiO$_2$ nanoparticles and irradiated under UV light of 550 W · m^{-2}. The results suggest that the degradation of the polymer matrix begins at the surface of the nanoparticles, implying that reactive oxygen species generated at the TiO$_2$ surface cause random breakage of the polymer chains.

Ali et al. (2016) have used TiO$_2$ nanotubes for the photocatalytic degradation of LDPE films. The degradation of pure and composite PE films was measured in terms of photo-induced weight loss and was confirmed surface roughness and tensile strength testing. The PE films with 10% dye sensitized TiO$_2$ nanotubes shows a degradation of around 50% under visible light over a short period of 45 days.

Kemp and McIntyre (2006) have studied the effects of adding dispersed powders of various forms of TiO$_2$ on the photodegradation of PE. The study describes the results for the photo-oxidative degradation of PE films pigmented with undoped and transition metal (Cr, Mn, V, Mo, and W) doped TiO$_2$ powders. The effect of doping TiO$_2$ with metals revealed that W and Mo-doped TiO$_2$ pigments are relatively aggressive toward the PE film degradation while rutile TiO$_2$ with Mn and V induces a protective effect toward the PE film. Even the most aggressive of the metal-doped pigments are less photoactive than the Degussa P-25 material, containing both rutile and anatase.

Shengying et al. (2010) have reported the photocatalytic degradation of PE plastic under the sunlight irradiation with polypyrrole/TiO$_2$ (PPy/TiO$_2$) nanocomposite, which was prepared by sol–gel and emulsion polymerization methods. The escape of volatile organics came from degradation of PE resulted in the formation of the cavities on PE plastic surface. The photocatalytic degradation was compared with pure TiO_2 and the results show that PPy/TiO_2 photocatalyst is much faster than that using PPy or TiO_2 which is due to the fact that nanocomposite has an enhanced visible light capturing ability.

Asghar et al. (2011) have reported comparative solid-phase photocatalytic degradation of PE films with metal (Fe, Ag, and Fe/Ag mix) doped and undoped TiO_2 nanoparticles under three different conditions such as UV radiation, artificial light, and darkness (Sathishkumar et al., 2013). Weight of PE films steadily decreased and led to a maximum of 14.34% reduction under UV irradiation with Fe/Ag-doped TiO_2 nanoparticles and a maximum of 14.28% reduction under artificial light with Ag-doped TiO_2 nanoparticles in 300 h. No weight reduction was observed under darkness. The results reveal that PE-TiO_2 compositing with metal doping has the potential to degrade the PE waste under irradiation without any pollution.

The photocatalytic degradation of the LDPE-TiO_2 was also studied where the rate of weight loss of the composite film was comparatively higher than the pure LDPE film after UV light irradiation for 400 h (Zan et al., 2006).

12.3.2 Photocatalytic Degradation of PP–photocatalyst Composites

Kamrannejad et al. (2014) studied the photocatalytic degradation of PP by UV irradiation of the composite film of PP and carbon-coated TiO_2, and PP and commercially available TiO_2 nanoparticles (Degussa P-25) for 500 h. The mechanical and morphological properties of PP composite films were investigated before and after the exposure of UV radiation and compared with the pure PP sample. The FTIR spectra showed that the carbonyl peak intensity for pure PP film is very low, while in nano-composite films, the intensity was greater with a decrease in the carbon amount of TiO_2 powders. The degradation of PP nano-composite films is observed higher than pure PP films, which is reduced while increasing the carbon content of the TiO_2.

12.3.3 Photocatalytic Degradation of PS–photocatalyst Composites

The solid-phase photocatalytic degradation of PS plastic with pure TiO_2 (Shang et al., 2003b) and CuPc-sensitized TiO_2 photocatalyst ($TiO_2/CuPc$) (Shang et al., 2003a) in the ambient air under fluorescent and UV light irradiation has been reported. CO_2 and volatile organic compounds were produced during the photocatalytic degradation through chain scission resulting into weight loss and decrease in average molecular weight.

Zan et al. (2004, 2006) have prepared modified TiO_2, PS-grafted TiO_2, PS-G-TiO_2 and PS-g-TiO_2. The titania nanoparticles were modified by grafting polymer on its surface (G-TiO_2) and dispersed into the styrene to prepared PS-G-TiO_2. In PS-g-TiO_2 formation, grafted titania (G-TiO_2) was suspended in the solvent in which PS particles were dissolved. For the both cases, the degradation process carried out in ambient air at room temperature under UV lamp. The weight loss and the cleavage of the benzene rings were observed in the both PS-matrix of composite films PS-G-TiO_2 (Zan et al., 2004) and PS-g-TiO_2 (Zan et al., 2006) under UV-light irradiation for 396 and 300 h, respectively. The modification by grafting polymer with the TiO_2 nanoparticles leads to obtain effective dispersion process for composite film preparation which would be hopeful new eco-friendly polymeric material. Bandyopadhyay and Basak (2007) have studied the photocatalytic degradation of PS by using ZnO under UV light irradiation. There was a net weight loss of 16% in only 2 h, recorded for PS on modification with ZnO, which slightly increased up to 18% with the help of a dye. The main chain cleavage of PS was observed from the viscosity average molecular weight data, which actually decreases with increasing extent of photo degradation with different ZnO concentrations.

Chakrabarti et al. (2011) have prepared composite films by dispersing ZnO semiconductor particles in the matrix of PVC and PS. Photocatalytic degradation of the composite films (PVC-ZnO and PS-ZnO) has been studied under tropical sunlight in the presence of air and water. As results, increase in ZnO loading increased the rate and extent of degradation in both cases.

Kemp and McIntyre (2006) have also reported the photocatalytic degradation of PS containing undoped and transition metal (Cr, Mn, V, Mo, and W) doped TiO_2 powders where the similar degradation results were observed for PS as observed for PE degradation.

Fa et al. (2013) have prepared a novel photodegradable TiO_2-Fe(St)$_3$-PS nanocomposite by embedding titania Fe(St)$_3$ (TiO_2-Fe(St)$_3$) into the commercial PS. The nanocomposite was irradiated in an ambient air at room temperature under UV light irradiation for 480 h. The results show that the photodegradation efficiency of TiO_2-Fe(St)$_3$-PS nanocomposite film was higher than that of the pure PS film and TiO_2-PS composite film. The TiO_2-Fe(St)$_3$-PS composite was found to be a promising new eco-friendly polymer material.

12.3.4 Photocatalytic Degradation of PVC-photocatalyst Composites

Hoffmann et al. (1995) reported the preparation of PVC/TiO$_2$ composite film and studied its photocatalytic degradation under UV light irradiation in the ambient air in order to assess the feasibility of developing photodegradable polymers. As a result, the PVC-TiO$_2$ films quickly whitened with visible light scattering from growing cavities. According to the spectral analysis, the surface chlorine content gradually decreased with irradiation and some chlorine-containing volatile products were formed. The process proves that the PVC-TiO$_2$ composite has a potential viability to be used as a photodegradable product.

The enhanced photocatalytic degradation of PVC (Fa et al., 2011) was observed for PVC composite film containing TiO$_2$ nanoparticles modified with FePcCl$_{16}$. The PVC-(FePcCl$_{16}$)-TiO$_2$ thin film irradiated under UV light indicated that the degradation efficiency of PVC-FePcCl$_{16}$-TiO$_2$ composite film was significantly higher than that of pure PVC film and PVC-TiO$_2$ composite film. The studies demonstrate that the modification of TiO$_2$ nanoparticles greatly promotes the photocatalytic degradation of the composite.

Kim et al. (2006) have investigated photodegradable PVC/TiO$_2$ nanohybrid as an eco-friendly alternative strategy to the current waste landfill and toxic byproduct-emitting incineration of PVC wastes. The HPCL with numerous –COOH groups and good miscibility with PVC as a binder for TiO$_2$ (PVC/HPCL-TiO$_2$) nanoparticles was prepared. The photocatalytic degradation of the samples was examined and verified from the change of surface morphology, chemical structure, molecular weight, and molecular-level structure after UV irradiation. The study presented an innovative approach to prepare TiO$_2$ well-dispersed PVC/TiO$_2$ nano-hybrid for the removal of PVC wastes from the environment.

Kemp and McIntyre (2006) have also studied photocatalytic degradation of PVC films pigmented with undoped and transition metal (Cr, Mn, V, Mo, and W)-doped TiO$_2$ powders. The observation correlates reasonably well with that reported results for the degradation of PS and PE (Kemp and McIntyre, 2006). Another interesting observation was that the levels of photodegradation recorded by the carbonyl development (by FTIR studies) closely matched that recorded by chain scission (molecular weight measurements) for the group-VI metal-doped TiO$_2$ suggesting a link between the rates of chain scission and of carbonyl formation (Kemp and McIntyre, 2006).

Gupta et al. (2017) have reported solid-phase photocatalytic degradation of PE and PVC with various photocatalysts such as CeO_2, ZnO, CuS, and TiO_2 under different light sources. In the FTIR spectra, the peak intensities corresponding to adsorbed CO_2 were significantly increased indicating the higher degradation rate of the PE in composites with increase in duration of exposure to light source. The presence of carbonyl bands proved the degradation of PVC-photocatalyst composite. The CeO_2 showed degradation efficiencies higher than 70% for PVC and 75% for PE under the fluorescent and solar radiation. The UV–visible spectroscopic studies show that the CeO_2, ZnO, and CuS photocatalysts are active in the visible spectrum resulting in enhanced degradation efficiency in fluorescent and solar radiation. TiO_2 showed lower degradation efficiency for both the polymers under all radiation types.

Yang et al. (2011) have prepared PVC-POM/TiO_2 nanocomposite film by embedding nano-TiO_2 photocatalyst modified by POM into the commercial PVC plastic. The comparative degradation study of PVC-POM/TiO_2 with the PVC-TiO_2 film and the pure PVC film shows that PVC-POM/TiO_2 film has a high photocatalytic degradation rate which is due to the presence of PMO that promotes the separation of photogenerated electron–hole pair and mediate photogenerated electron transferring from TiO_2 conduction band to dioxygen. The authors have prompted a novel process to produce the environment-friendly photocatalytic polymer degradation.

12.3.5 Photocatalytic Degradation of PAM-photocatalysts Composites

Vijayalakshmi and Madras (2006) have reported the photocatalytic degradation of PAM copolymers and PEO in the presence of CS-TiO_2 using two different UV lamps of 125 and 80 W. Degradation of PEO was observed in both the cases, whereas PAM degraded only when exposed to lamp of higher power, even in the presence of catalyst. The results show that, CS-TiO_2 exhibits a higher degradation rate coefficient for both the polymers compared to the systems without any catalyst as well as with commercially available TiO_2-Degussa P-25. The observed high photocatalytic activity of CS-TiO_2 was attributed to the nano-size, high specific surface area, and high surface hydroxyl content.

Wang et al. (2016) have studied the photocatalytic performance of tungsten (W^{6+}) doped TiO_2 nanoparticles and viscosity reduction properties of the polymer by the degradation of PAM in oilfield sewage. The results show that W^{6+} doped TiO_2 has higher photocatalytic activity than the

TiO_2 nanoparticles and its photocatalytic performance of viscosity reduction is very significant.

12.3.6 Photocatalytic Degradation of other Plastic Polymers

Liau et al. (2008) have studied the kinetics of degradation of PVB/ nano-TiO_2 films under UV irradiation and different operating conditions at room temperature. Samples of the films were illuminated for different times in a UV chamber. The results showed that PVB was photodegraded according to the analytical data of the residual mass, chemical structures, and surface images of the films. The composition of TiO_2 in the composite samples is the most influential factor to affect the photodegradation rate.

Eren and Okte (2007) have reported the solid-state photocatalytic degradation of the methacryl and alkoxysilane functionalized stearate using powdered TiO_2 (Degussa P-25) under medium pressure mercury lamp (40% UV lines are below 400 nm) illumination in air. The weight of the polymer was reduced by 25% of its initial value after irradiation for 40 h. The surface morphology of the irradiated polymer films was analyzed by SEM to monitor the polymer degradation. The SEM analysis of the film after irradiation showed the formation of cavities around the TiO_2 particles.

Miyauchi et al. (2008) have reported enhanced photocatalytic degradation of PBS by dispersing nanocomposites of TiO_2 particles. PBS has interesting properties such as biodegradability, melt processability, and thermal and chemical resistance. The UV irradiation onto the TiO_2/PBS exhibited the highest decomposition and biodegradability. The strategy in the present study is applicable to various eco-friendly film products with efficient degradability.

Yuan et al. (2013) have prepared Fe-TiO_2/PEG composite using modified anatase titania (Fe-TiO_2 prepared by using ferric acetylacetonate complex) for the photodegradation of PEG. The modification endowed nano-TiO_2 not only higher catalytic activity (the ferric complex acts as a photosensitizer), but also surface oleophylicity (the ferric complex also acts as a surfactant). Fe-TiO_2/PEG composite was irradiated under UV light (40 W) and degradation was evaluated by the weight loss, and the change of molecular weight and distribution. From the morphological analysis, the cavities were formed on the surface when the composites were irradiated for a certain time. With prolonging the UV irradiation, the surface of composite became coarse more and more, and the structure of composite tended to be destroyed.

12.3.7 Photocatalytic Degradation of Plastic Polymers Using Different Photocatalyst Suspension in Water

There are some reports on photocatalytic degradation of polymers in hetero-geneous catalysts suspension in water. Phonsy et al. (2015) have studied the photocatalytic degradation of PE using aqueous suspension of TiO_2 and ZnO under UV irradiation in a multi lamp photoreactor for 300 h. The degradation process was found to be pH dependent and favorable in acidic condition. The oxidizing agents such as H_2O_2 and PDS accelerated the degradation process. The results indicated that the higher efficiency of TiO_2 (20% more) than ZnO in degradation of PE in water. The study showed good potential for the safe removal of PE plastic by semiconductor photocatalysis. Recently, de Bandara et al. (2017) investigated photocatalytic behaviorof ZrO_2 nanoparticles (syn-thesized by a sono-chemical method) and TiO_2 (synthesized by a sol–gel method) for the degradation of PE and PP under the sun simulator as well under the real sun light. The results suggested that ZrO_2 nanoparticle sus-pension exhibited higher degradation of PE and PP the TiO_2 nanoparticles suspension.

Penot et al. (1983) have investigated the photocatalytic behavior of ZnO for degradation of isotactic PP in the presence of typical oxidants. As a result, UV excitation of the ZnO dispersed in isotactic PP induced a primary hydroperoxidation and the decomposition of the formed hydroperoxides into alcohols and ketones.

Marimuthu and Madras (2008) have also studied the degradation of PAA using $CS-TiO_2$ in the presence of three different oxidizers such as benzoyl peroxide, dicumyl peroxide, and azobisisobutyronitrile. The pho-tocatalyst in the UV-assisted oxidative degradation of PAA enhances the random chain scission of the polymer radical, and the rate of polymer degradation in the presence of $CS-TiO_2$ is significantly higher than that obtained with Degussa P-25. The effect of oxidizer concentration on the degradation rate was investigated, and found that the degradation rate increased significantly with oxidizer concentration. The present developed model can be used to predict the degradation rate of the polymer in the presence of any initiators by only knowing its dissociation rate constant values.

12.4 Mechanism of Photocatalytic Degradation of Polymers

The degradation of polymer macromolecules denotes all processes which lead to a decline of polymer properties like tensile strength, color, shape, etc. (Yousif and Haddad, 2013). Chemical processes related to degradation may lead to a reduction of average molar mass due to macromolecular chain bond scission or to an increase of molar mass due to cross linking rendering the polymer insoluble. Once the polymer chains are broken down into smaller fragments, photocatalytic degradation becomes favorable (Shang et al., 2003a; Yousif and Haddad, 2013).

The literature survey shows that the photocatalytic degradation mechanism has been extensively discussed (Crawford and Hughes, 1997; Kuzina and Mikhailov, 1998; Torikai et al., 1998; Bauer et al., 2001; Shang et al., 2003b; Imanishi et al., 2007; Tan et al., 2011; Barakat et al., 2013; Imanishi et al., 2014; Gnanaprakasam et al., 2015). The photocatalytic degradation process of plastic (PE, PP, PS, and PVC) under UV irradiation occurred via direct absorption of photons by plastic macromolecule to create excited states, and then undergo chain scission, branching cross-linking and oxidation reactions. When TiO_2 is irradiated with UV light of energy (3.2 eV) greater than or equal to its bandgap ($\lambda < 387$ nm), VB holes (h_{vb}^+) and conduction band electrons are produced (Figure 12.2). Subsequently, the electrons (e_{cb}^-)

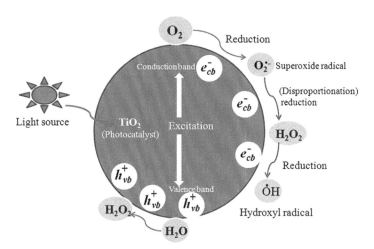

Figure 12.2 Mechanism of TiO_2 photocatalysis showing formation of •OH and H_2O_2 (Phonsy et al., 2015).

(conduction band) can be captured by O_2 that adsorbs on the surface of TiO_2 which lead to the formation of active oxygen species such as $O_2^{\bullet-}$ (Gnanaprakasam et al., 2015).

Consequently, $O_2^{\bullet-}$ could react with H_2O to produce surface hydroperoxy radicals ($HO_2^{\bullet}$) and hydroxyl ion (OH^-). At the same time, photogenerated holes are trapped by surface hydroxyl groups or H_2O adsorbing on the surface of TiO_2 to produce $HO^{\bullet}$ radical. The $HO^{\bullet}$ radicals are the most important oxidant in photocatalytic oxidation to attack the polymer chains, leading to the irregular bond scission and forming carbon-centered radicals like -($^{\bullet}CHCH_2$)- and -($^{\bullet}CHCHPh$)- (Gnanaprakasam et al., 2015). The following are the steps involved in formation of HO radical and polymer chain scission (see Scheme 12.1. where $h\nu$ is the energy essential to transfer the electron from VB to conduction band).

Once the carbon-centered radicals are introduced in the polymer chain, their successive reactions lead to the chain cleavage with the oxygen incorporation and species containing carbonyl intermediates (ethene, formaldehyde/acetaldehyde and ethanol) and carboxyl groups are produced. These intermediates can be further oxidized, finally forming CO_2 and H_2O by the aid of reactive oxygen species (Gnanaprakasam et al., 2015).

$$TiO_2 + h\nu \longrightarrow \bar{e}_{cb} + h_{vb}^+$$

$$\bar{e}_{cb} + O_2 \longrightarrow O_2^-$$

$$O_2^- + H_2O \longrightarrow HO_2^{\cdot} + OH^-$$

$$h_{vb}^+ + OH^- \longrightarrow \dot{O}H$$

$$h_{vb}^+ + H_2O_{ads} \longrightarrow \dot{O}H + H^+$$

$$2HO_2^{\cdot} \longrightarrow H_2O_2 + O_2$$

$$h_{vb}^+ + H_2O_2 \longrightarrow 2\dot{O}H$$

$$-(CH_2\text{-}CH_2)\text{-} + \dot{O}H \longrightarrow -(\dot{C}H\text{-}CH_2)\text{-} + H_2O$$

$$-(\dot{C}H\text{-}CH_2)\text{-} + O_2 \longrightarrow -(CH(\dot{O}O)CH_2)\text{-}$$

$$-(CH(\dot{O}O)CH_2)\text{-} + -(CH_2\text{-}CH_2)\text{-} \longrightarrow -(CH(OOH)CH_2)\text{-} + -(CH_2\dot{C}H)\text{-}$$

$$-(CH(OOH)CH_2)\text{-} + h\nu \longrightarrow -(CH\dot{O}\text{-}CH_2)\text{-} + \dot{O}H$$

$$-(CH\dot{O}\text{-}CH_2)\text{-} \longrightarrow -CHO + \dot{C}H_2\text{-}CH_2\text{-}$$

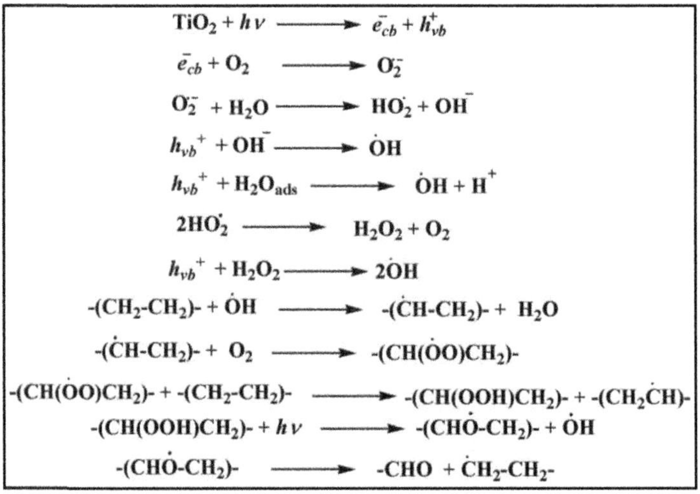

Scheme 12.1 Proposed mechanism for photocatalytic degradation through polymer chain scission.

12.5 Conclusions

In recent years, the uses of plastic materials have rapidly increased but it is well established that rapid photodegradation of these materials is possible when they are exposed to sunlight/UV light in the presence of photocatalysts. The present review was humble attempt to state out the photocatalytic degradation process for the plastic polymers. Over the past two decades, nano-TiO_2 has set the standard as a benchmark photocatalyst for the degradation. Various modified forms of TiO_2 like metal doped, polymer/TiO_2 composites, and core–shell TiO_2 have been developed to enhance the photoactivity of first generation of TiO_2 and compared to the commercial Degussa P-25 TiO_2, which is used as a yardstick for photoactivity assessment. This type of materials holds a huge potential of improvement, although in order to achieve this, further research efforts will be required. Therefore, the discussion in the review suggests that a cleaner and greener environment, and renewable and sustainable energy economy can be achieved through photocatalysis. The authors do believe that more effective results in the research area will be forthcoming in the next decades.

In summary, it can be concluded that although the particular directions that photocatalysis will take in the next years are still uncertain, they will surely lead to bright technological developments.

Acknowledgments

The authors are thankful to UGC-New Delhi, India, for financial assistant under Dr. D.S. Kothari Post-Doctoral Fellowship to carry out the research work.

References

Ali SS, Qazi IA, Arshad M, Khan Z, Voice, TC, Ch. Mehmood T. (2016). Photocatalytic degradation of low density polyethylene (LDPE) films using titania nanotubes. Environmental Nanotechnology Monitoring and Management. 5:44–53. DOI:org/10.1016/j.enmm.2016.01.001

Anpo M, Kamat PV. (2010). Environmentally benign photocatalysts: applications of titanium oxide-based materials. Springer Science & Business Media. London.

Asghar W, Qazi IA, Ilyas H, Khan AA, Awan MA, Aslam, MR. (2011). Comparative solid phase photocatalytic degradation of polythene films with doped and undoped TiO_2 nanoparticles. Journal of Nanomaterials. 1–8. DOI:10.1155/2011/461930

Bandara N, Dahanayake D, de Silva RM, Gunasekara S, de Silva KMN, Thanabalasingam K. (2017). Is nano ZrO_2 better photocatalyst than nano TiO_2 for degradation of plastics? RSC Advances. 7:46155–63. DOI: 10.1039/c7ra08324f

Bandyopadhyay A, Basak GC. (2007). Studies on photocatalytic degradation of polystyrene. Materials Science and Technology. 23(3):307–14. DOI: 10.1179/174328407X158640

Barakat NAM, Kanjwal MA, Chronakis IS, Kim HY. (2013). Influence of temperature on the photodegradation process using Ag-doped TiO_2 nanostructures: negative impact with the nanofibers. Journal of Molecular Catalysis A: Chemistry. 366:333–40. DOI: org/10.1016/j.molcata.2012.10.012

Barnes DKA, Galgani F, Thompson RC, Barlaz M. (2009). Accumulation and fragmentation of plastic debris in global environments. Philosophical Transactions of the Royal Society B. 364(1526):1985–98. DOI: 10.1098/rstb.2008.0205

Bauer C, Jacques P, Kalt A. (2001). Photooxidation of an azo dye induced by visible light incident on the surface of TiO_2. Journal of Photochemistry and Photobiology: Chemistry. 140:87–92. DOI: org/10.1016/S1010-6030(01)00391-4

Bhatkhande DS, Pangarkar VG, Beenackers AACM. (2002). Photocatalytic degradation for environmental applications – a review. Journal of Chemical Technology and Biotechnology. 77(1):102–16. doi.org/10.1002/jctb.532

Byrappa K, Subramani AK, Ananda S, Rai KML, Dinesh R, Yoshimura M. (2006). Photocatalytic degradation of rhodamine B dye using hydrothermally synthesized ZnO. Bulletins in Material Science. 29(5):433–8. DOI: org/10.1007/BF02914073

Carp O, Huisman CL, Reller A.(2004). Photoinduced reactivity of titanium dioxide. Progress in Solid State Chemistry. 32(1–2):33–177. DOI:org/10.1016/j.progsolidstchem.2004.08.001

Central Pollution Control Board. Plastic waste management. Available at: www.cpcb.nic.in/139-144.pdf [Accessed 20 Feb 2010].

Chakrabarti S, Bhattacharjee S, Sil D, Chaudhuri B. (2011). Comparison of the photocatalytic degradation of polyvinyl chloride and polystyrene with zinc oxide semiconductor catalyst under tropical sunlight. Chemical Technology: An Indian Journal. 6(1):58–64.

Chakrabarti S, Chaudhuri B, Bhattacharjee S, Das P, Dutta BK. (2008). Degradation mechanism and kinetic model for photocatalytic oxidation of PVC-ZnO composite film in presence of a sensitizing dye and UV radiation. Journal of Hazardous Materials. 154:230–6. DOI: org/10.1016/j.jhazmat.2007.10.015

Chen X, Mao SS. (2007). Titanium dioxide nanomaterials: synthesis, properties, modifications, and applications. Chemical Reviews 07: 2891–959. DOI: 10.1021/cr0500535

Cho S, Choi W. (2001). Solid-phase photocatalytic degradation of PVC-TiO_2 polymer composites. Journal of Photochemistry and Photobiology A: Chemistry. 143:221–8. DOI: org/10.1016/S1010-6030(01)00499-3

Crawford KD, Hughes KD. (1997). Rapid formation and spectroscopic observation of polystyrene conjugation in individual micron-diameter particles with visible radiation. Journal of Physical Chemistry B. 101(6):864–70. DOI: 10.1021/jp962040o

Daghrir R, Drogui P, Robert D. (2013). Modified TiO_2 for environmental photocatalytic applications: a review. Industrial and Engineering Chemistry Research. 52: 3581–99. DOI: 10.1021/ie303468t

Dearfield K, Douglas L, George R, Ehling UH, Moore MM, Sega GA. (1995). Acrylamide: A review of its genotoxicity and an assessment of heritable genetic risk. Mutation Research. 330:71–99.

Eren T, Okte AN. (2007). Polymerization of methacryl and triethoxysilane functionalized stearate ester: Titanium dioxide composite films and their photocatalytic degradations. Journal of Applied Polymer Science. 105:1426–36. DOI: org/10.1002/app.26307

Fa W, Gong C, Tian L, Peng T, Zan L. (2011). Enhancement of photocatalytic degradation of poly(vinyl chloride) with perchlorinated iron (II) phthalocyanine modified nano–TiO_2. Journal of Applied Polymer Science. 122:1823–8. DOI: org/10.1002/app.34230

Fa W, Guo L, Wang J, Guo R, Zheng Z, Yang F. (2013). Solid-phase photocatalytic degradation of polystyrene with TiO_2/Fe(St)$_3$ as catalyst. Journal of Applied Polymer Science. 128(5):2618–22. DOI: org/10.1002/app.37751

Fa W, Yang C, Gong C, Zan L, Peng T. (2010). Enhanced photodegradation efficiency of polyethylene-TiO_2 nanocomposite film with oxidized polyethylene wax. Journal of Applied Polymer Science. 118:378–84. DOI: org/10.1002/app.32413

Fujishima A, Rao TN, Tryk DA. (2000). Titanium dioxide photocatalysis. Journal of Photochemistry and Photobiology C: Photochemical Reviews. 1(1):1–21. DOI. org/10.1016/S1389- 5567(00)00002-2

Fujishima, K. Hashimoto, T. Watanabe (eds.), (1999). TiO_2 Photocatalysis fundamentals and applications, BKC, Inc. Tokyo.

Gewert B, Plassmann M, Sandblom O, Macleod M. (2018). Identification of chain scission products released to water by plastic exposed to ultraviolet light. Environmental Science and Technology Letters. 5(5):272–6. DOI: 10.1021/acs.estlett.8b00119

Gilani N. (2017). Uses of PVC plastic. Sciencing, Santa Monica, USA.

Gnanaprakasam A, Sivakumar VM, Thirumarimurugan M. (2015). Influencing parameters in the photocatalytic degradation of organic effluent via nanometal oxide catalyst: a review. Industrial Journal of Material Science. 1–16. DOI: org/10.1155/2015/601827

Gouvea CAK, Wypych F, Moraes SG, Duran N, Nagata N, Zamora P. (2000). Semiconductor-assisted photocatalytic degradation of reactive dyes in aqueous solution. Chemosphere. 40:433–40. DOI: org/10.1016/S0045-6535(99)00313-6

Grover A, Gupta A, Chandra S, Kumari A, Khurana SMP. (2015). Polythene and environment. International Journal of Environmental Science. 5(6):1091–105. DOI: 10.6088/ijes.2014050100103

Gupta A, Lakshmi YN, Manivannan R, Victoria SN. (2017). Visible range photocatalysts for solid phase photocatalytic degradation of Polyethylene and polyvinyl chloride Jourla of the Chilean Chemical Society. 62(1):3393–8.

Hagfeldt A, Graetzel M. (1995). Light-induced redox reactions in nanocrystalline systems. Chemical Reviews. 95:49–68. DOI: 10.1021/cr00033a003

Herrmann JM. (1999) Heterogeneous photocatalysis: fundamentals and applications to the removal of various types of aqueous pollutants. Catalysis Today. 53:115–29. DOI: org/10.1016/S0920-5861(99)00107-8

Hoffmann, MR, Martin ST, Choi W, Bahnemannt DW. (1995). Environmental applications of semiconductor photocatalysis. Chemical Reviews. 95:69–96. DOI: 10.1021/cr00033a004

Imanishi A, Fukui K. (2014). Atomic-scale surface local structure of TiO_2 and its influence on the water photooxidation process. Journal of Physical Chemistry Letters. 5:2108–17. DOI: 10.1021/jz5004704

Imanishi A, Okamura T, Ohashi N, Nakamura R, Nakato Y. (2007). Mechanism of water photooxidation reaction at atomically flat TiO_2

(Rutile) (110) and (100) Surfaces: Dependenceon solution pH. Journal of the American Chemical Society. 129 (37): 11569–78. DOI: 10.1021/ja073206+

IUPAC Recommendations. (1996). Glossary of basic terms in polymer science. Available at: http://goldbook.Iupac.org/Mo3667.html.

Kabra K, Chaudhary R, Sawhney RL. (2004). Treatment of hazardous organic and inorganic compounds through aqueous-phase photocatalysis: a review. Industrial and Engineering Chemistry Research. 43(24): 7683–96. DOI: 10.1021/ie0498551

Kamrannejad MM, Hasanzadeh A. (2014). Photocatalytic degradation of polypropylene/TiO_2 nano-composites. Materials Research. 17(4):1039–46.

Kemp TJ, McIntyre RA. (2006a). Influence of transition metal-doped titanium(IV) dioxide on the photodegradation of polystyrene. Polymer Degradation and Stability. 91:3020–5. DOI:10.1016/j.polymdegradstab. 2006.08.005

Kemp TJ, McIntyre RA. (2006). Transition metal-doped titanium(IV) dioxide: characterisation and influence on photodegradation of poly(vinyl chloride). Polymer Degradation and Stability. 91:165–94. DOI: 10.1016/j.polymdegradstab.2005.04.033

Kim SH, Kwak SY, Suzuki T. (2006). Photocatalytic degradation of flexible PVC/TiO_2 nanohybrid as an eco-friendly alternative to the current waste landfill and dioxin- emitting incineration of post-use PVC. Polymer. 47(9):3005–16. DOI:10.1016/j.polymer.2006.03.015

Kudo A, Miseki Y. (2009). Heterogeneous photocatalyst materials for water splitting. Chemical Society Reviews. 38:253–78. DOI: 10.1039/ B800489G

Kulkarni A, Dasari H. (2018). Current status of methods used in degradation of polymers: A review. In Proceedings of international conference on research in mechanical engineering sciences. MATEC Web of Conferences. 144: 02023. DOI: org/10.1051/matecconf/201814402023

Kuzina SI, Mikhailov AI. (1998). Photo-oxidation of polymers-2. Photochain reaction of peroxide radicals in polystyrene. European Polymer Journal. 34:291–9. DOI: org/10.1016/S0014-3057(97)00169-9

Leong S, Razmjou A, Wang K, Hapgood K, Zhang X, Wang H. (2014). TiO_2 based photocatalytic membranes: a review. Journal of Membrance Science. 472:167–84. DOI: 10.1016/j.memsci.2014.08.016

Liang W, Luo Y, Song S, Dong X, Yu X. (2013). High photocatalytic degradation activity of polyethylene containing polyacrylamide grafted TiO_2. Polymer Degradation and Stability. 98:1754–61. DOI: org/10.1016/j.polymdegradstab.2013.05.027

Liau LCK, Chou WW, Wu RK. (2008). Photocatalytic lithography processing via poly(vinyl butyral)/TiO_2 photoresists by ultraviolet (UV) Exposure. Industrial Engineering and Chemical Research. 47:2273–8. DOI: 10.1021/ie071331o

Linsebigler AL, Lu G, Yates JT. (1995). Photocatalysis on TiO_2 surfaces: principles, mechanisms, and selected results. Chemical Reviews 95:735–58. DOI: 10.1021/cr00035a013

Lizama C, Freer J, Baeza J, Mansilla HD. (2002). Optimized photodegradation of Reactive Blue 19 on TiO_2 and ZnO suspensions. Catalysis Today. 76:235. DOI:org/10.1016/S0920-5861(02)00222-5

Marimuthu A, Madras G. (2008). Photocatalytic oxidative degradation of poly(alkyl acrylates) with nano TiO_2. Industrial Engineering and Chemical Research. 47:2182–90. DOI: 10.1021/ie0712939 CCC: $40.75

McCrum NG, Buckley CV, Bucknall CB. (1997). Principles of Polymer Engineering. Oxford University Press. New York.

Mehmood CT, Qazi IA, Baig MA, Arshad M, Qudos A. (2015). Application of photodegraded polythene films for the treatment of drimarene brilliant red (DBR) dye. International Biodeterioration and Biodegradation. 1–9. DOI: org/10.1016/j.ibiod. 2014.12.014

Mills A, Hunte SL. (1997). An overview of semiconductor photocatalysis. Journal of Photochemistry and Photobiology A: Chemistry. 108:1–35. DOI:org/10.1016/S1010-6030(97)00118-4

Miyauchi M, Li Y, Shimizu H. (2008). Enhanced degradation in nanocomposites of TiO_2 and biodegradable polymer. Environmental Science and Technology. 42:4551–4. DOI: 10.1021/es800097n

Ning XH, Meng QL, Han YL, Zhou DY, Li L, Cao L, Weng JK, Ding R, Wang ZB. (2017). Flower-shaped TiO_2 clusters for highly efficient photocatalysis. RSC Advances. 7:34907–11. DOI: 10.1039/C7RA05949C

Owen ED. ed. (1984). Degradation and stabilization of PVC. Elsevier: London.

Painter PC, Coleman MM. (1997). Fundamentals of polymer science: An introductory text. Technomic Pub. Co. Lancaster.

Palmisano G, Augugliaro V, Pagliaro M, Palmisano L. (2007). Photocatalysis: A promising route for 21st century organic chemistry. Chemical Communications. 3425–37. DOI:10.1039/B700395C

Phonsy PD, Yesodharan S. Yesodharan EP. (2015). Enhancement of semiconductor mediated photocatalytic removal of polyethylene plastic wastes from the environment by oxidizers. Research Journal of Recent Sciences. 4(10):105–12.

Owen ED. ed. (1984). Degradation and stabilization of PVC. Elsevier: London.

Penot G, Arnaud R, Lemaire J. (1983). ZnO-photocatalyzed oxidation of isotactic polypropylene. Die Angewandte Makromolekulare Chemie. 117:71–84.

Piringer OG, Baner AL. (2008). Plastic packing: Interactions with food and pharmaceuticals (2nd ed.) Wiley-VCH.

Rajabi HR, Khani O, Shamsipur M, Vatanpour V. (2013). High performance pure and Fe^{3+}-ion doped ZnS quantum dots as green nanophotocatalysts for the removal of malachite green under UV-light irradiation. Journal of Hazardous Materials. 250–1:370–8.

Sakthivel S, Neppolian B, Shankar MV, Arabi-ndoo B, Palanichamy M, Murugesan V. (2003). Solar photocatalytic degradation of azo dye: comparison of photocatalytic efficiency of ZnO and TiO_2. Solar Energy Materials and Solar Cells. 77:65. DOI: org/10.1016/S0927-0248(02)00255-6

Sathishkumar P, Pugazhenthiran N, Mangalaraja RV, Asiri AM, Anandan S. (2013). ZnO supported $CoFe_2O_4$ nanophotcatalysts for the mineralization of Direct Blue 71 in aqueous environments. Journal of Hazardous Materials. 252–3:171–9. DOI: org/10.1016/j.jhazmat.2013.02.030

SCCNFP. (2000). Opinion of the scientific committee on cosmetic products and non-food products intended for consumer concerning titanium dioxide, Colipa No. S75, adapted by the SCCNFP during the 14th plenary meeting of 24 October 2000.

Schneider J, Anpo M, Matsuoka M, Takeuchi M, Bahnemann DW, Zhang J, Horiuchi Y. (2014). Understanding TiO_2 photocatalysis: Mechanisms and materials. Chemical Reviews. 114(19):9919–86. DOI: org/10.1021/cr5001892

Seymour RB. (1989). Polymer Science before and after 1899: notable developments during the lifetime of Maurits Dekker. Journal of Macromolecular Science A. 26(8):1023–32. DOI: org/10.1080/00222338908052032

Shah AA, Hasan F, Hameed A, Ahmed S. (2008). Biological degradation of plastics: a comprehensive review. Biotechnology Advances. 26(3): 246–65. DOI: 10.1016/j.biotechadv.2007.12.005

Shang J, Chai M, Zhu YJ. (2003a). Solid-phase photocatalytic degradation of polystyrene plastic with TiO_2 as photocatalyst. Solid State Chemistry. 174:104–10. DOI: org/10.1016/S0022-4596(03)00183-X

Shang J, Chai M, Zhu Y. (2003b). Photocatalytic degradation of polystyrene plastic under fluorescent light. Environmental Science and Technology. 37: 4494–9. DOI: 10.1021/es0209464

Shengying L, Shihong X, Lijun H, Fei X, Yonghong W, Li Z. (2010). Photocatalytic degradation of polyethylene plastic with polypyrrole/TiO_2 nanocomposite as photocatalyst. Polymer-Plastics Technology and Engineering. 49:400–6. DOI: org/10.1080/03602550903532166

Soltaninezhad M. Aminifar A. (2011). Study nanostructures of semiconductor zinc oxide (ZnO) as a photocatalyst for the degradation of organic pollutants. International Journal of Nano Dimension. 2(2): 137–45. DOI: 10.7508/IJND.2011.02.007

Sun RD, Irie H, Nishikawa T, Nakajima A, Watanabe T, Hashimoto K. (2003). Suppressing effect of $CaCO_3$ on the dioxins emission from poly(vinyl chloride) (PVC) incineration. Polymer Degradation and Stability. 79:253–6. DOI: org/10.1016/S0141-3910(02)00288-4

Tan YN, Wong CL, Mohamed AR. (2011). An overview on the photocatalytic activity of nano-doped-TiO_2 in the degradation of organic pollutants. ISRN Material Science. 1–18. DOI: org/10.5402/2011/261219

Thomas RT, Sandhyarani N. (2013). Enhancement in the photocatalytic degradation of low density polyethylene-TiO_2 nanocomposite films under solar irradiation. RSC Advances. 3:14080–7.

Torikai A, Kobatake T, Okisaki F. (1998). Photodegradation of polystyrene containing flame retardants: effect of chemical structure of the additives on the efficiency of degradation. Journal of Applied Polymer Science. 67:1293–300. DOI: org/10.1002/(SICI)1097-4628(19980214)67:7<1293::AID-APP20>3.0.CO;2-1

Verma R, Vinoda KS, Papireddy M, Gowda ANS. (2016). Toxic pollutants from plastic waste - a review. Procedia Environmental Sciences. 35:701–8. DOI: org/10.1016/j.proenv.2016.07.069

Vijayalakshmi SP, Madras G. (2006). Photocatalytic degradation of poly(ethylene oxide) and polyacrylamide. Journal of Applied Polymer Science. 100;3997–4003. DOI: 10.1002/app.23190

Vinu R, Madras G. (2010). Environmental remediation by photocatalysis. Journal of the Indian Institute of Science. 90:2:189–230.

Wada Y, Yin H, Yanagida S. (2002). Environmental remediation using catalysis driven under electromagnetic irradiation. Catalysis Surveys from Japan. 5:127–38.

Wahab R, Hwang IH, Kim YS. (2011). Non-hydrolytic synthesis and photo-catalytic studies of ZnO nanoparticles. Chemical Engineering Journal. 175:450–7. DOI:org/10.1016/j.cej.2011.09.055

Wang G, Zhuo X, Wang Y. (2016). Photocatalytic Degradation of Poly-acrylamide in Oilfield Sewage by Nano-sized TiO_2 Doped with W Ion. 2nd International Conference on Chemical and Material Engineering (ICCME 2015). MATEC Web of Conferences 39, 01013. DOI: org/10.1051/matecconf/20163901013

Wunsch JR. (2000). Polystyrene: synthesis, production and applications. Smithers Rapra Publishing, Shawbury, Shrewsbury, Shropshire, UK.

Xiao-Jing L, Guan-Jun Q, Jie-Rong C. (2008). The effect of surface modification by nitrogen plasma on photocatalytic degradation of polyvinyl chloride films. Applied Surface Science. 254:6568–74. DOI:org/10.1016/j.apsusc.2008.04.024

Yang C, Tian L, Ye L, Peng T, Deng K, Zan L. (2011). Enhancement of photocatalytic degradation activity of poly(vinyl chloride)$-TiO_2$ nanocomposite film with polyoxometalate. Journal of Applied Polymer Science. 120:2048–53. DOI: org/10.1002/app.33316

Yuan F, Li P, Qian H. (2013). Photocatalytic degradation of polyethylene glycol by nano-titanium dioxide modified with ferric acety-lacetonate. Synthesis and Reactivity in Inorganic Metal. 43:321–4. DOI: org/10.1080/15533174.2012.740729

Yousif E, Haddad R. (2013). Photodegradation and photostabilization of polymers, especially polystyrene: review. Springer Plus. 2:398–428. DOI: 10.1186/2193-1801-2-398

Zan L, Fa W, Wang S. (2006). Novel photodegradable low-density polyethylene-TiO_2 nanocomposite film. Environ. Science and Technology. 40:1681–5. DOI: 10.1021/es051173xCCC

Zan L, Tian L, Liu Z, Peng Z. (2004). A new polystyrene-TiO_2 nanocompos-ite film and its photocatalytic degradation. Applied Catalysis A: General. 264:237–42. DOI: org/10.1016/j.apcata.2003.12.046

Zan L, Wang S, Fa W, Hu Y, Tian L, Deng K. (2006). Solid-phase photo-catalytic degradation of polystyrene with modified nano-TiO_2 catalyst. Polymer. 47:8155–62. DOI:10.1016/j.polymer.2006.09.023

Zapata PA, Rabagliat FM, Lieberwirth I, Catalina F. (2014). Study of the photodegradation of nanocomposites containing TiO_2 nanoparticles dispersed in polyethylene and in poly(ethylene-co-octadecene). Poly-mer Degradation and Stability. 109:106–14. DOI: org/10.1016/j.polym degradstab.2014.06.020

Zhao J, Chen C, Ma W. (2005). Photocatalytic degradation of organic pollutants under visible light irradiation. Topics in Catalysis. 35:269–78.

Zhao X, Li Z, Chen Y, Shi L, Zhu Y. (2008). Enhancement of photocatalytic degradation of polyethylene plastic with CuPc modified TiO_2 photocatalyst under solar light irradiation. Applied Surface Science. 254:1825–9. DOI:10.1016/j.apsusc.2007.07.154

Zhao X, Zongwei L, Chen Y, Shi L, Zhu Y. (2007). Solid-phase photocatalytic degradation of polyethylene plastic under UV and solar light irradiation. Journal of Molecular Catalysis A: Chemical. 268:101–6. DOI:10.1016/j.molcata.2006.12.012

Zhu S, Wang D. (2017). Photocatalysis: basic principles, diverse forms of implementations and emerging scientific opportunities. Advanced Energy Materials. 1700841–65. DOI: 10.1002/aenm.201700841

13

Thermo-Mechanical Process Using for Recycling Polystyrene Waste

Ahmad K. Jassim

Production and Metallurgical Engineering, Basra, Iraq
E-mail: ahmadkj1966@yahoo.com

Plastic is one of the main materials that are commonly used in everyday life which becomes an integral part of our lives because it is lightweight, durable, and strong an organic polymer. The consumption rate of plastic material resource is increased due to overpopulation and rapid development of industries and lifestyle which lead to reduce their resource. The waste that is produced by using plastic material equals to 12% of the total solid waste material. It becomes the biggest problem today which affects our environment because it is not biodegradable waste. Humans have always produced waste and disposed it as a solid waste material. The growing of using plastic materials leads to increase their waste which needs large area of landfill. However, disposal of these wastes in environment is considered to be a big problem. Polystyrene waste is one of solid waste materials that need to be managed and reused due to it is very low biodegradability and existence large quantities. Therefore, recycling and reusing of these waste help to overcome the problem of solid waste. In this chapter, thermo-mechanical process was used for recycling polystyrene waste to produce useful parts that can be used in our lives. Therefore, this chapter will offer a short introduction about recycling, plastic waste management, sustainable manufacturing process, thermo-mechanical process, and case study as example for recycling plastic waste including polystyrene waste and styrene rubber waste using thermos-mechanical process.

13.1 Introduction

Plastic can be defined as macromolecules formed by polymerization and having the ability to be shaped by the application of reasonable amount of heat and pressure or another form of forces (El Newehy, 2016). It is the mostly used man-made material in the world which is still growing rapidly and becomes an integral part of our lives due to their low density and low cost (Tapkire et al., 2014; Muanza and Mbohwa, 2017; Seghiri et al., 2017). Therefore, the plastic waste was increasing in recent years and it becomes a big environmental problem because it is a very low biodegradable material (Tapkire et al., 2014). Polystyrene is the best packaging plastic material that is used in many applications which can be found in home, offices, restaurants, and industries. Therefore, a high amount of polystyrene waste is disposed after use, which creates a challenge because it is harmful to our environment. So, finding alternative methods of disposing waste is becoming a major research issue (Abbes et al., 2005; Samper et al., 2010; Aminudin et al., 2011; Tapkire et al., 2014; Muanza and Mbohwa, 2017; Seghiri et al., 2017; Gonzalez et al.).

Solid-state recycling process becomes an effective and powerful methodology to realize the green state forming from recyclable wastes to useful parts. It is one of the solutions for getting rid of the mountains of trash (Jassim, 2017). Sustainable recycling of post-consumer packaging plastic solid waste provides opportunities to reduce oil usage, carbon dioxide emissions, and quantities of waste requiring disposal. It is considered as a sustainable manufacturing process dealing with municipal solid waste to produce new products without utilizing new materials hence contributing to sustainable development (Muanza and Mbohwa, 2017).

Different strategies and methods have been invented by previous researchers in order to recycle polystyrene waste. Therefore, it becomes one of the most widely used plastics that can be recycled until infinity times. Generally, there are three methods used for recycling polystyrene which are mechanical, chemical, and thermos methods (Abbes et al., 2005; Muanza and Mbohwa, 2017). Modern recycling includes re-grinding, re-melting, and de-polymerization (Donati, 2008). In this research, recycling of polystyrene waste has been applied to produce new useful parts.

13.2 Plastic Waste Management

Solid wastes are generated by human activities. The problem has to do with the quantum of solid waste generated and effective ways of management.

Solid waste management is understood as supervised handling of waste materials from source through recovery processes to disposal. It involves control of generation, storage, collection, transportation, processing, and disposal of solid waste with the aim of protecting environmental quantity and human health and preserving natural resources (Peprah et al., 2015).

Currently, the world annual consumption of plastic materials is increasing to be 100 million tons because it becomes an integral part of our lives. The production of plastic waste is growing fast because plastic is used to produce packaging, shopping bags, PET bottles, and replaces heavier ferrous materials that are used in automotive industries. This is because it has lightweight, low density, user friendly design, and fabrication capabilities with low cost (Miller et al., 2014; El Newehy, 2016; Subramanian). Therefore, a high amount of plastic waste is being accumulated and becomes the major constitute of municipal and industrial waste in cities which accounts approximately 17% of the total waste. Plastic waste pollutes beaches and oceans as well as landscape. The plastic bags kill animals and the burning of plastic releases hazardous gases into the air and leaves toxic residues in the form of ash and fumes with carbon monoxide, chlorine, and hydrochloric acid because plastics include materials composed of various elements such as carbon, hydrogen, oxygen, nitrogen, chlorine, and sulfur (Rahman, 2014; El Newehy, 2016; Jassim, 2017).

Solid waste now is a global problem and must be addressed in order to solve the world's resources and energy challenges. The common plastics that can be recycled are polyethylene, polypropylene, and polystyrene (Plastic Waste Management Institute, 2009; El Newehy, 2016; Shehu, 2017). Recycling is one of the most conventional and easiest ways for everyone to participate in. However, recycling plastic to make new products is not easy as we think (Li et al., 2009).

13.3 Sustainable Manufacturing Process

Sustainable manufacturing process and solid waste management are used for conserving valuable natural resources, preventing the unnecessary emission of gas, and protecting public health. It is the goal to reduce environmental impacts and offer economic opportunities (Jassim, 2017).

Recycling has always been through to be a means to reduce emission and the consumption of virgin materials and the conservation of other natural resource such as water, air, land, and energy (Shehu, 2017).

3Rs model is considered as state-of-art philosophy as integrated solid waste management and 6Rs model as state for sustainable manufacturing process. 3Rs model means reduce, reuse, and recycle, while 6Rs model means reduce, reuse, recycle, redesign, recover, and remanufacture (Peprah et al., 2015; El Newehy, 2016). Therefore, in this chapter, we will explain methods that are used for recycling and remanufacturing polystyrene waste and styrene rubber waste that are generated from disposed food containers and car tires.

13.4 Thermo-Mechanical Process

Plastic recycling is defined as the process of recovering scrap or waste plastic and reprocessing the material into useful products. The developing technology allows plastic waste to be recycled by a number of methods which are landfilling, incineration, recycling, and biodegradation.

All plastic can be disposed in landfills; however, landfilling is considered highly wasteful as it requires a vast amount of space and the chemical constituents and energy contained in plastic is lost waste in this disposal route. On the other hand, incineration returns some of the energy from plastic production. However, it is tending to cause negative environment and health effects as hazardous gases may be released into the atmosphere in the process. Moreover, many plastics can be recycled and the materials recovered can be given a second life; however, this method is not fully utilized due to difficulties with the collection and sorting of plastic waste. Recycling is the effective way to deal with plastic waste. Furthermore, biodegradable of plastic is defined as the decomposition of plastic by the active of living organisms. It has the potential to solve a number of waste management. However, biodegradable plastics are not without controversy (Noren, 2018).

They can be grouped into four main categories—mechanical recycling, feedstock recycling, and thermal recycling. Therefore, it is important to select a method for recycling plastics to impose the least social cost and limit the environmental impact (Plastic Waste Management Institute, 2009; Rahman, 2014; Shehu, 2017).

Mechanical recycling is defined as a way of making new products out of unmodified plastic waste. The process includes cutting, shredding, sorting, contamination separation, floating and cleaning, extrusion, filtering, and pelletizing; 78% of mechanically recycled plastics are converted to recycle raw plastic intermediate, while the remaining 13% are converted directly into products (Plastic Waste Management Institute, 2009; Shehu, 2017).

Moreover, thermal recycling is defined as melting of plastic waste by using heating source.

13.5 Case Study

To study the possibility to recycle plastic waste by using the thermo-mechanical recycling method, polystyrene waste and rubber waste have been collected to use for producing new useful products.

13.5.1 Materials

Polystyrene is a versatile plastic that can be rigid or foamed which has a low melting point. It is often combined with rubber to make high impact polystyrene which is used for packing and durable applications requiring toughness. Polystyrene is one of the long known synthetic polymerized thermoplastic resins that are produced in different forms. Typical applications include protective packing, foodservice packing, bottles, and food containers. Therefore, polystyrene sheet or molded and disposable packaging for food has been collected from municipal and landfills as a waste of human activities as shown in Figure 13.1. They were used as a raw material to produce new materials that are used to produce rigid parts.

13.5.2 Production Procedure

Heat treatment process was used to produce rigid polystyrene waste powder. The procedure for recycling polystyrene waste is divided into three main stages. The first stage is melting the waste to reduce their size and transfer it to rigid materials by heating inside a furnace with a temperature between 150 and 250°C as shown in Table 13.1. The second stage includes quenching re-melting polystyrene waste by water media as shown in Table 13.2. The

Figure 13.1   Collected polystyrene wastes.

Table 13.1 Melting temperature for melting polystyrene wastes

Sample Number	Melting Temperature (°C)	Melting Time (Minutes)
1	150	5, 15, 30
2	175	5, 15, 30
3	200	5, 15, 30
4	225	5, 15, 30
5	250	3, 15, 30

Table 13.2 Quenching condition for melted polystyrene wastes

Sample	Melting Temperature (°C)	Melting Time (Minutes)	Quenching by
1	250	5	Water
2	250	15	Water

third stage is crushing and grinding the quenching waste to produce fine powder to be ready for forming.

The produced polystyrene waste powder was used without any materials and with recycling tire waste as rubber powder to produce rigid materials. Recycling rubber tire was added with 10% as a plasticizer and then mixed with polystyrene waste and heated at 250°C for 10 min.

13.6 Results

13.6.1 Shape of Products

The shape of melting and quenching polystyrene waste and the grinding powder that were prepared in this work are shown in Figures 13.2 and 13.3.

Figure 13.2 Polystyrene waste melted at 150–250°C for 15–30 min.

Figure 13.3 Polystyrene waste powder after melting at 150–250°C for 15–30 min.

The result shows that there is a possibility to produce white powder and with collar depending on the burning temperature and time of burning.

13.6.2 IR Inspection

Infrared analysis for polystyrene waste powder with and without recycling rubber is shown in Figures 13.4 and 13.5.

13.6.3 Hardness

Shore hardness device was used to measure the hardness of products according to ASTM D2240 and the results are shown in Table 13.3. The results show that hardness was decreased by using rubber with polystyrene waste. However, impact toughness was increased by adding rubber materials to polystyrene.

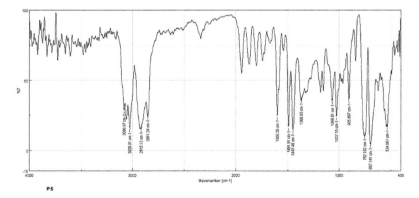

Figure 13.4 IR inspection for recycling polystyrene waste without rubber waste.

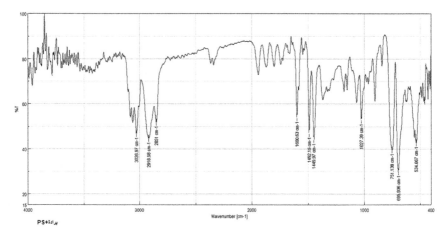

PS+رابر

Figure 13.5   IR inspection for recycling polystyrene waste with rubber waste.

Table 13.3   Shore hardness test

| Sample Number | Shore Hardness HSD for Recycling | |
	Polystyrene Waste	Polystyrene Waste with Rubber Waste
1	121	110
2	143	108
3	163	115
4	157	130
5	126	115
6	138	109
7	149	113
8	148	123
9	149	116
Mean value	144	115

Table 13.4   Impact toughness test

Sample Number	Sample Materials	Impact Toughness (J/m^2)
1	Polystyrene waste	2.2
2	Polystyrene waste with rubber waste	12

13.6.4 Impact Toughness

Table 13.4 shows the result of impact toughness test that was done for parts produced by using polystyrene waste with and without rubber waste which was prepared according to ASTM D-256. The result shows that the impact toughness was increased by added rubber waste to polystyrene waste.

The impact strength is increased to be six times compared with polystyrene waste.

13.6.5 Workability

The recycled polystyrene waste that was prepared in this work has a good workability because it is possible to form it and make a different shape as shown in Figures 13.6–13.8. There is a possibility to make hole through waste products with continuous chips and with less waste powder by using drill as shown in Figure 13.6. Moreover, it is possible to make marking or printing shape on the waste product with clear shape as shown in Figure 13.7. In addition, the surface of product is very good and has a smooth surface as shown in Figure 13.8 and the thickness can be managed as required.

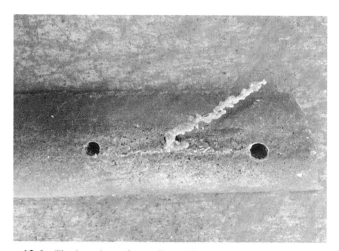

Figure 13.6   Final product of recycling polystyrene waste with rubber waste.

Figure 13.7   Final product of recycling polystyrene waste without rubber waste.

Figure 13.8   Final product of recycling polystyrene waste with rubber waste.

13.7 Conclusion

It is possible to produce useful parts from polystyrene waste by using thermos-mechanical process that includes heating the waste at 250°C for 30 min inside the furnace and quenching by water. The results show that impact toughness for recycling polystyrene waste is 2.2 J/m^2 which was increased to be 12 J/m^2 after adding rubber powder. In addition, the max hardness of recycling polystyrene waste is 163 HSD which was reduced to be 130 HSD after adding recycled rubber waste. Finally, the final products that were produced from recycling polystyrene waste have a good workability.

References

Abbes IB, Bayoudh S, Baklouti M, Papon E, Leclereq D. (2005). Converting waste polystyrene into adsorbent: Optimisation of reaction parameters and properties. Progress in Rubber, Plastic and Recycling Technology. 22(3):170–93.

Aminudin E, Md Din MF, Mohamad Z, Noor ZZ, Iwao K. (2011) A review on recycled expanded polystyrene waste as potential thermal reduction in builing materials, 2011 International Conference on Environment and Industrial Innovation, IACSIT press, 12:113–8.

Donati G. (2008). Innovations in recycled expanded polystyrene form for use in electronic protective packaging. California State Polytechnic, University San Luis Obispo. pp. 1–12.

El Newehy M. (2016). Arab Chemistry week, King Saud University, Petrochemical Research, Department of Chemistry, College of Science. Available at: http://fac.ksu.edu.sa,www.environment-expert.com/compa nies/keyword-plastic-recycling 30821/location-Saudi Arabia.

Gonzalez MTG, Romero JFR, Lucas A, Fernandez IG, Gutierrez GD. Waste expanded polystyrene recycling by cymene using liquid or supercritical Co_2 for solvent recovery.

Jassim AK. (2017). Recycling of polyethylene waste to produce plastic cement. Elsevier, Science Direct, Manufacturing Procedia. 8:635–42.

Jassim AK. (2017). Sustainable solid waste recycling, chapter 1, Skills Development for sustainable manufacturing book, intech open science publisher.

Li N, Mahat D, Park S. (2009). WPI, Reduce, Reuse and Replace: a study on solutions to plastic wastes.

Miller L, Soulliere K, Beaulien SS, Teseny S, and Tam E. (2014). Challenges and alternative to plastic recycling in the automotive sector. Materials Journal. 7:5883–902. Available at: www.mdpi.com/journal/materials.

Muanza BG, Mbohwa C. (2017). Drivers to sustainable plastic solid waste recycling: a review, Elsevier, Science Direct, Manufacturing Procedia. 8:649–56.

Noren A. (2018). Methods of plastic waste disposal and possible complication, plastic word, plastic nightmare.

Peprah K, Amoah ST, Achana GTW. (2015). Assesing 3Rs model in relation to municipal solid waste management in Wa, Chana. World Environment. 5(3):112–20.

Plastic Waste Management Institute. (2009). An introduction to plastic recycling.

Rahman FA. (2014). Reduce, reuse and recycle: alternatives for waste management, New Mexico state university. Available at: www.aces. nmsu.edu.

Samper M, Garcia-Sanoguera D, Parres F, Lopez J. (2010). Recycling of expanded polystyrene from packaging. Progress in Rubber Plastics Recycling Technology. 26(2):83–92.

Seghiri M, Boutoutaou D, Kroker A, Hachani MI. (2017). The possibility of making a composite material from waste plastic. Elsevier, Energy Procedia. 119:163–9.

Shehu SI. (2017). Separation of plastic waste from mixed waste: existing and emerging sorting technologies performance and possibilities of increased recycling rate with Finland as case study, Master thesis,

Lappeenranta university of technology, LUT School of energy technology, Faculty of technology.

Subramanian PM. Plastic recycling and waste management in the US, S. P. M technologies.

Tapkire G, Pariher S, Patil P, Kumarat HR. (2014). Recycling plastic used in concrete paver block. International Journal of Research in Engineering and Technology. 3(9):33–5.

14

Hydrogen and Methane Production Under Conditions of Dark Fermentation Process with Low Oxygen Concentration

Gaweł Sołowski[1], Bartosz Hrycak[1], Dariusz Czylkowski[1], Izabela Konkol[1], Krzysztof Pastuszak[2], and Adam Cenian[1]

[1]Instytut Maszyn Przepływowych im R. Szewalskiego, Polskiej Akademii Nauk
[2]Katedra Algorytmów i Modelowania; Wydział Informatyki, Telekomunikacji i Informatyki Politechniki Gdańskiej
E-mail: gsolowski@imp.gda.pl; dariusz.czylkowski@imp.gda.pl; bartosz.hrycak@imp.gda.pl; izabela.konkol@imp.gda.pl; krzpastu@pg.edu.pl; cenian@imp.gda.pl

In this paper, results of dark fermentation of sour cabbage with volatile suspended solids (VSS) 5 and 10 g/L in the presence of oxygen are presented. The oxygen flow rates were 0.56 up to 2.73 mL/h. The highest volume of hydrogen was obtained for flow rate 0.63 O_2 mL/h giving oxygen concentration in biogas in the range 0.2–6.2%.

14.1 Introduction

Dark fermentation is a type of anaerobic digestion process where sugar or glycerol is converted into hydrogen, carbon dioxide, and low organic acids (Chaganti et al., 2012). Anaerobic digestion is a minimum double-step process usually four-step process (Chasnyk et al., 2015). The organic raw material (fats, proteins, and sugars or a mixture of it) is hydrolyzed into organic acids, oligosaccharides or simpler sugars, and glycerol, then the feed is converted by bacteria like *Clostridia* or *Enterobacter* into hydrogen and simple organic acid (butyric or acetic acid), and finally by methanogenic bacteria, it is converted into methane (Skorek et al., 2003). The dark fermentation occurs in the third process. The hydrogen bacteria usually start to become also methanogenic and producing methane with other typically methane bacteria (Giovannini et al., 2016). The blocking of methane occurs by stressing the bacteria (Xing et al., 2010; Chaganti et al., 2012; Sinbuathong et al., 2015) and lowering the pH of inoculum to pH 5.0–6.0 (Hawkes et al., 2002). However, it can also occur in other conditions of pH (Han et al., 2015; Alibardi and Cossu, 2016; Chatellard et al., 2016; Najafpour et al., 2016) or without any pre-treatment of inoculum like in infection of wounds (dark fermentation process is known here as gas gangrene) (Chi et al., 1995). The reaction of dark fermentation can occur in three path ways (Hussy et al., 2005; Bartacek et al., 2007) (see Figure 14.1).

An is part of anaerobic digestion process, see in Figure 14.2.

Thus, to understand the process better, it is necessary to test different materials that can potentially produce hydrogen without pre-treatment. One of the known problems which must be solved before viable industrialization of the process is the need to inhibit methane production from the process as that removes hydrogen from biogas. Another problem is hydrogen sulfur

$$C_6H_{12}O_6 + 4H_2O \rightarrow 2CH_3COOH + 2CO_2 + 4H_2 \qquad \Delta^0G=\text{-48 kJ mol}^{-1} \qquad (1)$$

$$C_6H_{12}O_6 + 2H_2O \rightarrow CH_3CH_2CH_2COOH + 2CO_2 + 2H_2 \qquad \Delta^0G=\text{-137 kJ mol}^{-1} \qquad (2)$$

$$C_6H_{12}O_6 + 3H_2O \rightarrow CH_3COOH + 2CH_3CH_2OH + 3H^+ \qquad \Delta^0G=\text{-97 kJ mol}^{-1} \qquad (3)$$

Figure 14.1 Scheme of dark fermentation possible reactions (Sołowski et al., 2018).

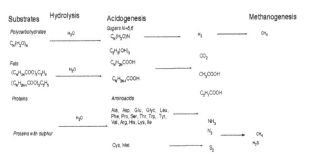

Figure 14.2 Scheme of anaerobic digestions (Chasnyk et al., 2015).

formation; this can be also prevented by the addition of small amounts of oxygen (Duangmanee, 2009; Terry, 2010; Abo-Hashesh and Hallenbeck, 2012; Khoshnevisan et al., 2017). Thus, it seems worth checking which process, methanogenesis or hydrogenesis, is more sensitive to oxygen presence. As the material for the chapter was chosen sour cabbage, the sour cabbage (cabbage that is preserved by lactic acid fermentation—popular in Middle Europe) was used as a substrate. After the promising results obtained for VSS 10 g/L (Sołowski et al., 2017), the research was continued for 5 g/L.

14.2 Materials and Methods

The fermentation process of sour cabbage was performed in reactors (jars) of volume 2 dm^3. As inoculum, sludge from a biogas plant in Lubań (Pomerania Region) was used. Either 5 or 10 g/L VSS were applied to each batch of sour cabbage for the process. The cabbage was prepared by milling and mixing it. The fermentation process was continued for up to 5 days (115 h) at temperature 38°C and initial pH $\sim$7.9 in the presence of oxygen and no oxygen. Various oxygen flow rates for reactors with VSS content equal 5 g/L were used: 0.4, 0.63, 1.7, and 2.7 mL/h, and no oxygen. The results were compared with previous one for 10 g/L and flow rates of 1.4, 2.4, and 4.5 mL/h and initial pH 7.9. The oxygen was added twice a day for 5 days through the top of the jars. Air was added to each reactor in such a way that corresponded to the averaged flow rates given above. As for comparison there was also the reactor only with sludge. The biogas produced was measured using the Owen method (Logan et al., 2002) and analyzed using TCD-GC.

14.3 Results and Discussion

GC analysis allowed the determination of methane, hydrogen, hydrogen sulfide, carbon dioxide, and nitrogen concentration. The biogas volumes obtained during 5 days (115 h) were: 2.52 dm^3 in anaerobic conditions; 2.19 dm^3 for oxygen flow rate 0.56 mL/h; 2.73 dm^3 for oxygen flow rate 0.63 mL/h; 3.19 dm^3 for oxygen flow rate 1.7 mL/h; and 3.98 dm^3 in oxygen flow rate 2.73 mL/h. In the reactor with only sludge, there was no biogas obtained. The total biogas production is shown in Figure 14.3 for load 5 g/L and Figure 14.4 for 10 g/L.

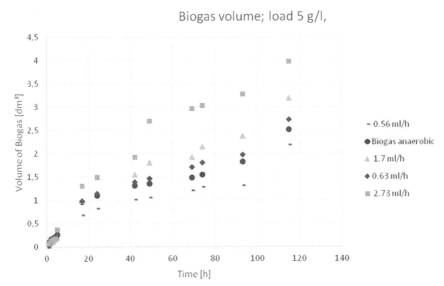

Figure 14.3   Time evolution of total biogas production under strict anaerobic conditions (red dots) and with addition of oxygen at flow rates: 0.56 (dark blue dash), 0.63 (violet rhombi), 1.7 (green triangles), and 2.7 mL/h (light blue squares).

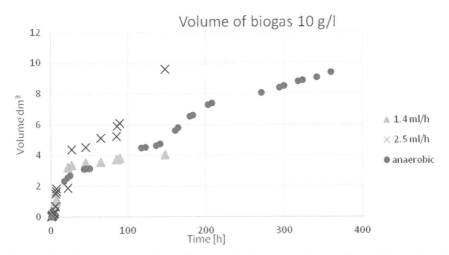

Figure 14.4   Time evolution of total biogas production: anaerobic conditions (blue dots) and with oxygen flow rates: 1.4 (green triangles) and 2.5 mL/h crosses.

Under strict anaerobic conditions, the volumes of methane and hydrogen produced during 115 h were 1.81 and 0.0021 dm^3, respectively. Under different oxygen flow rate conditions, the volumes of methane and hydrogen were:

- 0.52 dm^3 of methane and 0.001 dm^3 of hydrogen in oxygen flow rates 0.56 mL/h
- 0.97 dm^3 of methane and 0.0066 dm^3 of hydrogen in oxygen flow rates 0.62 mL/h
- 0.74 dm^3 of methane and 0.0026 dm^3 of hydrogen in oxygen flow rates 1.7 mL/h
- 0.71 dm^3 of methane and 0.0008 dm^3 of hydrogen in oxygen flow rates 2.7 mL/h.

In the case of substrate concentration 10 g/L, the volume of methane and hydrogen was:

- 0.21 dm^3 of methane and 0.176 dm^3 of hydrogen in oxygen flow rates of 2.5 mL/h
- 0.26 dm^3 of methane and 0.07 dm^3 of hydrogen in oxygen flow rates of 1.4 mL/h.

The details for loads 5 g/L are shown in Figures 14.5 (methane) and 14.6 (hydrogen) and for loads 10 g/L in Figures 14.7 (methane) and 14.8 (hydrogen).

It can be discerned that in the case of sour cabbage load 5 g/L, the highest hydrogen production was observed for flow rate 0.63 mL/h (hydrogen concentration reached 0.33% in biogas). For oxygen flow rate 1.7 mL/h, the hydrogen production is similar to anaerobic conditions. However, for oxygen flow rates 0.56 and 2.73 mL/h, the production is lower than that under anaerobic conditions. The methane production for oxygen flow rate 0.63 mL/h is almost half of that under anaerobic conditions. In the case of methane production, the maximum is reached under anaerobic conditions. As the oxygen flow rates increase, methane production declines.

Oxygen was added to the reactor 17 h after feeding, by injection of a given amount of air in order to achieve the requested averaged oxygen flows (calculated in mL/h). After some time (when biogas volume reached the level of at least 0.45 dm^3 necessary for gas chromatograph), the following measurements were performed, i.e., 49, 93, and 115 h after starting the experiment. In between, the following oxygen injection was performed to reach assumed oxygen flows.

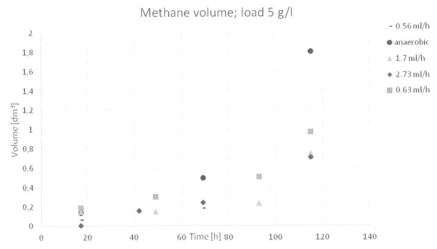

Figure 14.5   Time evolution of total methane production under strict anaerobic conditions (red dots) and with small oxygen flow rates: 0.56 (dark blue dashes), 0.63 (blue squares), 1.7 (green triangles), and 2.73 mL/h (violet rhombi).

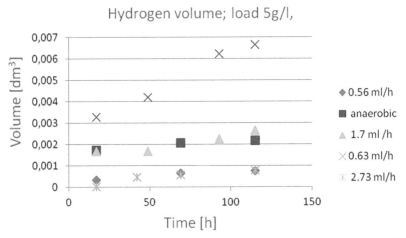

Figure 14.6   Time evolution of total hydrogen production under strict anaerobic conditions (red squares) and with small oxygen flow rates: 0.56 (dark blue diamonds), 0.63 (violet crosses), 1.7 (green triangles), and 2.73 mL/h (light blue stars).

In the case of load 5 g/L and flow rate:

- 2.73 mL/h, oxygen concentration in biogas was falling from above 20% (measured 12 h after first oxygen injection) to 1.1% after few injections

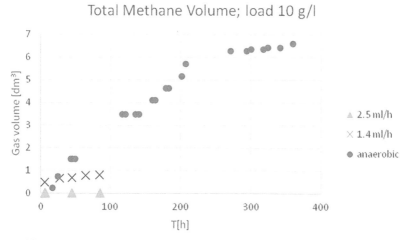

Figure 14.7 Time evolution of total methane production under strict anaerobic conditions (blue dots) and with small oxygen flow rates: 1.4 (crosses) and 2.5 mL/h (green triangles).

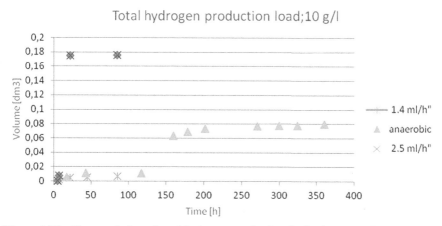

Figure 14.8 Time evolution of total hydrogen production for load 10 g/L of sour cabbage under strict anaerobic conditions (green triangles) and with small oxygen flow rates: 2.5 (violet crosses) and 1.4 mL/h (light blue stars).

- 0.63 mL/h, oxygen concentration in biogas was falling from above 6% (measured 12 h after first oxygen injection) to 0.2% after few injections.

In the case of load 10 g/L and flow rate:

- 2.5 mL/h, oxygen concentration in biogas was falling from above 12% (measured 5 h after first oxygen injection) to 0.5% after few injections

- 1.4 mL/h, oxygen concentration in biogas was falling from above 11% (measured 5 h after first oxygen injection) to 1.9% after few injections.

The highest hydrogen conversion of 1.3 mL H_2/g of sour cabbage was found for flow rate 0.63 mL/h and load 5 g/L. In comparison, in the case of anaerobic conditions, 0.4 mL H_2/g of the substrate was obtained. Hydrogen production under flow rate 0.63 mL/h is even six times higher in volume than in the case of flow rates 0.56 and 2.73 mL/h.

The highest conversion of hydrogen for load 10 g/L (4.8 mL H_2/g of sour cabbage) was obtained for oxygen flow 4.5 mL/h, while under anaerobic conditions, production 0.7 mL H_2/g of sour cabbage was measured. The hydrogen concentration in biogas was up to 13% for 10 g/L and about 8% in 5 g/L. The hydrogen content of biogas from sour cabbage was similar to that measured in gas gangrene (Chi et al., 1995). Generally, the hydrogen production for load 5 g/L is almost two times lower than for load 10 g/L for anaerobic conditions. This trend is reversed for other substrates like glucose (Pan et al., 2008). In the condition of reaction, the sour cabbage was better source of hydrogen than sunflower stalk or dungeon compost some source in optimal hydrogen condition (Fan et al., 2006; Monlau et al., 2013). On the other hand, the hydrogen production by dark fermentation from cabbage waste is 50 times less than from cassava (Su et al., 2009). The comparison of hydrogen production from different sources is presented in Table 14.1.

Table 14.1 Comparison of hydrogen production from different substrates

Type of Substrate/Pre-Treatment	Hydrogen Yield (mL H_2/g Substrate)	Type of Inoculum	Type of Process	Process Conditions (g/L/°C/pH)	References
Sour cabbage/no	1.3	Untreated sludge	Batch	(5/38/7.9 microaeration)	This chapter
Sour cabbage/no	4.8	Untreated sludge	Batch	(10/38/7.9 microaeration)	Sołowski et al., 2017
Reed canary grass /3% of HCl, 121°C, 90 min	36	Untreated sludge	Batch	(5/38/3.4–6	Lakaniemi et al., 2011
Mixture of hydrothermally pretreated asbestos to glucose of proportion 1:6/heat shock	1.3	Activated Sludge	Batch	(5/35/5.45)	Spasiano, 2018

Table 14.1 (Continued)

Type of Substrate/Pre-Treatment	Hydrogen Yield (mL H$_2$/g Substrate)	Type of Inoculum	Type of Process	Process Conditions (g/L/°C/pH)	References
Grass/ *comminuted 20 mesh +3 h 105°C + 0.5 % NaOH*	4.39	Cracked cereal	Batch	(5/35/7.0)	Cui et al., 2012
Cassava/α- *amylase+ glucoamylase – heat gelatinizing*	240	Activated sludge	Batch	(10/35/7.0)	Su et al., 2009
Cow dung compost/2 %HCl *+ 8-min microwaves*	0.5	Dung compost	Batch	(25/36/7.0)	Fan et al., 2006
Sunflower cornstalk/diluted HCl	2.3	Activated sludge	Batch	(5/35/5.5)	Monlau et al., 2013
Microalgal biomass/heat shock	20.9	Activated Sewage Sludge	Batch	(10/30/7)	Batista et al., 2018
Depackaging wastes/none	1.4	Untreated sludge	Batch	(4/37/7)	Noblecourt et al., 2018

14.4 Conclusions

Sour cabbage is a potential source for the production of hydrogen by dark fermentation under either strict anaerobic conditions, but with a small addition of oxygen, which inhibits methanogenesis and improves hydrogenesis.

After 115 h of dark fermentation process under strict anaerobic conditions and substrate load 5 g/L, methane production is twice as high and hydrogen production is three times as low compared with a system with oxygen flow rate 0.63 mL/h. The flow rate that can improve hydrogen production in the case of 5 g/L feed rate must be higher than 0.56 mL/h and lower than 1.7 mL/h. The optimal oxygen flow rate for highest hydrogen production in system with feed 5 g/L was 0.63 mL/h.

For feed 10 g/L, the oxygen flow rate should be minimum 2.5 mL/h in order to increase hydrogen production. The highest hydrogen production was for flow rate 4.5 mL/h (Sołowski et al., 2017). The phenomena need further research.

Acknowledgments

The research was funded from grant of Institute of Fluid-Flow Machinery, Polish Accademy of Science in Gdansk FBW-44 – Solowski and Strategic Project NCBR, "New ecoeneregetics technologies for sustainable and low-emission development of arable area" Biostateg 3/344128/12/NCBR/2017.

References

Abo-Hashesh M, Hallenbeck PC. (2012). Microaerobic dark fermentative hydrogen production by the photosynthetic bacterium, Rhodobacter capsulatus JP91. International Journal of Low-Carbon Technologies. 7:97–103. doi:10.1093/ijlct/cts011

Alibardi L, Cossu R. (2016). Effects of carbohydrate, protein and lipid content of organic waste on hydrogen production and fermentation products. Waste Management. 47:69–77. doi:10.1016/j.wasman.2015.07.049

Bartacek J, Zabranska J, Lens PNL. (2007). Developments and constraints in fermentative hydrogen production. Biofuels, Bioproducts and Biorefining. 1:201–14.

Batista AP, Gouveia L, Marques PASS. (2018). Fermentative hydrogen production from microalgal biomass by a single strain of bacterium Enterobacter aerogenes – Effect of operational conditions and fermentation kinetics. Renewable Energy. 119:203–9. doi:10.1016/j.renene.2017.12.017

Chaganti SR, Kim DH, Lalman JA. (2012). Dark fermentative hydrogen production by mixed anaerobic cultures: effect of inoculum treatment methods on hydrogen yield. Renewable Energy. 48:117–21. doi:10.1016/j.renene.2012.04.015

Chasnyk O, Sołowski G, Shkarupa O. (2015). Historical, technical and economic aspects of biogas development: case of Poland and Ukraine. Renewable and Sustainable Energy Reviews. 52:227–39. doi:10.1016/j.rser.2015.07.122

Chatellard L, Trably É, Carrère H. (2016). Cross-impact of initial sugars type and microbial community origin and history on fermentative production of biohydrogen and biomolecules from lignocellulosic biomass. 21st World Hydrogen Energy Conference. 2016:497–8.

Chi CH, Chen KW, Huang JJ, Chuang YC, Wu MH. (1995). Gas composition in Clostridium septicum gas gangrene. Journal of the Formosan Medical Association Taiwan Yi Zhi. 94:757–9.

Cui M, Shen J. (2012). Effects of acid and alkaline pretreatments on the biohydrogen production from grass by anaerobic dark fermentation. International Journal of Hydrogen Energy. 37:1120–4. doi:10.1016/j.ijhydene.2011.02.078

Duangmanee T. (2009). Micro-aeration for hydrogen sulfide removal from biogas. Iowa State University.

Fan YT, Zhang YH, Zhang SF, Hou HW, Ren BZ. (2006). Efficient conversion of wheat straw wastes into biohydrogen gas by cow dung compost. Bioresource Technology. 97:500–5. doi:10.1016/j.biortech.2005.02.049

Giovannini G, Donoso-Bravo A, Jeison D, Chamy R, Ruíz-Filippi G, Vande Wouver A. (2016). A review of the role of hydrogen in past and current modelling approaches to anaerobic digestion processes. International Journal of Hydrogen Energy. 1. doi:10.1016/j.ijhydene.2016.07.012

Han W, Na D, Wen Y, Hong J, Feng Y, Qi N. (2015). Biohydrogen production from food waste hydrolysate using continuous mixed immobilized sludge reactors. Bioresource Technology. 180:54–8. doi:10.1016/j.biortech.2014.12.067

Hawkes FR, Dinsdale R, Hawkes DL, Hussy I. (2002). Sustainable fermentative hydrogen production: Challenges for process optimisation. International Journal of Hydrogen Energy. 27:1339–47. doi:10.1016/S0360-3199(02)00090-3

Hussy I, Hawkes FR, Dinsdale R, Hawkes DL. (2005). Continuous fermentative hydrogen production from sucrose and sugarbeet. International Journal of Hydrogen Energy. 30:471–83. doi:DOI 10.1016/j.ijhydene.2004.04.003

Khoshnevisan B, Tsapekos P, Alfaro N, Díaz I, Fdz-Polanco M, Rafiee S, et al. (2017). A review on prospects and challenges of biological H2S removal from biogas with focus on biotrickling filtration and microaerobic desulfurization. Biofuel Research Journal. 4:741–50. doi:10.18331/BRJ2017.4.4.6

Lakaniemi AM, Koskinen PEP, Nevatalo LM, Kaksonen AH, Puhakka JA. (2011). Biogenic hydrogen and methane production from reed canary grass. Biomass and Bioenergy. 35:773–80. doi:10.1016/j.biombioe. 2010.10.032

Logan BE, Oh SE, Kim IS, Van Ginkel S. (2002). Biological hydrogen production measured in batch anaerobic respirometers. Environmental Science and Technology. 36:2530–5. doi:10.1021/es015783i

Monlau F, Aemig Q, Trably E, Hamelin J, Steyer JP, Carrere H. (2013). Specific inhibition of biohydrogen-producing Clostridium sp. after dilute-acid pretreatment of sunflower stalks. International Journal of Hydrogen Energy. 38:12273–82. doi:10.1016/j.ijhydene.2013.07.018

Najafpour GD, Shahavi MH, Neshat SA. (2016). Assessment of biological Hydrogen production processes: a review. IOP Conference Series: Earth and Environmental Science. 36:012068. doi:10.1088/1755-1315/36/1/012068

Noblecourt A, Christophe G, Larroche C, Fontanille P. (2018). Hydrogen production by dark fermentation from pre-fermented depackaging food wastes. Bioresource Technology. 247:864–70. doi:10.1016/j.biortech. 2017.09.199

Pan CM, Fan YT, Zhao P, Hou HW. Fermentative hydrogen production by the newly isolated Clostridium beijerinckii Fanp3. International Journal of Hydrogen Energy. 33:5383–91. doi:10.1016/j.ijhydene.2008.05.037

Sinbuathong N, Kanchanakhan B, Suchat L. (2015). Biohydrogen production from normal starch wastewater with heat-treated mixed microorganisms from a starch factory. International Journal of Global Warminglobal Warming. 7:293–306.

Sołowski G, Hrycak B, Czylkowski D, Cenian A, Pastuszak K. (2017). Oxygen sensitivity of hydrogenesis and methanogenesis. Contemporary Problems of Power Engineering and Environmental Protection. 2018:157–9.

Sołowski G, Shalaby MS, Abdallah H, Shaban AM, Cenian A. (2018). Production of hydrogen from biomass and its separation using membrane technology. Renewable and Sustainable Energy Reviews. 82:3152–67. doi:10.1016/j.rser.2017.10.027

Skorek J, Cebula J, Latocha L, Kalina J. (2003). Pozyskiwanie i energetyczne wykorzystanie biogazu z biogazowni rolniczych. Gospodarka Paliwami i Energia. 12:15–9.

Spasiano D. (2018). Dark fermentation process as pretreatment for a sustainable denaturation of asbestos containing wastes. Journal of Hazardous Materials. 349:45–50. doi:10.1016/j.jhazmat.2018.01.049

Su H, Cheng J, Zhou J, Song W, Cen K. (2009). Improving hydrogen production from cassava starch by combination of dark and photo fermentation. International Journal of Hydrogen Energy. 34:1780–6. doi:10.1016/j.ijhydene.2008.12.045

Terry PA. (2010). Application of ozone and oxygen to reduce chemical oxygen processing plant. International Journal of Chemical Engineering. 2010:1–6. doi:10.1155/2010/250235

Xing Y, Li Z, Fan Y, Hou H. (2010). Biohydrogen production from dairy manures with acidification pretreatment by anaerobic fermentation. Environmental Science and Pollution Research. 17:392–9. doi:10.1007/s11356-009-0187-4

15

Oxidation of Lignin from Wood Dust to Vanillin Using Ionic Liquid Medium and Study of Its Antioxidant Activity

Gyanashree Bora and Jyotirekha G. Handique*

Department of Chemistry, Dibrugarh University, Dibrugarh, Assam, India
E-mail: jghandique@rediffmail.com
*Corresponding Author

Vanillin can be considered as one of the world's most popular flavoring materials and has wide-ranging applications in food, beverages, cosmetics, perfumery, and pharmaceutical industry due to its antimicrobial and antioxidant properties. But the production of natural vanillin is very low and only about 1% of the global demand of vanillin can be fulfilled by actual Vanilla orchid. On the other hand, lignin is the second most abundant constituent of wood after cellulose and can be a potential substrate for the production of vanillin. Due to the increasing demand and decreasing supply of natural vanilla day by day, research on oxidation of lignin to produce vanillin is of very much importance nowadays as the conventional method for this oxidation is very much toxic and harmful to the environment.

So, in our present study, we have used wood dust for the production of vanillin by the isolation of lignin from the wood dust using an ionic liquid at a temperature of 100°C followed by the oxidation of the lignin using oxygen and chloroform in a hydrophilic ionic liquid medium. The Fourier transformation infrared spectroscopy, mass spectroscopy, and UV–Visible spectroscopy are used to characterize the product. Also, we have studied the antioxidant activity of the synthesized compound and compared the results with the standard vanillin.

15.1 Introduction

In the recent years, there has been an escalating interest for the research of natural and healthy foods, especially regarding the ingredients used in food items such as flavoring agents and preservatives. Among the assortment of natural flavors that are in use, vanilla occupies a well-known market place and is widely used for the preparation of ice creams, cakes, chocolates, liquors, soft drinks, and perfumery (Ranadive, 1994); also vanilla is well known as antioxidant additive, antifoaming agent, vulcanization inhibitor, and chemical precursor for pharmaceutical and agrochemicals industries (da Silva et al., 2009). Vanilla is a complex combination of various flavor components which can be extracted from the cured pods of different species of plant genus Vanilla: *Vanillus planifolia* and *Vanillus tahitensis* (Rao and Ravishankar, 2000). The flavor profile of vanilla contains more than 200 components, out of which only 26 components have concentrations more than 1 mg/kg. The fragrance and flavor of vanilla extract is accredited mostly due to the presence of vanillin (4-hydroxy-3-methoxybenzaldehyde) which occurs with a concentration of 1.0–2.0% w/w in cured vanilla (Westcott et al., 1994; Bettazzi et al., 2006; Sharma et al., 2006). Thus, extraction of vanillin from the vanilla beans or seed pods of the tropical vanilla orchid is very inadequate due to the existence of very less amount of Vanillin in Vanillin orchid, which covers only 0.2% of the market place requirement and also, the production cost is extremely high (Ibrahim et al., 2009).

On the other hand, lignin is an amorphous reticulated network polymer, composed of phenyl propane units, synthesized by radical coupling of mostly three hydroxypropanoids; coniferyl alcohol, coumaryl alcohol, and sinapyl alcohol (Figure 15.1), where the monomers are linked by a number of ether (which are non-condensed) as well as and C–C (which are condensed) bonds (Christiernin, 2006). Along with cellulose and hemicelluloses, it is one of the three essential components of wood, which forms a special supramolecular architecture (Sarkanen and Ludwig, 1971; Salmé and Burgert, 2008). It is considered as the second largest source of organic raw materials, constituting about 16–25 wt% of hardwoods, 23–35 wt% of softwoods, and 4–35 wt% of most biomass (Bridgwater, 2004; Gosselink et al., 2004). This polymer plays two major functions in the tree; provides rigidity and helps in binding the fibers to each other. Softwood lignin exclusively contains guiacyl type lignins which are derived from coniferyl alcohol. Hardwood lignin contains syringyl units, which are derived from sinapyl alcohol having two methoxy-groups,

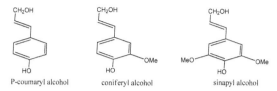

Figure 15.1   Monomers of lignin.

and it results in lignin with a larger number of ether bonds. Hardwood lignin also has an average of 30% guaiacyl units (Christiernin, 2006).

It is a precious resource for various chemicals, energy, and can be used as a potential substrate for the production of vanillin. Lignin-based vanillin is appropriate as a flavoring agent, as it has a more affluent flavor profile than oil-based flavoring agents (Esposito et al., 1997). Oxidation of softwood lignin in basic medium yields up to 23–28 wt.% of vanillin depending on the wood species and oxidation of hardwood lignin yields up to 30–51% of vanillin and syringaldehyde mixture (Heitner et al., 2010). A complex mixture of three aldehydes—vanillin, syringaldehyde, and 4-hydroxybenzaldehyde can be obtained by the oxidation of grass lignins (wheat and rice straw, etc.) and their total yield is lesser in comparison to wood lignin oxidation (Klinke et al., 2002).

The process of isolation of lignocellulosic materials by using the Kraft pulping or sulfate process uses concentrated and strong bases, particularly a mixture of sodium hydroxide and sodium sulfide, known as white liquor (Wallberg and Jönsson, 2006). After processing the lignocellulosic materials by this isolation process, black liquor is obtained which contains lignin fragments and can be used to produce vanillin by means of oxidation (Fargues et al., 1996). However, additional steps are required after the oxidation in order to neutralize the high pH value of black liquor for the vanillin recovery. This process to produce vanillin is highly corrosive, hazardous, and can cause harm to the environment (Hocking, 1997). Hence, it is no longer accepted. Organosolvent process has also been developed in order to isolate lignin (Botello et al., 1999). Nevertheless, because of the use of organic solvents those are highly volatile, highly flammable, and toxic, this process is also not widely accepted. Moreover, the process also usually requires high pressure condition and various advanced equipment for oxidation of lignin due to the low boiling point of organic solvents with high vapor pressure (Barbera et al., 2011).

Ionic liquids can be called green solvents as they have some attractive properties like negligible vapor pressure (so they are non-volatile), environmentally non-toxic, non-flammable, benign, recyclable, and chemically inert (Shamsuri and Abdullah, 2010a,b; Shamsuri and Abdullah, 2011; Shamsuri and Daik, 2012). Ionic liquids are also of very much importance as they are capable of dissolving and blending biopolymers with high efficiency. Also, there is no occurrence of severe side reactions and recovery of the products is very much simple as well (Xie et al., 2005; Shamsuri et al., 2012).

Since ionic liquids offer a potentially clean medium for carrying out chemical reactions or such processes, so much attention has been paid on using these as reaction media for lignocellulosic materials, viz., cellulose and lignin (Pu et al., 2007; Feng and Chen, 2008). In this study, wood dust is used as the raw material. The lignin was isolated by using an ionic liquid from wood dust and the isolated lignin is then oxidized to produce vanillin. We have used a hydrophilic ionic liquid medium and no catalyst is used to avoid possible metal contamination in the product. The product has been characterized by various spectroscopic analyses, viz., Fourier transform infrared, UV–Visible, and mass spectrometry. The antioxidant activity of the vanillin produced is studied using ABTS, DPPH, and hydroxyl radical assays, and the results are compared with standard vanillin.

15.2 Experimental

15.2.1 Materials

1-Butyl-3-methylimidazolium chloride and ABTS were purchased from TCI. Methanol (LR grade), H_2O_2, and $K_2S_2O_8$ were purchased from "Rankem." $K_2H_2PO_4$, chloroform, sodium hydroxide, and potassium hydroxide were purchased from Merck. Deoxyribose and ascorbic acid were purchased from Acros-organic. DPPH, TBA, trichloroacetic acid, and EDTA were purchased from Sigma–Aldrich. $FeCl_3$ was purchased from Fisher Scientific. Wood dust, used in the experiment, was supplied by local sawmill in Dibrugarh. Standard vanillin was purchased from SRL.

15.2.2 Preparation of Ionic Liquid

1-Butyl-3-methylimidazolium hydroxide is prepared by mixing 1-butyl-3-methylimidazolium chloride with KOH in acetone; 5 g of 1-butyl-3-methylimidazolium chloride and 1.6 g of KOH (1:1 equivalent) are dissolved

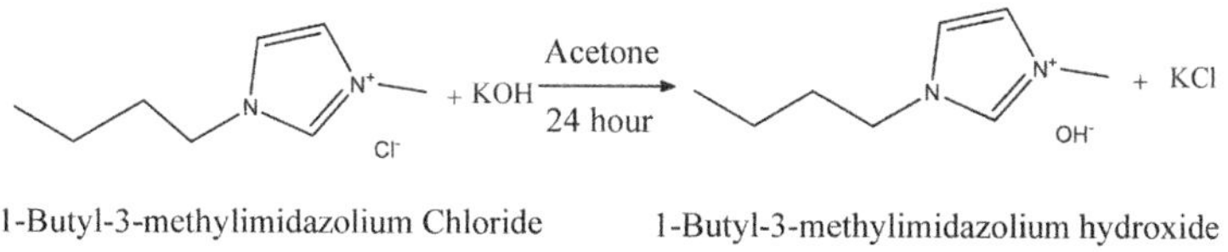

1-Butyl-3-methylimidazolium Chloride   1-Butyl-3-methylimidazolium hydroxide

Scheme 15.1 Preparation of ionic liquid.

in acetone (Scheme 15.1) and allowed to react in a 100 mL round bottom flask placed on a magnetic stirrer for 24 h. After that, the reaction mixture is filtered and the filtrate is evaporated in vacuum to get the pure dark brown ionic liquid.

15.2.3 Isolation of Lignin from Wood Dust

The isolation of lignin from wood dust by the use of an ionic liquid medium is done according to a procedure described by Shamsuri and Abdullah (2012) with slight modification. For the isolation process, the wood dust is grinded and kept in an oven for overnight in order to remove the moisture present. After that, an oven dry weight of 5.2 g of wood dust is placed in a glass flask, and 0.1 moles concentration of 1-butyl-3-methylimidazolium hydroxide is poured into the flask by moles increasing to 0.2, 0.3, 0.4, and 0.5 separately. The flask is allowed to stir with 500 rpm at 25°C for 30 min. After that, the insoluble components in the solution are separated by filtration and soluble lignin is precipitated by adding methanol and thus soluble lignin is separated from the ionic liquid. The isolated compound from the ionic liquid is then filtered off and washed with distilled water for several times. After the filtration, the lignin is collected and dried in an oven at 85°C for at least 24 h until dryness and then weighted. Further adjustments in the duration of isolation process are made based on the higher yield of lignin by changing the time for 60, 90, 120, and 150 min. The temperature for isolation is kept nearly constant for 25°C in that condition and then elevated slowly to 40, 60, 80, and finally to 100°C by immersing in oil bath. The isolation processes are repeated for five times and the average yields are calculated. The dried lignin after isolation is kept in desiccant before further use.

15.2.4 Oxidation of Isolated Lignin to Vanillin

Two grams of the isolated lignin is dissolved into 0.1 moles concentration of 1-butyl-3-methylimidazolium hydroxide in a beaker by moles increasing

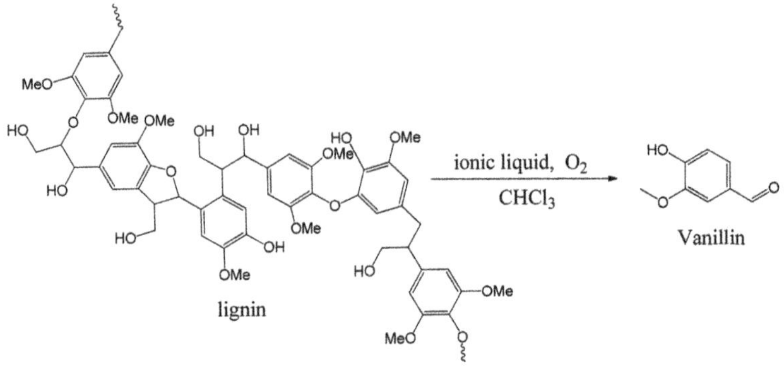

Scheme 15.2   Oxidation of lignin to vanillin.

to 0.2, 0.3, 0.4, and 0.5 separately to find out the maximum concentration of ionic liquid. Then, the solution is transferred into a round-bottom flask and kept on a magnetic stirrer and oxygen gas is passed through the flask for 1 h to oxidize the lignin. Then, the oxidative mixture is mixed with 25 mL chloroform and allowed to react for 30 min. Then the mixture is filtered and the filtrate is transferred to a separating funnel in order to extract the product and to separate any un-oxidized lignin. The solution present at the bottom of the funnel is then collected. After finding out the maximum concentration of ionic liquid, the time period of passing the oxygen through the flask is increased to 2, 3, 4, and 5 h to find out the optimum time period for the oxidation. The oxidation processes (Scheme 15.2) are also repeated for five times and the average yields are calculated. The product is soluble in chloroform and is used for characterization and further experiments.

15.2.5 Determination of Antioxidant Activity

The antioxidant activity of the vanillin produced by the oxidation of lignin is evaluated by DPPH, ABTS, and hydroxyl radical scavenging assay, and the results are compared with the activity of standard vanillin by using their UV absorbance spectra.

15.2.5.1 Measurement of DPPH radical scavenging

Study of the DPPH free radical scavenging activity is a very uncomplicated and widespread method to evaluate the antioxidant activity of an antioxidant. DPPH is easier to deal with as it is much stable compared to oxygen free radicals. Antioxidants provide hydrogen or electrons to free radicals and thereby scavenge them. When DPPH, a stable, purple-colored free radical

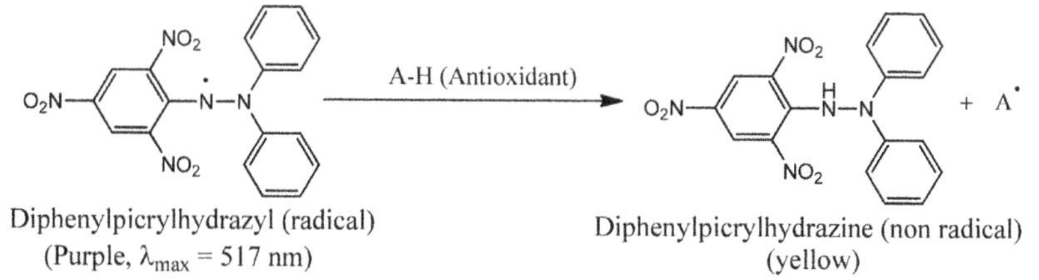

Diphenylpicrylhydrazyl (radical)
(Purple, λ_{max} = 517 nm)

Diphenylpicrylhydrazine (non radical)
(yellow)

Scheme 15.3 Principle of DPPH assay.

reacts with an antioxidant, it gets reduced to yellow-colored 1,1-diphenyl-2-picrylhydrazine. The principle of DPPH Radical Scavenging Assay is given in reaction Scheme 15.3. The reduction of the DPPH free radical is calculated by evaluation of the absorbance at 517 nm.

In DPPH assay, the antioxidant activity of the produced vanillin is measured on the basis of the scavenging activity of vanillin toward stable DPPH free radical according to the method described by Brand-Williams et al. (1995) with slight modification. For the DPPH assay, the stock solution of DPPH free radical is prepared by dissolving 0.004 g in 10 mL methanol in dark. Blank solutions are prepared by adding 200 µL DPPH free radical stock solution in 3 mL of methanol for different concentrations of the vanillin and their absorbances are measured at 517 nm. Then in each blank solution, 100 µL of the produced vanillin as well as standard vanillin with different concentrations (2, 4, 6, 8, and 10 mmol/L of methanol) is added and kept in dark for 30 min. After 30 min, their absorbances are measured at 517 nm.

15.2.5.2 Measurement of ABTS radical scavenging

The ABTS assay is based on the generation of 2, 2'-azinobis-(3-ethylbenzothiazoline-6-sulfonic acid) radical cation. The reaction between ABTS and potassium persulfate produces the blue/green ABTS$^{\cdot+}$ chromophore which has the absorption maxima at wavelengths 645, 734, and 815 nm. Interaction of this pre-formed radical cation with an antioxidant (Principle is given in Scheme 15.4) decreases the amount of radical to an extent, the concentration of the antioxidant, and the time period of the reaction depending on the activity of the antioxidant. Thus, the extent of decolorization of the free radical as percentage inhibition of the ABTS$^{\cdot+}$ radical is calculated as a function of time and concentration and is determined by comparing with the reactivity of the standard vanillin under the same conditions.

To determine the antioxidant activity of the produced vanillin by ABTS method, we followed the method of Arnao et al. (2001) with a little

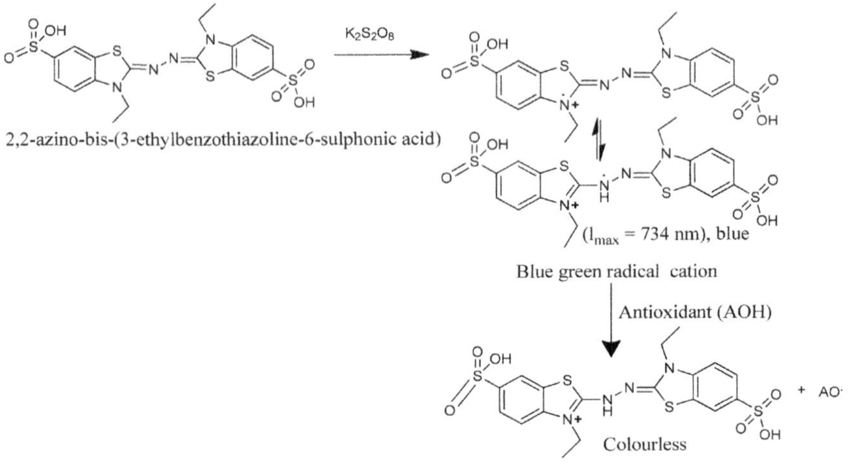

2,2-azino-bis-(3-ethylbenzothiazoline-6-sulphonic acid)

$(\lambda_{max} = 734$ nm), blue

Blue green radical cation

Antioxidant (AOH)

Colourless   + AO·

Scheme 15.4   Principle of ABTS assay.

amendment. For the ABTS assay, the stock solution is prepared by incorporation of 7 mM of ABTS solution as well as 2.4 mM of potassium persulfate solution. Then the working solution of ABTS is prepared by mixing both the stock solutions of ABTS and potassium persulfate in equal quantities and allowed them to react for almost 14–16 h at room temperature in dark. After that, 100 μL ABTS solution is mixed with 6 mL methanol in order to dilute the solution so that we can get an absorbance around 0.706 ± 0.01 units at 734 nm by using a UV spectrophotometer. Then, 2.5 mL ABTS solution is allowed to react with 100 μL of the solutions (prepared in methanol) of the vanillin produced from lignin as well as standard vanillin with four different concentrations (2, 4, 6, 8, and 10 mmol/L of methanol). The absorbance is observed at 734 nm after 6 min for each concentration.

15.2.5.3 Measurement of hydroxyl radical scavenging

Hydroxyl radical is one of the powerful reactive oxygen species in the biological system which reacts with polyunsaturated fatty acid moieties of cell membrane phospholipids and causes harm to cell. Hydroxyl radical scavenging assay is used to find the scavenging activity of free hydroxyl radicals that damage the body cells in the presence of different concentrations of antioxidants. Hydroxyl radical is formed by Fenton's reaction in the presence of Fe^{3+} and H^2O^2.

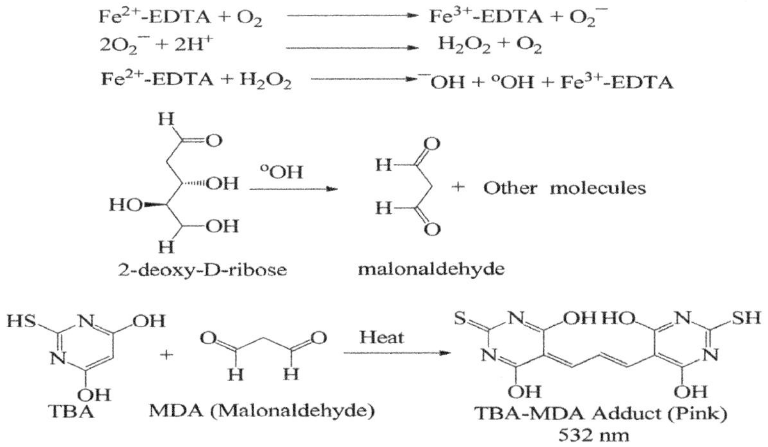

Scheme 15.5 Principle of hydroxyl radical assay.

The hydroxyl radical (·OH) is detected by its capacity to degrade 2-deoxy-d-ribose into fragments that on heating with TBA at low pH forms a pink chromogen with an absorbance at 532 nm. The complete principle of the hydroxyl radical scavenging assay starting from the Fenton's reaction to the formation of TBA-MDA Adduct is given in Scheme 15.5. Decrease of the absorbance at 532 nm is observed in the presence of any antioxidant.

Hydroxyl radical scavenging activity is measured in aqueous medium by using deoxyribose assay (Halliwell et al., 1987). Stock solutions of 10 mM deoxyribose, 10 mM $FeCl_3$, 1 mM EDTA, 1 mM ascorbic acid, and 10 mM H_2O_2 are prepared in distilled water; 1 mL of solution of produced vanillin is taken in various concentrations (2, 4, 6, 8, and 10 mmol/L) with 0.1 mL of EDTA, 0.1 mL of H_2O_2, 0.01 mL of $FeCl_3$, 0.36 mL of deoxyribose, 0.33 mL of phosphate buffer (pH -7.4), and 0.1 mL of ascorbic acid are added to it. The mixture is kept in an incubator for 1 h at 37°C. After that, the above reaction mixture is added to 1 mL of 10% trichloroacetic acid and 1 mL of TBA (0.025 M of NaOH), and heated in a boiling water bath at 80°C for 1 h and a pink chromogen is formed. The absorbance of the sample is measured at 532 nm. The same process is repeated for standard vanillin also.

15.2.5.4 Calculation of percentage inhibition

The percentage inhibition of vanillin by ABTS/DPPH/hydroxyl radical scavenging activity is calculated by using the formula:

$$\text{Percentage inhibition} = A_b - A_s / A_b \times 100\%$$

where

A_b = absorbance of the DPPH/ABTS/hydroxyl radical without sample
A_s = absorbance of DPPH/ABTS/hydroxyl radical with sample

By plotting the values of percentage inhibition as abscissa and the concentrations of vanillin as ordinate, the IC_{50} values of the produced vanillin and standard vanillin are calculated. This quantitative measure indicates how much vanillin is needed to inhibit the growth of a free radical reaction.

15.2.6 Recyclability Test

Ionic liquid is recycled for at least three times by following the same isolation and oxidation procedure thoroughly. The recyclability is measured by comparing the percentage of isolated lignin as well as vanillin with first time isolation and oxidation by intact ionic liquid. The recycled ionic liquid is filtered properly and dried prior to measurement. The ionic liquid thus obtained is utilized for the estimation of mass balances of recycled ionic liquid. The measurement is done as previously, and, if the results are not found to be satisfactory, purification, isolation, as well as oxidation are repeated in quintuplicate. The average yields of isolated lignin and vanillin are being used as comparison.

15.3 Results and Discussion

15.3.1 Optimum Condition for Isolation of Lignin

The weight of the vanillin produced by this method is calculated on an oven-dried (85°C, 24 h) basis. The condition of lignin isolation to get its highest solubility is optimized by altering the concentration of the ionic liquid. The total yield of soluble lignin increases from 0.128 to 0.227 g as the concentration of ionic liquid increases from 0.1 to 0.5 moles at 25°C for 30 min. The highest lignin yield is found at 0.5 moles of the ionic liquid with lignin solubility value of 0.227 g. So, it is found from the study that higher concentration of ionic liquid enhances the solubility of wood due to decrease in the amount of lignin saturation.

After finding out the maximum concentration of ionic liquid required for the highest solubility of lignin, the time optimization is done and found that increase of the duration of isolation increases the yields of soluble lignin. Higher yield of soluble lignin is found to be 0.305 g, for the duration of 120 min in room temperature. After this time period, solubility of the lignin wood samples is found to be almost similar and significant differences are not observed in the yields even after prolong durations. This implies that under the appropriate condition, the release of lignin from the cell walls of wood increases as the duration of isolation is increased.

We have also studied the effect of temperature on optimized reaction conditions and found that the lignin isolation is higher with higher temperature than in lower temperature. Solubility of lignin is increased from 0.305 (25°C) to 0.582 g (60°C), and up to 0.693 g for 100°C, at 120 min with 0.5 moles of ionic liquid. These results signifies that the presence of high temperature may enhance the interaction between lignin and ionic liquid and thus there is an increase in lignin solubility in the ionic liquid as well as the isolation of lignin. As can be seen, isolation of the wood dust (wt.% dry starting material) with 0.5 moles ionic liquid at 100°C for 120 min could be isolated only 13.3wt.% lignin of the wood.

15.3.2 Optimum Conditions for Oxidation of Lignin to Vanillin

During the process of oxidation, lignin undergoes depolymerization and as a result, the rate of depolymerization becomes dependent on the conditions employed for oxidation. During the dissolution process of lignin in ionic liquid, the primary chemical structure of lignin is not changed. However, its primary structure changes as soon as the formation of vanillin takes place with greater oxygen concentration. The reaction time of the oxidative mixture has to be maximized to extend the depolymerization. The temperature has also indicated similar effect on oxidation reaction, whereby high temperature increases the solubility of oxygen in the ionic liquid and enhances the lignin depolymerization rate. In contrast, increase in ionic liquid concentration does not result in improvement of the vanillin production for all investigated parameters; the resultant product is almost the same at the end of the oxidation reaction. On the other hand, this preliminary study also showed that the product acquired from these optimum reaction conditions released a scent identical to the standard vanillin, The percentage yields of isolated lignin from wood dust and produced vanillin by our method using the ionic liquid 1-Butyl-3-methylimidazolium hydroxide are given up to three cycles in the Table 15.1.

Table 15.1   Percentage yield of isolated lignin and produced vanillin up to three cycles

Product	First Cycle (% Yield in wt. %)	Second Cycle (% Yield in wt. %)	Third Cycle (% Yield in wt. %)
Lignin (isolated from wood dust)	13.3	12.6	10.4
Vanillin (yield is calculated compared to the amount of isolated lignin used for oxidation)	23.2	20.8	15.6

15.3.3 Characterization of Vanillin

The production of vanillin is monitored by TLC and the identity of the compound is confirmed by FT-IR, UV–Vis, and mass spectral analyses.

15.3.3.1 Infrared spectrum

The infrared spectrum of the vanillin produced by oxidation of lignin is recorded (Figure 15.2) within the range 400–4000 cm^{-1} using Shimadzu FT-IR spectrophotometer, Model: Prestige 21. In the FT-IR spectrum, a broad band at 3200 cm^{-1} and a sharp peak at 1668 cm^{-1} reveal the presence of a phenolic hydroxyl (O−H stretching) and an aldehyde carbonyl functional groups (C=O stretching), respectively, in the compound. In addition to this,

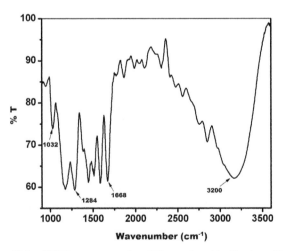

Figure 15.2   FT-IR spectrum of vanillin produced by lignin oxidation.

Table 15.2 FTIR bands of the vanillin produced by oxidation of lignin

Product (cm^{-1})	Band Origin	Functional Group
3200 cm^{-1}	O−H stretching	Phenol
1668 cm^{-1}	C=O stretching	Aldehyde
1284 cm^{-1}	C−O stretching	Ether
1032 cm^{-1}	C−O stretching	Phenol

sharp peaks at 1284 and 1032 cm^{-1} reveal the presence of C−O stretching of ether and C−O stretching of phenolic group. The FTIR bands of the vanillin produced by oxidation of wood dust lignin using ionic liquid medium by our method are given in Table 15.2 along with their band origins and functional groups for which these bands are observed.

15.3.3.2 UV–Vis spectrum

The electronic spectrum of the compound (Figure 15.3) is recorded in methanol within the range of 200–800 nm using Mak-Shimadzu, UV-1700 spectrometer and absorptions are observed at 275 and 309 nm owing to $\pi \rightarrow \pi^*$ and $n \rightarrow \pi^*$ electronic transitions, respectively. The $n \rightarrow \pi^*$ transition is observed due to the involvement of carbonyl group of the aldehyde substituent. The values confirm the formation of vanillin as these absorption values are in concordance with the previously reported values (Oliveira et al., 2014).

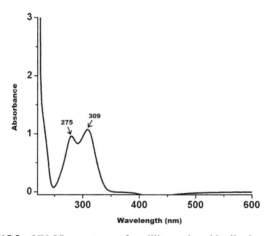

Figure 15.3 UV–Vis spectrum of vanillin produced by lignin oxidation.

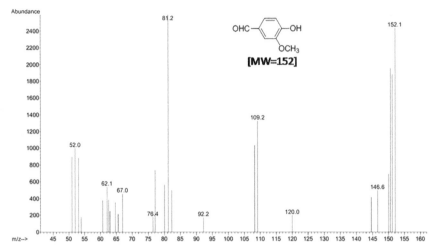

Figure 15.4   Mass spectra of the compound.

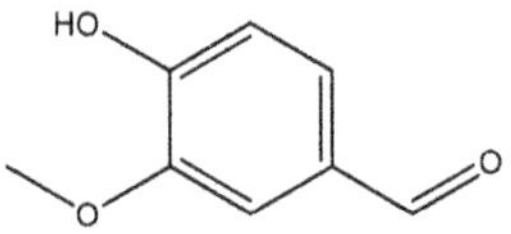

Figure 15.5   Structure of vanillin.

15.3.3.3 Mass spectrum

The identity of the compound was also confirmed by using Agilent GC-7820A/MS5975 analyzer. An m/z value of 152 corresponding to the molecular ion, $[M^+]$ of the desired compound reveals the formation of vanillin (Figure 15.4).

Thus, the produced compound from oxidation of isolated lignin from wood dust by our work is confirmed to be vanillin by IR, UV–Vis, and mass spectral analysis with the following chemical structure (Figure 15.5).

15.3.4 Determination of Antioxidant Activity

Since vanillin is substituted with the phenolic–OH groups, it shows considerable radical scavenging activity. The working out of the radical scavenging capacity is based on the UV absorbance spectrum of the vanillin after the reaction with the DPPH, ABTS, and hydroxyl radical. The radical absorptions

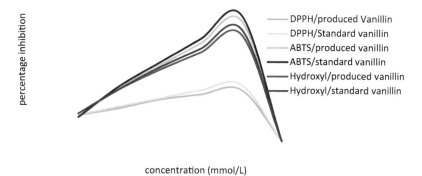

Figure 15.6 Percentage inhibition versus concentration curves of standard vanillin as well as produced vanillin produced by using DPPH, ABTS, and hydroxyl radical scavenging assays.

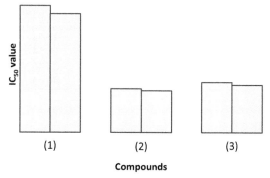

Figure 15.7 IC $_0$ values of produced vanillin and standard vanillin by DPPH (1), ABTS (2), and hydroxyl radical scavenging (3) assays.

are recorded using UV–Visible spectrophotometer, Hitachi model: U-3900H, wavelength range of between 200 and 600 nm. The observations of the experiments conducted to calculate the antioxidant activity of the produced vanillin and standard vanillin by DPPH, ABTS, and hydroxyl radical assays are presented in Figures 15.6 and 15.7 and Table 15.3.

IC$_{50}$ value of the vanillin produced from lignin oxidation by using ionic liquid in our process is very much comparable to that of standard vanillin and thus by using all the three assays, it is found that the antioxidant activity of the vanillin produced by the oxidation of lignin using ionic liquid is almost similar to the antioxidant activity shown by standard vanillin.

Table 15.3   Percentage inhibition and IC_0 values of produced vanillin and standard vanillin

Assays	Concentration (mmol/L)	% Inhibition of Produced Vanillin	IC_0 Values of Produced Vanillin	Percentage Inhibition of Standard Vanillin	IC_0 Values of Standard Vanillin
DPPH assay	2	9.32		9.46	
	4	11.87		12.47	
	6	15.03	32.84	15.48	30.75
	8	16.94		18.47	
	10	19.97		20.49	
ABTS assay	2	8.53		8.54	
	4	19.01		19.59	
	6	27.29	11.35	28.29	10.83
	8	35.45		36.99	
	10	43.65		45.69	
Hydroxyl radical assay	2	9.87		9.87	
	4	18.54		18.54	
	6	25.67	13.03	26.54	12.32
	8	32.08		33.46	
	10	38.69		40.65	

15.4 Conclusion

In conclusion, vanillin has been successfully obtained by the oxidation of lignin from wood dust using an ionic liquid medium. Both the processes, isolation of lignin from wood dust as well as oxidation of lignin to vanillin, are found to be dependent on the reaction conditions. The solubility of the lignin in ionic liquid is found to be enhanced by increase in the concentration of ionic liquid, duration of isolation, and high temperature whereas the oxidation of lignin to vanillin was favored by increased temperature and duration of oxidation, but is independent of ionic liquid concentration. The produced vanillin by our work is characterized by FTIR, UV–Visible, and mass spectrometry. Ionic liquid is recycled for at least three times by following the same isolation and oxidation procedure thoroughly. The ionic liquid used for both the isolation and oxidation process is recycled and the recyclability is calculated by comparing the percentage of isolated lignin and produced vanillin with first time isolation and oxidation product. The antioxidant activity of the produced vanillin is determined by using DPPH, ABTS, and hydroxyl radical scavenging assays and their percentage inhibition and IC_{50} values are determined and compared with standard vanillin. The

IC_{50} values of the produced vanillin are found to be comparable to that of standard vanillin by using all the three assays. Thus, it can be concluded that the produced vanillin by this method has almost equal antioxidant activity to that of standard vanillin.

Acknowledgments

The authors gratefully acknowledge the University Grant Commission-Basic Scientific Research program for providing the financial assistance for the research through the Research fellowship for Meritorious Students (UGC-BSR-RFSMS fellowship) and UGC-SAP-DRS and DST-FIST for analytical and technical support provided for the necessary research work. The authors are also grateful to Mahatma Gandhi University and International and Inter-University Center for Nanoscience and Nanotechnology for organizing the Fourth International Conference on Reuse and Recycling of Materials (ICRM-2018) and giving the opportunity to publish the research work.

References

Arnao MB, Cano A, Acosta M. (2001). The hydrophilic and lipophilic contribution to total antioxidant activity. Food Chemistry. 73:239–44.

Barbera L, Pelach MA, Perez I, Puig J,. Mutje P (2011). Upgrading of hemp core for papermaking purposes by means of organosolvent process. Industrial Crops and Production. 34:865–72.

Bettazzi F, Palchetti I, Sisalli S, Mascini M. (2006). A disposable electrochemical sensor for vanillin detection. Analytica Chimica Acta. 555:134–8.

Botello JI, Gilarranz MA, Rodriguez F, Oliet M. (1999). Preliminary study on products distribution in alcohol pulping of eucalyptus globules. Journal of Chemical Technology and Biotechnology. 74:14–148.

Bridgwater AV. (2004). Biomass fast pyrolysis. Thermal Science. 8:21–49.

Christiernin M. (2006). Lignin composition in cambial tissues of poplar. Plant Physiology and Biochemistry. 44:700–6.

da Silva EAB, Zabkova M, Araojo JD, Cateto CA, Barreiro MF, Belgacem MN, Ranadive AS. (1994). Vanilla* cultivation, curing, chemistry, technology and commercial products. Spices, Herbs and Edible Fungi. 517–6.

Esposito LJ, Formanek K, Kientz G, Mauger F, Maureaux V, Robert G, Truchet F. (1997). 'Vanillin. Vol. 24', Kirk-Othmer Encyclopedia of Chemical Technology, John Wiley & Sons, New York. 24(4):812–5.

Fargues C, Mathias A, Rodrigues A. (1996). Kinetics of vanillin production from kraft lignin oxidation. Industrial Engineering and Chemistry Research. 35:28–36.

Feng L, Chen ZI. (2008). Research progress on dissolution and functional modification of cellulose in ionic liquids. Journal of Molecular Liquids. 142:1–5.

Gosselink RJA, de Jong E, Guran B, Abächerli A. (2004). Co-ordination network for lignin—standardisation, production and applications adapted to market requirements (Eurolignin). Induatrial Crops and Production. 20:121–9.

Halliwell B, Gutteridge JMC, Aruoma OI. (1987). The deoxyribose method: a simple "test-tube" assay for determination of rate constants for reactions of hydroxyl radicals. Analytical. Biochemistry. 165:215–9.

Heitner C, Dimmel D, Schmidt JA. (2010). Lignins and Lignans: Advances in Chemistry. CRC Press: Boca Raton, FL. p. 14.

Hocking MB. (1997). Vanillin: synthetic flavouring from spent sulfite liquor. Journal of Chemical Education. 74:1055–9.

Ibrahim MNM, Sipaut CS, Yusof NNM. (2009). Purification of vanillin by a molecular imprinting polymer technique. Separation and Purification Technology. 66:450–6.

Klinke HB, Ahring BK, Schmidt AS, Thomsen AB. (2002). Characterization of degradation products from alkaline wet oxidation of wheat straw. Bioresource Technology. 82(1):15–26.

Oliveira CBS, Meurer YSR, Oliveira MG, Medeiros WMTQ, Silva FON, Brito ACF, Pontes DL, Neto VFA. (2014). Comparative study on the antioxidant and anti-toxoplasma activities of vanillin and its resorcinarene derivative. Molecules. 19:5898–912.

Pu Y, Jiang N, Ragauskas AJ. (2007). Ionic liquid as a green solvent for lignin. Journal of Wood Chemistry and Technology. 27:23–33.

Rao SR, Ravishankar GA. (2000). Vanilla flavor: production by conventional and biotechnological routes. Journal of Science of Food and Agriculture. 80:289–304.

Rodrigues AE. (2009). An integrated process to produce vanillin and lignin-based polyurethanes from Kraft lignin. Chemical Engineering. Research and Design. 87:1276–92.

Salmé L, Burgert I. (2008). Cell wall features with regard to mechanical performance. a review. COST Action E35 2004 – 2008. Wood Machining – Micromechanics and Fracture, Holzforschung. 63:121–9.

Sarkanen KV, Ludwig CH. (1971). Lignins – Occurence, Formation, Structure and Reactions' Polymer Letters, Wiley-Interscience, New York. pp. 228–30.

Shamsuri AA, Abdullah DK. (2010a). Protonation and complexation approaches for production of protic eutectic ionic liquids. Journal of Physical Science. 21:15–28.

Shamsuri AA, Abdullah DK. (2010b). Isolation and characterization of lignin from rubber wood in ionic liquid medium. Modern Applied Science. 4:19–27.

Shamsuri AA, Abdullah DK. (2011). Synthesizing of ionic liquids from different chemical reactions. Singapore Journal of Science and Research. 1:246–775.

Shamsuri AA, Abdullah DK. (2012). A preliminary study of oxidation of lignin from rubber wood to vanillin in ionic liquid medium. Oxidation Communications. 35(3):767–75.

Shamsuri AA, Abdullah DK, Daik R. (2012). Fabrication of agar/biopolymer blend aerogels in ionic liquid and co-solvent mixture. Cellulose Chemistry and Technology. 46:45–52.

Shamsuri AA, Daik R. (2012). Plasticising effect of choline chloride/urea eutectic-based ionic liquid on physicochemical properties of agarose films. Bioresource. 7:4760–55.

Sharma A, Verma SC, Saxena N, Chadda N, Singh NP, Sinha AK. (2006). Microwave and ultrasound assisted extraction of vanillin and its quantification by high-performance liquid chromatography in Vanilla planifolia. Journal of Separation Science. 29:613–9.

Wallberg O, Jönsson AS. (2006). Separation of lignin in kraft cooking liquor from a continuous digester by ultrafiltration at temperatures above 100°C. Desalination. 195(1):187–200.

Westcott RJ, Cheetham PSJ, Arraclough AJB. (1994). Use of organized viable vanilla plant aerial roots for the production of natural vanillin. Phytochemistry. 35:135–8.

Williams WB, Cuvelier ME, Berset C. (1995). Use of a free radical method to evaluate antioxidant activity. Lebensm.-Wiss. u.-Technology. 28:25–30.

Xie H, Li S, Zhang S. (2005). Ionic liquids as novel solvents for the dissolution and blending of wool keratin fibers. Green Chemistry. 7:606–8.

16

Thermochemical Recycling of Carbon-Based Solid Waste

Juma Haydary

Institute of Chemical and Environmental Engineering,
Faculty of Chemical and Food Technology
Slovak University of Technology, Radlinského 9, Bratislava
E-mail: juma.haydary@stuba.sk

A short review of research done in the field of thermochemical recycling of carbon-based solid waste at the Institute of Chemical and Environmental Engineering of the Slovak University of Technology in Bratislava, from 2005 to 2017, is presented in this work. During these years, thermochemical processing of different types of waste materials using methods such as pyrolysis and gasification has been studied. Characteristics of different waste materials including the behavior of their thermal decomposition, proximate and elemental composition, and their heating values have been provided. Experimental methods and laboratory techniques used for solid waste pyrolysis and gasification as well as analytical techniques used for raw materials and products characterization have been described. As the raw material, waste biomass from agricultural and forestry production, refused derived fuel (RDF), automobile shredder residue (ASR), waste tires, and multilayer packaging materials were used. The work is focused on both pyrolysis and gasification of solid waste. Kinetic parameters of thermal decomposition of different waste materials have been presented. Influence of process conditions on the amount and quality of products and optimization of process conditions were reviewed. Products were characterized by gas chromatography (GC), GC/mass spectrometry (GC/MS), elemental analysis, thermogravimetric analysis, and calorimetric analysis. Mathematical modeling of pyrolysis and gasification processes and computer design of industrial scale pyrolysis and gasification were studied. Tar content of the producer gas and the use of low-cost catalysts for tar reduction have been the subject of many experiments during these years. The effect of low cost catalysts such as dolomite, red clay, and char on the tar content of producer gas was estimated. A novel configuration of a two-stage pyrolysis/gasification process for the production of tar-free high hydrogen content gas has been presented.

16.1 Introduction

Carbon-based solid waste such as MSW, ISW, agricultural solid waste, WB, etc., represents not only a huge environmental challenge worldwide but also a considerable source of useful chemicals and energy. In EU countries, the European Directive on Waste (Directive 2008/98 / EC or Waste Framework Directive) determines the rules and tasks related to solid waste disposal. Many tasks resulting from this directive predetermine the necessity of developing new technologies for waste processing and recovery.

The scheme shown in Figure 16.1 explains the so-called thermochemical recycling. Raw materials and energy are used in industries to produce desired products. After using the products by consumers, large quantities of waste are produced. A part of this waste can be recycled via the mechanical recycling loop, where the addition of energy leads to new products. Unfortunately, the major part of his waste is landfilled or simply dumped in the environment.

An alternative to landfilling is the thermochemical recycling loop. By thermochemical recycling, we can transfer waste again to chemicals and energy, which can be used to produce new products. This includes processes such as pyrolysis, gasification, liquefaction, depolymerization, torrefaction, as well as anaerobic digestion and combustion.

Our focus in this work is on pyrolysis and gasification. These processes of thermochemical conversion have been the subject of study in many research projects (Juma et al., 2006; Ruiz et al., 2013). From 2005 to 2017, thermochemical recycling of different types of waste materials using methods such as pyrolysis and gasification was also the subject of research at the Institute of Chemical and Environmental Engineering of the Slovak University of Technology. In this work, a short review of the results achieved in this field is presented.

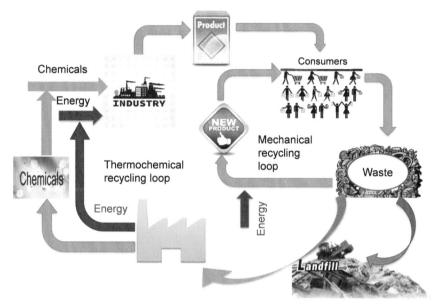

Figure 16.1   Thermochemical recycling loop.

16.2 Raw Materials and Their Characterization

Different types of wastes including MSW and its components, RDF and its components, ASR and its components, waste tires, and different types of WB from forestry and agriculture were the subject of study. We focused on heterogeneous complex waste mixtures and their components with the aim to study the existing challenges in thermochemical recycling of complex mixtures of waste.

All waste types were characterized by thermogravimetric analysis, elemental analysis, calorimetric analysis, and differential scanning calorimetry. Following some selected results of the characterization, studied waste materials are presented. More details can be found in Haydary et al. (2013), Haydary and Susa (2015), Haydary (2016), and Haydary et al. (2016).

16.2.1 Proximate and Elemental Composition

Elemental composition was estimated by a Vario Macro Cube ELEMENTAR elemental analyzer. A CHNS (carbon, hydrogen, nitrogen, sulfur) module with the combustion tube temperature of 1150°C and the reduction tube temperature of 850°C was used. Proximate composition was estimated using

Table 16.1   Proximate and elemental analysis of some studied waste materials

Waste	M	VM	FC	C	H	N	S	O	ASH
Wheat straw	1.91	79.92	15.18	42.8	5.44	0.6	0	46.26	4.9
Barley straw	1.94	85	8.1	49.7	6.9	1	0.7	34.8	6.9
Corn leaves and stalks	6.05	81.92	16	50.88	5.26	0.74	0.19	40.85	2.08
Sunflowers	14.6	82.19	10.5	44.77	5.58	0.26	2.19	38.64	8.56
Wood chips	7.04	85.89	13.15	52.7	5.34	0.5	0	40.5	0.96
ASR-rubber fraction	0.4	46.8	15.6	47.99	4.82	0.31	1.66	8.2	37.2
Polyurethane	1.4	91.4	5.8	61.75	8.43	6.38	0.19	21.84	1.4
ASR-plastic fraction	2.6	91.9	4	68.27	8.88	2.39	0.24	17.6	3.2
ASR-textile fraction	0.9	83.2	10	57.70	4.39	0.98	0.20	30.83	5.9
RDF-white paper	1.5	80.5	1.5	45.58	7.00	0.13	0.05	30.44	16.8
RDF-recycled paper	2.0	76.0	10.8	41.57	5.92	0.70	0.17	41.84	9.8
RDF-textile	4.0	85.0	6.9	51.05	4.92	0.71	0.21	37.46	5.65
RDF-rigid plastics	0.0	70.7	6.3	55.60	13.6	1.24	0.00	6.49	23.6
RDF-foil plastics	0.0	99.2	0.0	79.38	13.63	0.93	0.07	5.42	0.57
Polystyrene	0.0	98.4	0.0	88.66	7.6	0.65	0.06	1.18	1.86

thermal decomposition behavior of waste measured by a thermogravimetric device. Results of proximate and elemental composition for the studied waste types are presented in Table 16.1. All values are dry basis and the content of oxygen is calculated to 100 mass%.

16.2.2 Thermogravimetric Analysis and Kinetics of Thermal Decomposition

A simultaneous TG/DSC analyzer (Netzsch STA 409 PC Luxx. Germany) was used to determine the thermal decomposition kinetics and the content of volatile matter, fixed carbon, and ash. Experimental conditions were met at the linear heating rate of 5–20°C min^{-1} in the nitrogen flow of 60 mL h^{-1}. The samples were heated from 20 to 800°C; at this temperature, they were maintained for around 30 min and then they were combusted by introducing oxygen to the system. An example of the TG/DSC curve recorded for wheat

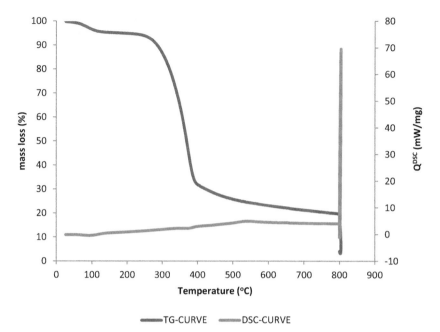

Figure 16.2 TG/DSC record for wheat straw.

straw is shown in Figure 16.2. DTG curves recorded at different heating rates are presented in Figure 16.3. From the thermogravimetric measurements, it results that all components of organic solid waste decompose at temperatures ranging from 250 to 600°C.

Thermogravimetric data were used to estimate kinetic parameters of thermal decomposition using different methods. A simple one-step one-curve method (Equation 16.1), multi-step one-curve method (Equation 16.2), and Friedman multi-curve isoconversional method (Equation 16.3) were applied. It was found that with many waste mixtures, an additive rule based on Equation (16.4) can be applied. Details on thermal decomposition kinetics were published in Koreňová et al. (2006), Koreňová et al. (2007), Haydary et al. (2012), and Haydary and Susa (2013).

$$\frac{d\alpha}{d} = A \exp\left(-\frac{E}{RT}\right)(1-\alpha)^n \tag{16.1}$$

$$\frac{d\alpha}{d} = \sum_{i=1}^{n} A_i \exp\left(-\frac{E_i}{RT}\right)(1-\alpha_i)^{n_i} \tag{16.2}$$

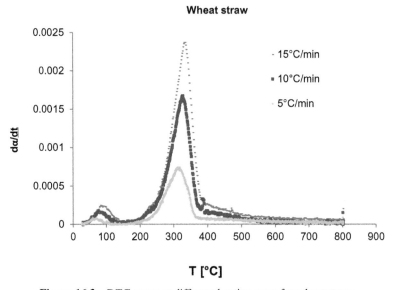

Figure 16.3 DTG curve at different heating rates for wheat straw.

$$\int_0^\alpha \frac{d\alpha}{f(\alpha)} = A \int_0^{t\alpha} \exp\left[-\frac{E}{RT}\right] d \quad\quad (16.3)$$

$$\frac{d\alpha_{ASR}}{d} = \sum_j w_j \frac{d\alpha_j}{d} = \sum_j w_j \left(\sum_{i=1}^n A_i \exp\left(-\frac{E_i}{RT}\right)(1-\alpha_i)^{n_i}\right)_j \quad\quad (16.4)$$

In these equations, A is the apparent pre-exponential factor, E is the apparent activation energy, T is the temperature, n is the reaction order, t is the time, α is the conversion, is the mass fraction, and R is the gas constant. Subscription i represents the reaction step and j the component.

16.2.3 Heating Value

The HHV of all waste components was measured using an FTT isoperibolic calorimetric bomb (Fire Testing Technology Limited). Results for the selected waste types are shown in Table 16.2.

Table 16.2 HHV of selected waste types

RDF Components	HHV (MJ/kg)	Waste Biomass	HHV (MJ/kg)	ASR	HHV (MJ/kg)
RDF-paper	15.24	Wheat straw	16.80	ASR-rubber	21.66
RDF-foil plastics	39.78	Barley straw	16.96	ASR-foam	29.39
RDF-rigid plastics	38.38	Corn leaves and stalks	16.97	ASR-plastics	35.00
RDF-textile	23.20	Sunflowers	14.68	ASR-textile	

16.3 Laboratory Scale Pyrolysis and Gasification Units

Pyrolysis of waste was carried out in a laboratory unit with a screw-type reactor. Two different modifications of the laboratory pyrolysis apparatus were developed; the first one, without a secondary catalytic stage, was used when maximization of the oil product was required, while the second one, with a secondary catalytic reactor, was applied for cases when maximization of gas product was required. Schemes of both laboratory scale pyrolysis units are shown in Figure 16.4. Inert atmosphere in the pyrolysis system was ensured by the nitrogen flow. Details of the pyrolysis units and experimental conditions used can be found in Juma et al. (2007), Haydary et al. (2009), and Susa and Haydary (2014, 2015).

Gasification experiments were realized in a two-stage laboratory batch gasification unit and in a two-stage gasification unit with a screw reactor and one or two fixed bed catalytic reactors. As gasification agents, air, pure oxygen, and oxygen enriched air were used. The units enabled estimation of

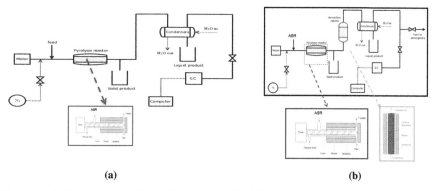

(a)   **(b)**

Figure 16.4 Laboratory scale pyrolysis units: (a) without a secondary catalytic reactor and (b) with a secondary catalytic reactor.

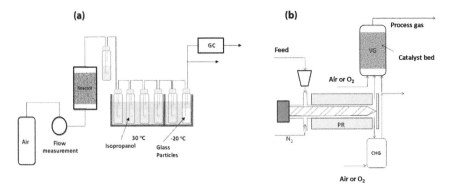

Figure 16.5   Laboratory gasification units: (a) batch reactor unit with a tar condensation system and (b) two-stage gasification unit with a secondary catalytic reactor for volatile fraction gasification and a secondary reactor for solid fraction gasification.

the producer gas tar content and gas composition. In Figure 16.5, schemes of both types of laboratory scale gasification units are presented. More details can be found in Haydary and Jelemenský (2014), Haydary (2016), and Haydary and Šuhaj (2017).

16.4 Catalysts and Their Characterization

Low-cost catalysts were prepared from materials such as dolomite, red clay, and pyrolytic char (Figure 16.6) by calcination of minerals or carbonization of char and impregnation with metals, such as Ni, or by acid activation. The prepared catalysts were characterized by pore structure and specific surface measurements, X-ray diffraction analysis, elemental composition, and thermogravimetric measurements.

Figure 16.6   Low-cost catalysts used in waste gasification.

Specific surface of dolomite-based catalyst varied between 10 and 15 m^2/g. A disadvantage of dolomite is the loss of mechanical strength after calcination. Raw clay catalyst has a specific surface area above 100 m^2/g which is, however, reduced to 20–30 m^2/g after calcination. Char catalyst has a specific surface area of around 40 m^2/g which increased by carbonization and activation to 90 m^2/g.

16.5 Pyrolysis Yields and Products Composition

Products of the pyrolysis and gasification experiments were characterized by different analytical methods. GC and an online gas micro-chromatograph were used for the determination of gas composition. GC/MS was used for the analysis of the pyrolytic liquid product. Elemental composition and heating value of both solid and liquid products were estimated by elemental analysis and bomb calorimetric analysis, respectively. A standard gravimetric method was applied to determine the tar content in gas.

Pyrolysis products were obtained in all three phases: solid (15–40%), liquid (10–50%), and gaseous (20–60%), in dependence on the waste type and process conditions. For many types of solid wastes, the optimum pyrolysis temperature has been found to be 550°C (Juma et al., 2006; Haydary and Susa, 2013; Haydary, 2016; Haydary et al., 2016). Reactor type and process conditions such as temperature, residence time, particle size, and inert gas flow rate can significantly influence the pyrolysis yields and product composition. Particle size has a crucial effect on the time required for reaching complete conversion (Haydary et al., 2012). The use of catalysts has significantly reduced the liquid yield and increased the gas yield. A comparison of catalytic and non-catalytic pyrolysis of wood chips biomass is shown in Figure 16.7. The gas produced contained mainly H_2, CO, CO_2, CH_4, and other hydrocarbons. Figure 16.8 shows the composition of gas obtained from the ASR pyrolysis. For other types of wastes, the trends are similar; however, gas from the pyrolysis of oxygenated waste such as lignocellulosic materials contains more oxides of carbon.

The oil product of waste pyrolysis is a very complex mixture of organic chemicals; its amount and composition are significantly affected by the process conditions such as pyrolysis temperature, feedstock size, inert gas flow rate, heating rate, and vapor cooling temperature. Optimization of oil yield was the subject of study in Mandal et al. (2018).

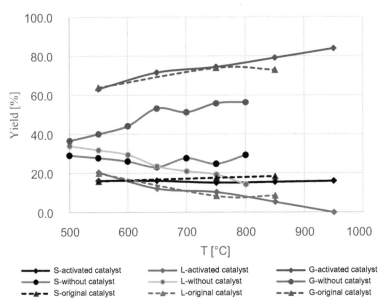

Figure 16.7   Product yields versus temperature for catalytic and non-catalytic pyrolysis of wood chips.

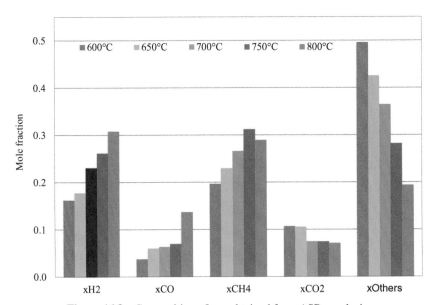

Figure 16.8   Composition of gas obtained from ASR pyrolysis.

Pyrolysis char from different types of wastes and biomasses has a great potential to be applied in different applications. Tire pyrolysis char has been studied for the use as reinforcement in new rubber products, as a low-cost adsorbent or catalyst (Korenova et al., 2006). Sulfur distribution into the products of waste tire pyrolysis was studied in Susa and Haydary (2013). Sulfur present in tire is distributed in all pyrolysis products; however, the sulfur concentration in pyrolysis char is higher than in raw tire. Biochar from WB can be used as bio-fuel or as an acid soil ameliorant. The average lower heating value of WB char is around 25 MJ/kg while tire pyrolysis char has a lower heating value, 30 MJ/kg.

16.6 Gasification Products and Their Composition

Raw gas produced by the gasification of all types of waste contains large amounts of tar. Therefore, a two-stage gasification system enabling secondary catalytic tar decomposition has been developed. Secondary catalysis effectively reduces tar content in the produced gas. Table 16.3 shows a comparison of tar yields in gas produced from the RDF gasification at different temperatures and the catalyst types used in the secondary reactor. In these experiments, red clay catalyst without Ni (RC) and Ni impregnated red clay (RC/Ni) were used. A similar catalytic effect was observed also for other types of waste being gasified.

Impregnation of catalysts with Ni significantly increased the content of H_2 in the produced gas and reduced the content of hydrocarbons. Table 16.4 shows the content of the major components of gas from the RDF gasification. When using Ni impregnated clay catalyst in the secondary reactor at 750°C, the content of hydrogen increased four times and that of CO more than five times compared to no catalyst introduced in the secondary reactor; all other conditions were the same.

Table 16.3 Tar yields in mg/g of RDF for non-catalytic and catalytic process

Temperature (°C)	700	750	800	850	900
Non-catalytic	9.04	9.39	7.32	8.77	8.48
RC	1.87	1.96	1.56	1.3	1.89
RC/Ni	0.81	0.62	0.76	0.63	0.71

Table 16.4   Content of major components of gas from RDF gasification

Component	Experiment	700 (°C)	750 (°C)	800 (°C)
Hydrogen	Non-catalytic	NA	6.84	10.3
[vol. %]	RC	20.8	17.8	15.6
	RC/Ni	36.1	27.7	20
Methane	Non-catalytic	NA	4.07	6.18
[vol. %]	RC	12.1	12.1	9.56
	RC/Ni	5.85	4.49	4.61
Hydrocarbons	Non-catalytic	NA	8.27	12.4
[vol. %]	RC	14.7	15.1	13.9
	RC/Ni	4.16	5.31	6.76
Carbon monoxide	Non-catalytic	NA	3.12	8.2
[vol. %]	RC	4.39	6.18	7.9
	RC/Ni	16.4	17.6	14.8
Carbon dioxide	Non-catalytic	NA	17.2	17.2
[vol. %]	RC	15.1	13.8	13.7
	RC/Ni	7.25	7.06	10.8
Nitrogen	Non-catalytic	NA	60.5	45.8
[vol. %]	RC	35.2	35.1	39.7
	RC/Ni	30.3	37.8	43.1

16.7 Computer-Aided Modeling of Waste Pyrolysis and Gasification

Modeling of thermochemical recycling processes enables predicting optimal process conditions and reducing the number of experiments in the process of design and operation. In modeling industrial scale pyrolysis and gasification units, in home developed models and experimental data obtained in the laboratory were combined with commercial simulators. The pyrolysis stage was modeled by the lump kinetic scheme described in Haydary (2017). The gasification process configuration shown in Figure 16.5b enables considering chemical equilibrium in the gasification step and thus calculating the equilibrium constant calculated based on the standard Gibbs free energy. However, the equilibrium model was combined with some empirical correlations based on laboratory experiments to predict the tar content of the gas. The proposed model is capable of predicting mass and energy balances of both steps, producer gas composition, gasification reactor temperature, gas heating value, and gas tar content using different types and amounts of the gasification agents.

The user model of pyrolysis stage was integrated into the Aspen Plus environment to carry out different case studies to optimize the process conditions.

Figure 16.9 shows gas composition versus air to RDF mass ratio ($m_{(Air)}/m_{(RDF)}$). It was found that optimum oxidizing agent to waste mass ratio is dependent on the feed elemental composition and type of the oxidizing agent. For RDF used in this work and air as the oxidizing agent, the optimum air to RDF mass ratio was 2.1.

The content of H_2 showed a maximum at $m_{(Air)}/m_{(RDF)} = 1$. However, from Figure 16.10, it results that at these values of $m_{(Air)}/m_{(RDF)}$, the

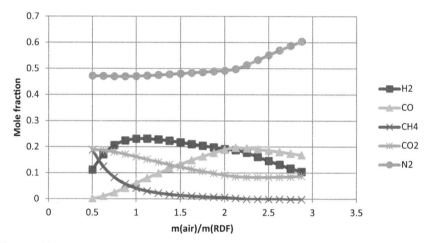

Figure 16.9 Gas composition at different $m_{(Air)}/m_{(RDF)}$ values using air as the gasification agent.

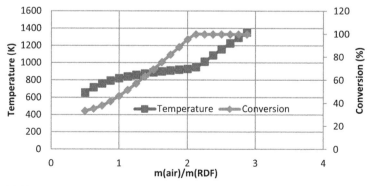

Figure 16.10 Conversion and reactor temperature at different $m_{(Air)}/m_{(RDF)}$ values using air as the gasification agent.

adiabatic reactor temperature is around 850 K and the conversion of RDF is only 50–60%. A 100% conversion was achieved at $m_{(Air)}/m_{(RDF)} = 2.1$ at a reactor temperature of around 950 K. More details on computer-aided modeling of pyrolysis and gasification have been published in Haydarý and Jelemenský (2014) and Haydary (2018).

16.8 Conclusion

Thermochemical recycling has a great potential to convert organic solid waste into valuable products and energy. All solid, liquid, and gaseous pyrolysis products can be used as a source of chemicals or as fuels. Both amount and composition of the pyrolysis and gasification products are strongly influenced by process conditions and the technology used. The use of catalysts in the secondary reactor can significantly reduce the tar content of the produced gas and increase its hydrogen content. Impregnation of catalysts with Ni significantly increased the content of H_2 in the produced gas and reduced the content of hydrocarbons. The two-stage pyrolysis/gasification system shows an incredible potential to produce low tar content gas with high content of H_2 and CO. Optimum oxidizing agent to waste mass ratio is dependent on the feed elemental composition and the type of the oxidizing agent and it decreases with the increasing concentration of oxygen in the oxidizing agent and increases with the increasing H_2 and C content in the feed.

Acknowledgment

This work was supported by Grant APVV-15-0148 provided by the Slovak Research and Development Agency.

References

Haydary J. (2016). Gasification of refuse-derived fuel (RDF). GeoScience Engineering 1. 62(1):37–44.

Haydary J. (2017). Modelling of two stage gasification of waste biomassm. Chemical Engineering Transactions 1. 61:1465–70.

Haydary J. (2018). Aspen simulation of two-stage pyrolysis/gasification of carbon based solid waste. Chemical Engineering Transactions. 70:1033–8.

Haydary J, Jelemenský Ľ. (2014). Design of biomass gasification and combined heat and power plant based on laboratory experiments. In Vol. 155 Springer proceeding in Physica ISBN 9783319055206, DIO:10.1007/978331905521-3

Haydary J, Jelemenský Ľ, Gašparovič L, Markoš J. (2012). Influence of particle size and kinetic parameters on tire pyrolysis. Journal of Analytical and Applied Pyrolysis. 97:73–9.

Haydary J, Jelemenský Ľ, Markoš J, Annus J. (2009). A laboratory set-up with a flow reactor for waste tire pyrolysis. KGK-KAUTSCHUK GUMMI KUNSTSTOFFE. 62(09):661–5.

Haydary J, Šuhaj P. (2017). Two stage pyrolysis/gasification of solid waste for tar free high hydrogen content syngas production. In 2017 Sustainable Industrial Processing Summit And Exhibition, Volume 5: New and Advanced Materials & Technologies for Energy, Environment and Sustainable, 135–44.

Haydary J, Susa D. (2013). Kinetics of thermal decomposition of aseptic packages. Chemical Papers. 67(12):1514–20.

Haydary J, Susa D. (2015). Characterization of automobile shredder residue for purpose of its thermal conversion. Journal of Solid Waste Technology and Management. 41:41–9.

Haydary J, Susa D, Dudáš J. (2013). Pyrolysis of aseptic packages (tetrapak) in a laboratory screw type reactor and secondary thermal/catalytic tar decomposition. Waste Management. 33:1136–41.

Haydary J, Susa D, Gelinger V, Čacho F. (2016). Pyrolysis of automobile shredder residue in a laboratory scale screw type reactor. Journal of Environmental Chemical Engineering. 4(1):965–72.

Juma M, Koreňová Z, Markoš J, Annus J, Jelemenský Ľ. (2006). Pyrolysis and combustion of scrap tire. Petroleum and Coal. 48(1):15–26.

Juma M, Koreňová Z, Markoš J, Jelemenský L, Bafrnec M. (2007). Experimental study of pyrolysis and combustion of scrap tire. Polymers for Advanced Technologies. 18(2):144–8.

Koreňová Z, Haydary J, Annus J, Markoš J, Jelemenský Ľ. (2007). Pore structure of pyrolyzed scrap tires. Chemical Papers. 62(1):86–91.

Koreňová Z, Juma M, Annus J, Markos J, Jelemeneský Ľ. (2006). Kinetics of pyrolysis and properties of carbon black from a scrap tire, Chemical papers-Chemicke zvesti. 60(6):422–6.

Mandal S, Haydary J, Bhattacharya TK, Tanna HR, Husar J, Haz A. (2018). Valorization of pine needles by thermal conversion to solid, liquid and gaseous fuels in a screw reactor. Waste and Biomass Valorization. 1–3.

Ruiz JA, Juárez MC, Morales MP, Muñoz P, Mendívil MA. (2013). Biomass gasification for electricity generation: review of current technology barriers. Renewable and Sustainable Energy Reviews. 18:174–83.

Susa D, Haydary J. (2013). Sulphur distribution in the products of waste tire pyrolysis. Chemical Papers. 67(12):1521–6.

Susa D, Haydary J. (2014). Tire pyrolysis in a continuous screw reactor. KGK-Kautschuk Gummi Kunststoffe. 67(1–2):53–6.

Susa D, Haydary J. (2015). Pyrolysis of biomass in a laboratory pyrolysis unit with a screw type reactor and a secondary decomposition reactor. Sustainable Environment Research. 25(5).

17

Fabrication and Characterization of Hair Keratin–Chitosan-Based Porous Scaffolds for Biomedical Application

Keshaw Ram Aadil* and Pratima Gupta*

Department of Biotechnology, National Institute of Technology Raipur, Raipur, 492010, Chhattisgarh, India
E-mail: kaadil7@gmail.com (KRA); pgupta.bt@nitrr.ac.in (PG)
*Corresponding Authors

Keratin extracted from human hair signifies a valuable source of biocompatible and biodegradable biopolymers. The aim of the present study was to fabricate keratin-Chitosan-based porous scaffolds using non-toxic cross-linkers and investigate its physico-chemical properties for biomedical and tissue engineering applications. Porous scaffolds were fabricated via freeze extraction methods. Keratin extracted from the human hair was blended with chitosan in different ratios using two different cross-linkers (glyoxal and 1-ethyl-3-(3-dimethylaminopropyl)-3-ethylcarbodiimide hydrochloride (EDC.HCL). Various physico-chemical properties (water solubility, swelling, and degradations) of the scaffolds were studied. The porosity of scaffolds was confirmed by scanning electron microscopy (SEM). FTIR analysis confirmed the amine/carboxyl groups of chitosan reacted with amine group of keratin in the presence of cross-linkers. Water solubility ranged from 9 to 69% and the least solubility was shown by glyoxal cross-linked chitosan–keratin (50:50) scaffolds (9.4%). Degradation study confirmed that the fabricated scaffolds were stable in the aqueous medium without any significant weight loss. It was concluded that the glyoxal cross-linked chitosan–keratin scaffold exhibited better physical properties compared to EDC cross-linked, and hence, would be a suitable biomaterial for biomedical and soft tissue engineering applications.

Images of chitosan-keratin blend   SEM Images of chitosan-keratin
                                                              scaffolds

Chitosan

17.1 Introduction

Presently, more attention has been given to naturally derived biologically active materials that display better biocompatibility and biodegradability for biomedical and tissue engineering applications. These materials mimic ECM and support growth and proliferation of new cells or tissues. Several abundant natural biopolymers such as alginate, cellulose, collagen, chitosan, gelatin, and starch are extensively used for the preparation of various value-added products for biomedical and tissue engineering applications (Croisier and Jérôme, 2013; Kakkar et al., 2014).

Among the natural biopolymers, keratin (45–60 kDa) is a family of fibrous proteins, which is found abundantly in nature. It forms the main constituent of hair, wool, nail, horn, and hooves of mammals, birds, and reptiles. Keratin can be easily obtained as a byproduct from the textile industries, poultry farms, hair salon, and feather from butchery (Li and Yang, 2014). Additionally, India alone exported approximately 1 million kilograms of hair keratin (Kakkar et al., 2014). Thus, it is of great interest to study the recycling of keratin for value-added products.

Keratin also contains cysteine amino acid residues (7–20%), which would be very useful for cell attachment and growth. The oxidation of these cysteine residues leads to inter- and intra-molecular covalent bonds, which are respon-sible for the toughness of the keratin fibers (Dowling et al., 1986; Kakkar et al., 2014). Moreover, it has cell adhesion sequences (RGD) (Arg-Gly-ASP) and LVD (Leu-Asp-Val) and cell binding motifs, which are also found in ECM proteins (Marshall et al., 1991; Kakkar et al., 2014).

Previously, keratin has been blended with different polymers (collagen, chitosan, bacterial cellulose, poly(vinyl alcohol), and poly(lactic acid)

(Kakkar et al., 2014; Li and Yang, 2014; Tanase and Spiridon, 2014; Keskin et al., 2017) for making improved scaffolds. However, the major drawback of this construct is its low physical properties and high brittleness (Li and Yang, 2014; Tanase and Spiridon, 2014). Such drawbacks can be overcome by cross-linking the biopolymeric components while blending and fabricating scaffolds, which has been reported to significantly improve physical and chemical properties of the scaffolds (Hua et al., 2016; Sanchez Ramirez et al., 2017). But most of the cross-linking agents are toxic to humans, and thus the use of non-toxic cross-linking agent has gained much attention in the cross-linking of biopolymers-based scaffolds particularly for biomedical applications. Glyoxal and EDC are both non-toxic and biocompatible cross-linking agents which offer a safe and effective method to cross-link amino linkage (Hua et al., 2016).

Chitosan is a polysaccharide obtained by the deacetylation of chitin and has many applications due to its low price, good oxygen barrier properties, biodegradability, biocompatibility, antimicrobial activity, and non-toxicity (Ahmed and Ikram, 2016; Sen et al., 2017; Xu et al., 2018).

With these backgrounds, in the present work, attempts were made to develop hair keratin–chitosan-based porous scaffolds for its biomedical applications. Glyoxal and EDC were used as cross-linking agents to improve their properties. The physicochemical and structural properties of the developed scaffolds were investigated.

17.2 Materials and Methods

17.2.1 Materials

Keratin (K) was extracted from human hair, collected from local salon shop via Shindai extraction method as described by Nakamura et al. (2002). Chitosan (CH), glyoxal, and EDC were purchased from Sigma–Aldrich, USA. Urea, Thiourea, β-mercaptoethanol, Tris HCl, and SDS were purchased from SRL Chemicals, India. Ultra-pure Milli Q water was used throughout the experiment. All other chemicals used were of analytical grade.

17.2.2 Extraction of Human Hair Keratin

Human hair was collected from a local hair salon of Rajnandgaon district of Chhattisgarh, India. Hair was washed extensively with 0.5% SDS (w/v) to remove dust and air dried. Further, dried hair was treated with hexane to remove surface grease and lipid for 3 h under shaking conditions.

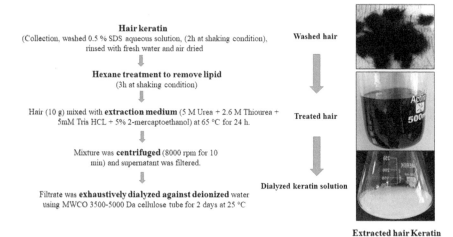

Figure 17.1 Schematic diagram for the extraction of keratin from human hair.

The sample was dried at room temperature for 12 h and cut into 4–5 mm length. To extract keratin, an amount of 10 g of cleaned hair was treated with 300 mL of a solution containing urea (5 M), 2.6 M Thiourea, 5 mM Tris HCl, and 5% 2-mercaptoethanol under shaking for 12 h at 65°C. The mixture was filtered with a muslin cloth and centrifuged at 8000 rpm for 10 min to remove insoluble matrix. Keratin aqueous solution obtained was dialyzed against distilled water with cellulose tube (3500 Dalton molecular weight cut off) for 2 days at room temperature, changing the distilled water frequently. To the end, keratin solution was lyophilized to obtain soluble pure keratin (Nakamura et al., 2002). The schematic diagram for the extraction of keratin is shown in Figure 17.1.

17.2.3 Preparation of Chitosan–Keratin Blend

In order to prepare chitosan–keratin blend, chitosan solution of 1% (w/v) was prepared by dissolving chitosan in 1% aqueous acetic acid solution. After that, pre-extracted hair keratin solution was blended with chitosan solution in different ratios: 25:75, 50:50, and 75:25. The resulting blend was homogenized in magnetic stirrer at 60°C for 1 h. Further, two different cross-linkers EDC (5 mM) and glyoxal (2% w/v) were used as cross-linking agents. After homogenization, the blend solution was placed at −20°C for 24 h. The frozen scaffolds were removed from the vials and immersed in pre-chilled acetone and kept at −20°C for 2 days. The scaffolds were taken out from the vials and dried at 60°C for 4 h. Finally, the samples were stored at room temperature for further characterization.

17.2.4 Characterization of Chitosan–Keratin Scaffolds

17.2.4.1 SEM analysis

Morphology characterization of scaffolds was studied by SEM analysis (JEOL JSM-6390LV). Samples were mounted on a metal stub and gold coated by using sputter coating technique for 20 s to make them conducting. Images of scaffolds were taken at 20 kV accelerating voltage at different magnifications. The pore size was calculated using Image J software.

17.2.4.2 FTIR analysis

Structure and chemical characterization of fabricated keratin–chitosan scaffolds were investigated by FTIR analysis (Perkin Elmer Spectrum One FTIR spectrophotometer) at a resolution of 4 cm^{-1} in the wavenumber region 400–4000 cm^{-1}. Spectra of samples were obtained from discs containing 1.0 mg sample in approximately 100 mg potassium bromide (KBr). The samples were conditioned at 55% RH and 25 $\pm$ 2°C.

17.2.4.3 XRD analysis

Wide-angle XRD study of keratin and keratin–chitosan scaffolds was carried out in order to examine the crystallinity of the samples. The samples were taken in powdered form and analyzed by using a PANalytical 3kW X'pert diffractometer system using Cu kα as X-ray source.

17.2.4.4 Water solubility test

The preliminary dry matter content of scaffolds was determined by drying to constant weight in an oven at 105°C. The scaffolds (1 $\times$ 1 cm^2), weighed (M_i), and immersed in 50 mL of water for 24 h at room temperature (25 $\pm$ 2°C) with agitation (60 rpm). The solution was then filtered through Whatman No. 1 filter paper to recover the remaining scaffolds, which was desiccated at 105°C for 24 h and weighted (M_f). The solubility of the mat was calculated as follows:

$$\text{Water solubility}(\%) = (W_i - W_f)/W_i \times 100$$

where W_i is the initial mass of the mat expressed as dry matter and W_f is the final mass of the remained desiccated mat (Aadil et al., 2016). All experiments were carried out in triplicate.

17.2.4.5 Swelling properties test

Swelling property of the scaffolds was studied as described by Aadil et al. (2016). Samples (1×1 cm^2 in size) were immersed in 25 mL of deionized water at room temperature ($25 \pm 2°C$) under shaking conditions. The weight gain of swollen films (W_s) was measured intermittently, after blotting the surface with Whatman No. 1 filter paper. The SR was calculated using the following equation:

$$\text{SR} = (W_s - W_d)/W_d \times 100$$

where W_s is the weight of swollen samples (g) and W_d is the weight of dry samples (g) (Aadil et al., 2016). The measurements were repeated three times and an average was taken as the result.

17.2.4.6 Biodegradability assay

The biodegradation of the scaffolds was studied by incubating with phosphate buffer for 1, 3, 7, and 14 days, as described by Aadil et al. (2018). Briefly, the chitosan–keratin scaffolds were vacuum dried overnight and the initial dry weight (W_i) of each sample (1×1 cm^2) was measured. The scaffolds were then immersed in PBS with shaking at 37°C. After 1, 3, 7, and 14 days of incubation, the scaffolds were vacuum dried overnight again and the residual dry weight (W_f) was measured. The remaining weight percentage was used as the index of the stability of the mats in an aqueous environment and calculated using the following formula:

$$\text{Remaining weight percentage } (\%) = W_f/W_i \times 100$$

17.3 Results and Discussion

Porous keratin–chitosan scaffolds via freeze-drying method were successfully fabricated. The prepared scaffolds were uniform in shape as shown in Figure 17.2. Scaffolds were flexible or spongy in nature after drying. Further characterization of keratin–chitosan scaffolds is discussed in the following section.

17.3.1 SEM Analysis

The surface morphology of keratin–chitosan scaffolds is shown in Figure 17.3. SEM images presented the non-uniform porous structure of scaffolds. The average diameter of pores ranged from 200 to 300 μm.

[A]

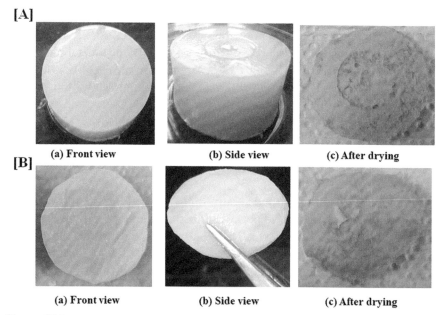

(a) Front view　　　　(b) Side view　　　　(c) After drying

[B]

(a) Front view　　　　(b) Side view　　　　(c) After drying

Figure 17.2　Photograph of prepared keratin–chitosan scaffolds: (A) glyoxal cross-linked and (B) EDC cross-linked scaffolds.

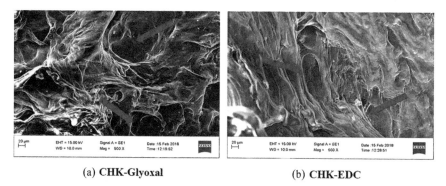

(a) **CHK-Glyoxal**　　　　　　　　(b) **CHK-EDC**

Figure 17.3　SEM micrograph of keratin–chitosan scaffolds: (a) CH-K-glyoxal and (b) CH-K-EDC cross-linked scaffolds.

The non-uniform pores of scaffolds might be due to the effect of drying, which also reduced the pore diameter. Compared to CH-K-EDC scaffold, CH-K-glyoxal shows partially spherical well-interconnected pores, which would be useful for cell growth, proliferation and migration during cell culture, and the wound healing process. The previous study of collagen–chitosan scaffolds also suggests that porous scaffolds would be important for wound healing and tissue engineering applications (Balajia et al., 2012; Kakkar et al., 2014). Porous scaffolds secure the diffusion of nutrients, metabolites, and biochemical cues for cell growth and differentiation (Xu et al., 2013). The porosity of previously reported different keratin-based scaffolds were found between 70 and 90% and the least porosity was observed in wool keratin scaffolds (69.5%) (Xu et al., 2013; Gupta and Nayak, 2015; Esparza et al., 2018).

17.3.2 FTIR Analysis

FTIR spectra of keratin, chitosan, and keratin–chitosan scaffolds are shown in Figure 17.4. The keratin distinctive peaks ascribed to Amide A, Amide I, II, and III were observed at 3337, 1652, 1540, and 1220–1300 cm^{-1}, respectively (Figure 17.4A) (Zeng et al., 2004; Kakkar et al., 2014). The peak of amide A is due to stretching vibrations of N–H bonds, which observed at 3337 cm^{-1}. Further, the peak of amide I appeared at 1652 cm^{-1} is because of the stretching vibration of C–O bonds. Amide II bonds were

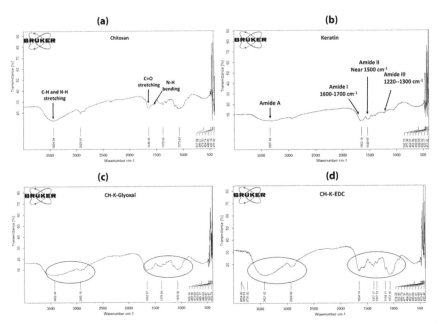

Figure 17.4 FTIR spectra of chitosan (a), keratin (b), CH-K-glyoxal (c), and CH-K-EDC (d) cross-linked scaffolds.

observed at 1540 cm^{-1} due to N–H bending and C–H stretching vibration. The characteristic peak of chitosan was observed at 3434 cm^{-1} ascribed to C–H and N–H group of chitosan. Peaks at 2923, 1648, and 1378 cm^{-1} were attributed to C–H, C=O, and N–H bending of chitosan, respectively (Kakkar et al., 2014; Sanchez Ramirez et al., 2017). After blending of keratin with chitosan in the presence of cross-linkers (glyoxal and EDC), the change of peaks intensity and shifting of the peaks were observed. In the presence of glyoxal, the major changes were observed at 3422, 2925, and 1600–1000 cm^{-1} suggesting the hydrogen bonding and amide linkage between keratin and chitosan. In the presence of EDC, a broad peak at 3421 cm^{-1} was appeared, which represented to C–H and N–H groups of chitosan. Further, changes were observed from 1600 to 1000 cm^{-1}, attributed to C=O and Amide II and III groups of chitosan and keratin. At 1317, 1154, and 1072 cm^{-1}, new peaks are appearing on the addition of EDC suggesting the hydrophobic interaction of keratin and chitosan in the presence of EDC (Figure 17.4D). The major changes and interaction were observed in glyoxal cross-linked scaffolds compared to EDC, suggesting the higher degree of interaction between keratin and chitosan molecules.

17.3.3 XRD Analysis

Figure 17.5 shows the XRD spectra of keratin and keratin–chitosan scaffolds. The keratin powder presented a small peak at about 23° and 26° (2θ), corresponding to the β-sheet structure (Li and Yang, 2014). These peaks

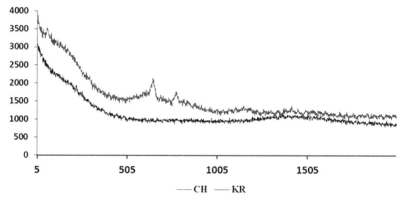

Figure 17.5 XRD spectra of keratin (A) and CH-K-glyoxal (B) cross-linked scaffolds.

were disappeared in the keratin–chitosan scaffolds cross-linked with gly-oxal (Figure 17.5). This evidence suggested that the strong interaction occurred between keratin and chitosan molecules in the blend in the presence of glyoxal.

17.3.4 Physical Characterization of Keratin–Chitosan Scaffolds

Under the physical characterization water solubility, swelling properties and degradation assay were performed. Table 17.1 presents the physical proper-ties of keratin–chitosan scaffolds. The least water solubility ($9.4 \pm 1.9\%$) was observed for CH + K + glyoxal (50:50) scaffolds whereas the solubility of CH + K + EDC (50:50) scaffolds was $69.5 \pm 1.6\%$. The decrease in solubility percentage of CH + K + glyoxal (50:50) is possibly due to the high degree of cross-linking of keratin with chitosan in the presence of glyoxal, which subsequently reduces the surface hydroxyl group present in the molecule. These results are consistent with the FTIR data suggesting the higher degree of cross-linking between keratin and chitosan in the presence of glyoxal. The swelling percentages were 192.4 ± 9.6 and 410 ± 2.7 for CH + K + glyoxal (50:50) and CH + K + EDC (50:50), respectively. To the best of our knowledge, there is no report on the water solubility and swelling properties of hair keratin–chitosan scaffolds. Kakkar et al. (2014) reported the water uptake capacity of keratin–chitosan–gelatin composite scaffolds ($>1700\%$). In contrast to this, swelling properties of hair keratin–chitosan scaffolds were very low (410% for CH+K+EDC). Xu et al. (2013) reported that on increasing the keratin concentration, water uptake capacity of the scaffold was also increased. This is consistent with our results as displayed in Table 17.1. The swelling capacity of 25% keratin containing scaffolds was 93.9 ± 4.2, whereas higher keratin content (75%) scaffolds was only 202 ± 7.9. The higher swelling capacity of scaffolds could be used for biomedical mainly wound healing applications.

The degradation study suggested that the keratin–chitosan scaffolds were stable in aqueous media. In the case of CH + K + glyoxal, only 9% weight loss was observed (Figure 17.6) after 14 days of incubation. While in the case of CH + K + EDC, 30% weight loss was observed on the 14th day. Degradation an study suggested that glyoxal cross-linked scaffolds were more stable in an aqueous medium in contrast to EDC cross-linked. The more stability of glyoxal cross-lined keratin–chitosan scaffolds is possibly due to the higher degree of cross-linking of keratin with chitosan in the presence of glyoxal. The results are also consistent with the FTIR result. The stability

Table 17.1 Physical properties of keratin-chitosan scaffolds

Samples	Water Solubility (%)	Swelling Ratio (%)
CH+K+ Gly (75:25)	46.7 ± 2.7	93.9 ± 4.2
CH+K+ Gly (50:50)	9.4 ± 1.9	192.4 ± 9.6
CH+K+ Gly (25:75)	14.9 ± 1.3	202 ± 7.9
CH+K+ EDC (50:50)	69.5 ± 1.6	410 ± 2.7

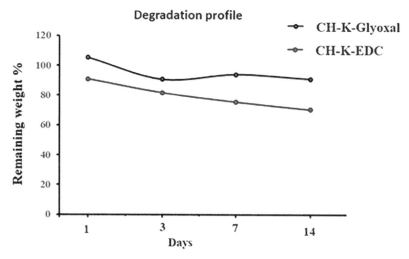

Figure 17.6 Degradation profile of (A) CH-K-glyoxal and (B) CH-K-EDC cross-linked scaffolds.

of scaffolds in aqueous media is possible due to the hydrophobic in nature of keratin and chitosan, which enhances the stability of scaffolds in the aqueous medium. In addition, the interaction between keratin and chitosan molecules in the presence of cross-linkers also improved its stability (Xu et al., 2013).

17.4 Conclusion

The present study shows the successful fabrication of hair keratin–chitosan-based porous scaffolds via freeze-drying method. Keratin–chitosan scaffolds have a non-uniform porous structure. FTIR analysis confirmed the amine/carboxyl groups of chitosan reacted with the amine group of keratin in the presence of cross-linkers. XRD data also suggest that a strong interaction occurred between keratin and chitosan molecules in the blend in the presence of glyoxal. Due to intermolecular interaction, physicochemical properties particularly water solubility and degradation rate of the scaffolds were improved. Degradation study confirmed that the fabricate scaffolds were stable in the aqueous medium without any significant weight loss. It was observed that the glyoxal cross-linked chitosan–keratin scaffold exhibited better physical properties compared to EDC cross-linked, and hence, would be a suitable biomaterial for biomedical and soft tissue engineering applications.

Acknowledgment

This study was funded by Science and Engineering Research Board (SERB), New Delhi, India, under the National Post-Doctoral Fellowship (NPDF) scheme (File No. PDF/2016/001156) to KRA.

References

Aadil KR, Jha H. (2016). Physico-chemical properties of lignin-alginate based films in the presence of different plasticizers. Iranian Polymer Journals. 25:661–70.

Aadil KR, Nathani A, Sharma CS, Lenka N, Gupta P. (2018). Fabrication of biocompatible alginate-poly(vinyl alcohol) nanofibers scaffolds for tissue engineering applications. Material Technology: Advanced Performance Material. 33(8):507–12.

Ahmed S, Ikram S. (2016). Chitosan based scaffolds and their applications in wound healing. Achievements in the Life Sciences. 10:27–37.

Balajia S, Kumara R, Sripriya R, Rao U, Mandal A, Kakkara P, Reddy NK, Sehgal PK. (2012). Characterization of keratin–collagen 3D scaffold for biomedical applications. Polymer Advance Technology. 23:500–7.

Croisier F, Jérôme C. (2013). Chitosan-based biomaterials for tissue engineering. European Polymer Journal. 49:780–92.

Dowling LM, Crewther WG, Parry DA. (1986). Secondary structure of component 8c-1 of α-keratin. An analysis of the amino acid sequence. Biochemistry Journal. 236:705–12.

Esparza Y, Bandara N, Ullah A, Wu J. (2018). Hydrogels from feather keratin show higher viscoelastic properties and cell proliferation than those from hair and wool keratins. Materials Science and Engineering C. 90:446–53.

Gupta P, Nayak KK. (2015). Optimization of Keratin/Alginate scaffold using RSM and its characterization for tissue engineering. International Journal of Biological Macromolecules. 85:141–9.

Hua J, Li Z, Xia W, Yang N, Gong J, Zhang J, Qiao C. (2016). Preparation and properties of EDC/NHS mediated crosslinking poly (gamma-glutamic acid)/epsilon-polylysine hydrogels. Materials Science and Engineering C. 61:879–92.

Kakkar P, Verma S, Manjubala I, Madhan B. (2014). Development of keratin–chitosan–gelatin composite scaffold for soft tissue engineering. Materials Science and Engineering C. 45:343–7.

Keskin Z, Urkmez AS, Hames EE. (2017). Novel keratin modified bacterial cellulose nanocomposite production and characterization for skin tissue engineering. Materials Science and Engineering C. 75:1144–53.

Li S, Yang X-H. (2014). Fabrication and characterization of electrospun wool keratin/poly(vinyl alcohol) blend nanofibers. Advances in Materials Science and Engineering. Article ID 163678, 1–7.

Marshall RC, Orwin DFG, Gillespie JM. (1991). Structure and biochemistry of mammalian hard keratin. Electron Microscopy Review. 4:47–83.

Nakamura A, Arimoto M, Takeuchi K, Fujii T. (2002). A rapid extraction procedure of human hair proteins and identification of phosphorylated species. Biological Pharmaceutical Bulletin. 25:569–72.

Sanchez Ramirez DO, Carletto RA, Tonetti C, Giachet FT, Varesano A, Vineis C. (2017). Wool keratin film plasticized by citric acid for food packaging. Food Packaging and Shelf Life. 12:100–6.

Sen T, Ozcelik B, Qiao GG, Ozmen MM. (2017). Hierarchical porous hybrid chitosan scaffolds with tailorable mechanical properties. Materials Letters. 209:528–31.

Tanase CD, Spiridon I. (2014). PLA/chitosan/keratin composites for biomedical applications. Materials Science and Engineering C. 40:242–7.

Xu S, Sang L, Zhang Y, Wang X, Li X. (2013). Biological evaluation of human hair keratin scaffolds for skin wound repair and regeneration. Materials Science and Engineering C. 33:648–55.

Xu Y, Han J, Chai Y, Yuan S, Lin H, Zhang X. (2018). Development of porous chitosan/tripolyphosphate scaffolds with tunable uncross-linking primary amine content for bone tissue engineering. Materials Science and Engineering C. 85:182–90.

Zeng MF, Fang ZP, Xu CW. (2004). Effect of compatibility on the structure of the microporous membrane prepared by selective dissolution of chitosan/synthetic polymer blend membrane. Membrane Science. 230:175–81.

18

Waste Paper: A Potential Source for Cellulose Nanofiber and Bio-nanocomposite Applications

Le Van Hai[1,2,*], Sunanda Roy[1], Ruth M. Muthoka[1], Jung Ho Park[1], Hyun-Chan Kim[1], and Jaehwan Kim[1,*]

[1]CRC for Nanocellulose Future Composite, Inha University, Incheon city, Republic of Korea
[2]Department of Pulp and Paper Technology, Phutho College of Industry and Trade, Phutho, Vietnam
E-mail: levanhai121978@gmail.com; jaehwan@inha.ac.kr
*Corresponding Author

Cellulose is a polysaccharide of a linear chain of β (1-4) glycosidic bonds. It is a renewable, recyclable, and environmentally friendly material. Wood cellulose has been in close contact with human since life on earth. Cellulose has been using for multifunctional purposes such as papers, fuel, construction, building, food, and cosmetics, for many decades. Most popular cellulose products are paper products (printing, newspaper, packaging, tissues, etc.) for a long time. Cellulose fibers can be extracted from different cellulose sources such as wood (hardwood and softwood) and non-wood sources (bamboo, rice straw, algae, and cattail) and waste paper as well. In recent years, cellulose nanofibers (CNFs) have been a trendy research subject, due to its lightweight and high mechanical properties. Converting waste paper to CNFs could be a great way of resources saving. In this chapter, waste paper and native cellulose sources will be discussed in terms of CNF extraction, their properties, and potential applications. CNFs can be extracted by methods like mechanical treatment, enzyme extraction, chemical, oxidation, and growing bacterial cellulose. To produce NCC, a different type of acid (HCl, H_2SO_4, or HNO_3) can be used. NCC can also be referred to as cellulose nanowhisker or nanocrystalline cellulose. By TEMPO-oxidation treatment, CNF extraction is usually named TOC-NFs. By mechanical methods, namely, grinders, high-pressure homogenizers, and ball-milling, nanocellulose extraction is usually called nanofibrillated or micro-fibrillated cellulose (NFC/MFC). In this chapter, NCC, TOCNFs, and NFC are called CNFs. Several potential applications of CNFs such as composites, smart materials, food packaging, coating, drug delivery, and other are also going to be discussed. From relevant publications and research resources, it has been found that the properties of CNFs that are extracted from waste cellulose source are quite similar to that of the CNFs extracted from native cellulose sources. Thus, waste paper is a potential cellulose source of CNF isolation, composites, and smart material applications.

18.1 Introduction

Cellulose is the main component of paper products. In general, to produce papers (copy, NP, packaging, tissues, etc.), cellulose combines with binders, pigments, additives, starch, $CaCO_3$, and other components. Cellulose is extracted from wood and non-wood sources by a different pulping method such as sulfite pulp, kraft sulfate pulp, semi-chemical pulp, mechanical pulp, and others (Roberts, 1996). After pulping, washing, and bleaching, cellulose can be used for different products and purposes such as paper printing, copy, packaging, and so on. According to Klemm et al. (2005), cellulose was first named by Anselme Payen in 1938. Waste paper can be recycled to papers, NP, copy, and so on. After being used, waste papers can be a source of fuel, landfilling, or recycling for new products. However, landfilling process leads to environmental pollution, and it is entirely expensive. To reduce the environmental impact, waste paper resources can be reproduced to a different type of paper or even converted to high-end value products. Recycled cellulose fiber can be used for many productions such as copy-printing, packaging, tissues, and so on. In this chapter, the waste paper will be discussed in the stages of recycled cellulose converted to CNFs. Properties and applications of extracted CNFs will also be portrayed.

CNF is a nanomaterial with at least one dimension less than 100 nm. Thus, to observe the morphology of CNFs, a high-quality electron microscopy is needed to determine the fibers such as TEM, FE-SEM, AFM, and DLS, and a nanoparticle analyzer is used to analyze the width and length of CNFs. From the literature, CNFs extracted from the native source and waste paper have almost the same morphology. However, the size depends on the isolation methods being adapted. From previous publications, it is shown that the width of recycled CNFs varies from 3 to 40 nm and the length from hundred nm to several mm long (Takagi et al., 2013; Wang et al., 2013; Danial et al., 2015; Mohamed et al., 2015; Hai et al., 2018). While the native CNFs have several 2–100 nm width and few micrometer length (Turbak et al., 1983; Moon et al., 2011; Hai et al., 2015; Hai, 2017; Hai and Seo, 2017). Primarily, the dimension of native CNFs is solely dependent on the cellulose sources and extraction process.

Every material selected for any purpose needs to have well-known properties such as mechanical, thermal, crystallinity, and so on. CNF is not out of the criteria for determining its properties for any used purposes. To determine the thermal stability of cellulose nanomaterials, the thermogravimetric analysis is usually conducted. In general, NCC, another form of CNFs, has lower thermal stability compared to their native cellulose sources. It has been reported that NCC decomposes between 200 and 300°C (Moon et al., 2011; Hai et al., 2015; Hai and Seo, 2017), whereas CNFs extracted by mechanical treatment (grinder, blender, etc.) usually have almost equal thermal stability with the native cellulose sources (Lee et al., 2009; Moon et al., 2011; Hai, 2017). While NCC decomposes at a lower temperature, the thermal stability of NFC obtained by mechanical method varies from around 300 to 330°C. The higher the thermal stability, the better the thermal resistance of materials. Hence, this review paper is also aimed at investigating the thermal stability of recycled CNFs from NCC and NFC for the comparison with CNFs extracted from native sources.

To evaluate the crystallinity index (CrI%) of cellulose or CNFs, there are several methods. In this context, XRD is a useful tool. A large number of publications have concerned to the CrI% of cellulose and CNFs. It is essential to evaluate the CrI% of cellulose and CNFs to understand the mechanical properties. Crystallinity index of NCC is often higher than the CrI% of native cellulose sources, because the amorphous parts of cellulose are released and destroyed during acid hydrolysis of NCC. This leads NCC with higher CrI%. Additionally, NFC or also sometimes called MFC is also lower than CrI% of native cellulose sources produced by the grinding or high-pressure homogenization process. The amorphous part remains with crystal areas in NFC. Thus, NFC is more flexible than NCC.

There are extensive applications of CNFs such as reinforcing composite materials. Many research papers have mentioned that CNFs can be used as reinforcing materials due to its lightweight, high tensile strength, Young's modulus, and it is also bio-degradable. CNFs has been used for different types of composites such as polymer lattices (Favier et al., 1995), POE (Azizi Samir et al., 2004), PE, LDPE, PVC, PVA, and PLA (Dufresne, 2012; Li, 2012), all-cellulose nanocomposite (Pullawan et al., 2014; Hai et al., 2015, 2017), and smart materials (Kim et al., 2015). There are significant potential applications of CNFs that are indicated in a number of research on different fields: heavy metal adsorption (Stephen et al., 2011), drug delivery (Kolakovic et al., 2012), nanocellulose-PEG bio-nanocomposite wound dressings (Sun et al., 2017), composite membrane (Kizltas et al., 2013; Ma et al., 2014), supercapacitor materials (Yang et al., 2015), electro-conductive composites (Shi et al., 2013), as well as in organic solar cells (Zhou et al., 2013). CNFs are also indicated in some additional research areas such as coating, lubricants, colorants, and others (Lavoine et al., 2012). CNFs have also been reported in paper coating (Lavoine et al., 2014; Afra et al., 2016). By using CNFs for coating, the surface porosity is decreased, while air resistance, strength, and stiffness are improved (Lavoine et al., 2014; Afra et al., 2016).

The CNFs extracted from the waste paper have also been used for different types of composite materials (Takagi et al., 2013; Hai et al., 2017). It has been noticed that CNFs from waste papers offered the same capability of reinforcement as produced by native CNFs. The extracted CNFs from waste papers were tested, for producing several bio-degradable composites, green composites, and PET packing materials where improvement in the mechanical properties of the composites was observed (Takagi et al., 2013; Hai et al., 2017; Lei et al., 2018).

18.2 Methods

18.2.1 Conversion of CNF from Waste Paper

In general, used papers can be recycled to new paper products. The waste paper goes through several treatment stages including sorting, de-inking, bleaching, pulping followed by papermaking process. In the case of CNF extraction from waste paper, it first needs sorting and removing contamination like binders, staples, inks, and so on. The waste paper then transfers to several stages of CNF extraction. The extraction of CNFs is depended on the selection methods such as oxidation, acid hydrolysis, or mechanical treatments. The selection process that converts cellulose to the CNFs effects on the properties of CNFs such as thermal stability, crystallinity, morphology, and so on. The process of converting waste paper to CNFs is simplified

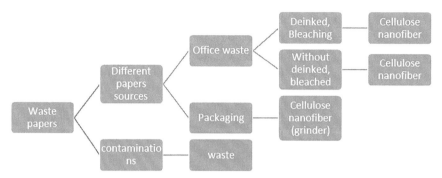

Figure 18.1 The process to convert CNFs from waste papers.

in Figure 18.1. A detail of processing methods such as acid hydrolysis, mechanical, or oxidation process is discussed in the following section.

18.2.2 NCC from Native and Waste Paper

NCC was first isolated from cellulose using sulfuric acid and named micelles by Ranby in 1949 (Rånby and Ribi, 1950). Later in 1953, NCC was also reported by another group Mukherjee and Woods (1953). There are different approaches to produce NCC using sulfuric acid, hydrochloric acid, and phosphotungstic acid (Ranby, 1949; Rånby and Ribi, 1950; Mukherjee and Woods, 1953; Gilkes et al., 1992; Revol et al., 1992; Dong et al., 1996; Edgar and Gray, 2001; Fahma et al., 2010; Winter et al., 2010; Yu et al., 2013; Gaspar et al., 2014; George et al., 2014; Gaspar et al., 2014; Liu et al., 2014; Santos et al., 2014). It was found that factors such as acid concentration, isolation time, and temperature affect the properties and yielding of NCC (Yu et al., 2013). Ideally, 55–65 wt% of concentrated sulfuric acid, temperature between 50 and 70°C, and reaction time from 15 min to several hours are used to isolate NCC. Briefly, cellulose source was mixed with acid (H_2SO_4, HCl, etc.) and stirred from 15 min to several hours and temperature from 50 to 70°C. The acid was removed by centrifugation and further treatment with dialysis with DI water or NaOH. Many potential applications of NCC in the area of coating, lubricants, pigments, additives, and composite reinforcement (Winter et al., 2013; George et al., 2014; Gaspar et al., 2014) have also been reported.

The above paragraph basically discussed the isolation process of NCC from native cellulose sources. Now the question arises how NCC can be extracted from waste paper. Typically, NCC, NFC, or TEMPO-oxidation

CNFs extracted from waste paper source have not much difference with CNFs extracted from native cellulose sources. However, some different activities should be taken into account during collection, sorting, de-inking, and bleaching stages while using waste papers. For native cellulose extraction process, wood and non-wood materials need to convert from trees to cellulose pulp through the pulping process. In this chapter, we will basically emphasize only on the isolation process of CNFs from waste paper. Takagi et al. (2013) isolated NCC from waste newspaper using 64% H_2SO_4, 70°C, and 1 h—in brief, waste newspaper (NP) through alkaline treatment, bleaching treatment, and acid hydrolysis process. Treated waste newspapers were treated with acid hydrolysis 64% for an hour at 70°C. The acid hydrolysis samples were stopped reaction by cold water and washing until pH becomes neutral. Mohamed et al. (2015) also isolated NCC from waste newspaper using 65 wt% H_2SO_4. From Mohamed et al. (2015), NP was ground with a grinder, then alkali treatment with 5 wt% NaOH, boiled at 125°C for 2 h, and washed with DI water until neutral. Waste NP was bleached treatment used HNO_3 and 2 w/v% $NaClO_2$ until a white sample was gained. Then NCC was extracted from waste NP using 65 wt% H_2SO_4 solution at 45°C and 60 min reaction time. From Danial et al. (2015), nanocrystal cellulose was extracted from waste paper (ONP) source using 60% (v/v) H_2SO_4, 45°C, the reaction for 1 h. Also from Campano et al. (2017), different types of waste papers such as ONP, recycled NP, waste NP plus two-step of treatment include alkali and bleaching treatment (NP-B), and MCC were used for NCC extraction. The authors compared the properties of NCC extraction from waste paper with NCC extraction from MCC source. NCC was extracted from waste papers using 60% wt H_2SO_4, 45°C, and reaction time was 90 min (Campano et al., 2017). Tang et al. (2015), NCC extracted from waste paper using 60% (w/w) H_3PO_4, 6 h, room temperature. The overall isolation process of NCC is briefly illustrated in Figure 18.2.

18.2.3 TOCNF from Native and Waste Paper

TEMPO-oxidation is the most convenient and popular process for extraction of TOCNFs where chemical oxidation followed by mechanical treatment

Figure 18.2   Illustration of nanocrystal cellulose isolation process.

are applied together. Recent years, a huge number of research publications concerned with TOCNFs. The TEMPO-oxidation treatment for cellulose fiber was first mentioned by Isogai and Kato (1998). They studied a wide range of processing parameters including chemical dosage, time, and temperature to obtain the best-oxidized cellulose. Later, TEMPO-oxidation treatment was also adapted for the isolation of TOCNFs. Some relevant articles on the same are referred here for more information (Saito et al., 2006, 2007; Okita et al., 2009). Since the landmark paper published by Isogai & Kato, TEMPO-oxidation of cellulose has gained large attention for fiber extraction. In this context, Saito et al. (2006), used never dried cotton and never dried bleached sulfite pulp from *Pinus pinaster* to isolate TOCNFs. The dosage of the NaClO (sodium hypochlorite, a chemical agent used for TEMPO oxidation) was 0–6.3 mmol/g and the reaction time was 5–120 min. Okita et al. (2009) produced TOCNFs from never dried softwood TMP; 5 g of TMP was treated with TEMPO (0.0625 g), NaBr (0.625 g), and NaClO (0.8–26 mmol), at a pH 10 to obtain the CNFs.

Likewise, TEMPO-oxidation was also used to extract TOCNFs from waste paper sources. In our recent publications (Hai et al., 2017, 2018), we demonstrated the preparation of TOCNFs by TEMPO-oxidation followed by homogenizer treatment. The TOCNFs from this TEMPO-oxidation were then used in green all-cellulose composite (Hai et al., 2017), where matrix and filler are both cellulose based. In another recent article Hai et al. (2018), we have discussed the isolation of TOCNFs from waste paper and compared the properties with native hardwood and cotton sources. Figure 18.3 demonstrates TEMPO-oxidized TOCNF isolation from waste papers.

18.2.4 NFC from Native and Waste Paper

NFC is produced by mechanical treatments using grinders, a high-pressure homogenizer, and others. According to Turbak et al. (1983) and Charreau et al. (2013), NFC was first produced in 1977. NFC is more flexible than NCC due to the fact that it retains both crystalline and amorphous areas in its structure. NFC is also very long compared to NCC. To produce NFC, high energy consumption is needed to extract nanosized fiber from microsized

Figure 18.3 Illustration of TOCNF isolation from waste paper.

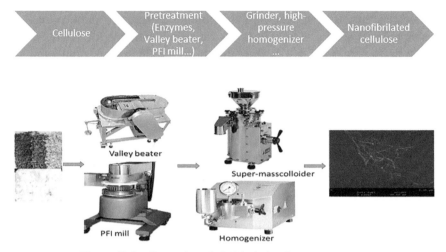

Figure 18.4 Illustration of NFC isolation from waste paper.

cellulose fibers. However, NFC is environmental friendly compared to acid hydrolysis or TEMPO-oxidized methods. NFC can be produced by several approaches such as using super mass collider, a high-pressure homogenizer, grinder, ball mill, blender, and so on. NFC produced by super mass collider is very common and carried out by many research groups. To produce NFC, cellulose fibers are passed through a grinding machine with a certain number of passes. The gap between grinding disks can be adjusted to vary the shear force to break down cellulose into nanosized fibers.

NFC was also produced from waste paper sources. Wang et al. (2013) described the preparation of NFC where 1% of the waste paper was passed 20 times through the supper mass collider (MKCA6-2) at 1600 rpm. The reported width of the NFC was found to 30–100 nm. Figure 18.4 shows the general process to extract NFC by mechanical method.

18.3 Results and Discussion

18.3.1 Morphology by TEM, SEM, and AFM

To determine the morphology of CNFs, different equipment such as AFM, FE-SEM, TEM, and DSL can be used. Typically, a diluted suspension of CNFs is dropped on the silicon wafer, glass slide for AFM determination or dropped on the copper grid for TEM measurement. The specimen has to be dried thoroughly before the measurement. CNFs can also be sprayed on a glass slide, or carbon tape and then the morphology of CNFs is determined by AFM or FE-SEM. The spraying–drying process allows one to attain uniformly dispersed and less dense samples to capture individual fiber compared to the dropped casting process. TEM samples can be coated or non-coated for morphology study. In the case of SEM and FE-SEM, deposited samples are always required coatings such as carbon, gold, or platinum for morphology evaluation.

Tagaki et al. (2013) obtained CNFs extracted from waste cellulose source that had a width ranging from 10 to 40 nm. Mohamed et al. (2014) mentioned 5.8 nm wide and 120 nm long nanocrystalline cellulose that was obtained from waste paper. Danial et al. (2015), was also able to produce nanocrystal cellulose with 3–10 nm diameter and 100–300 nm length. We also produced CNFs from recycled cellulose and native cellulose and found fibers with a diameter of 20 nm and length between 500 and 800 nm. In a recent paper, Lei et al. (2018) produced cellulose nanocrystals with 10–25 nm diameter and 220–770 nm length. Only in a specific case, Wang et al. (2013) claimed of producing very long CNFs of 3 mm length and 30–100 nm diameter. Based on another recent report by Thambiraj and Ravi Shankaran (2017), the width of CNFs was found to be 10 and 180 $\pm$ 60 nm in width and length, respectively, which they isolated from waste paper too. Figure 18.5 shows the morphologies of different types of CNFs that we captured by AFM, FE-SEM, and TEM.

18.3.2 Thermal Degradation Properties of Cellulose (TGA, TG-DTA)

Cellulose material has thermal stability up to 330–360°C and starts to decompose above that. Thus, it is necessary to know the right condition for cellulose material to be used in high-temperature applications. However, CNFs (NCC, NFC, and TOCNF) show low thermal stability compared to cellulose. Moreover, CNF from different sources has different thermal stability. NFC has almost the same thermal stability as native cellulose sources while NCC displays lower thermal stability than the native one. Several research groups mentioned that the low thermal stability of NCC is due to sulfate groups present in NCC. Kumar et al. (2014) studied the decomposition behavior of NCC and recorded the decomposition started at 220–270, 270–330 (with 40–50% char residual), and 330–470°C (with around 20–40% char residual). According to the report (Hai et al., 2015; Hai and Seo, 2017), the degradation of different NCC starts from 185 to 320°C. However, to follow the

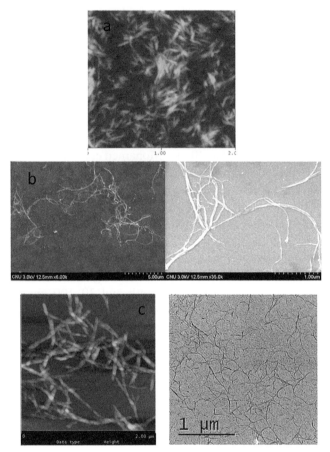

Figure 18.5   CNFs by AFM, FE-SEM, and TEM, (a) nanocrystals, (b) nanofibrillated, and (c) TOCNFs.

high thermal stability as like cellulose (330–360°C), several methods were employed including NaOH treatment to replace the sulfate groups from NCC. Another method to retain the high thermal stability is also reported by Haafiz et al. (2014) where NCC is subjected to a swelling agent, namely, N,N-dimethylacetamide with 0.5% lithium chloride. Haafiz et al. (2014) found that the thermal stability of NCC treated with swelling agent had decomposition max reach to 490°C. They observed that the treated NCC has higher thermal stability than that of the MCC, starting materials. This reflects the importance of chemical treatment in the enhancement of thermal stability of CNC. According to Abraham et al. (2013), the thermal stability of CNFs extracted by steam explosion is lower (327°C) than that of native cellulose (335°C). However, the thermal stability of NFC produced by mechanical methods such as (grinding, high-pressure homogenizer. . .) is almost similar to the native cellulose sources. A report by Hai (2017) states that the degradation temperature of cellulose and NFC was started around 300°C.

The above paragraphs were basically focused on the thermal stability of native CNFs. In the following paragraph, we will be discussing on the thermal stability of nanocellulose (CNFs, NCC, etc.) extracted from waste papers. Danial et al. (2015) found that the CNFs which were isolated from waste paper had a decomposition temperature between 187 and 320°C. Mohamed et al. (2015) also noticed that the NCC from waster paper started to degrade at around 180°C. In our case, the CNFs extracted from the waste paper (Hai et al., 2018) had a slightly higher decomposition temperature (209–311°C) than the above reported values. The difference in temperature could be the cause of different cellulose sources and extraction processes as mentioned before. Wang et al. (2017) reported that the decomposition temperature for their waste paper extracted NCC was in between 349 and 353°C which was slightly lower (363–369°C) than the native NCC. Lei et al. (2018) noticed a relatively longer decomposition temperature period (180–400°C) for the extracted NCC. From the above results, it is apparent that recycled CNF and CNF from a native source almost have a similar thermal stability.

18.3.3 Crystallinity Index (XRD)

The crystallinity index of cellulose and nanocrystal cellulose can be measured by the XRD method. Table 18.1 indicates the crystallinity index of CNF isolated from different cellulose sources and different research groups. As can be seen, the crystallinity index of cellulose nanocrystals improved significantly after the acid hydrolysis process regardless of the starting materials/sources.

Table 18.1 shows the results of CrI% of different cellulose sources including native cellulose and waste paper sources. To the best of our knowledge, there is only one paper concerned on NFC until now. There is no information on crystallinity index of NFC that is extracted from the waste paper (Wang et al., 2013).

18.3.4 Potential Applications of CNFs and NCC from Waste Paper

The potential of CNFs is immense, and it has proven its capability in the area of reinforcing composites, green composites, coatings, lubricants, and

Table 18.1   Crystallinity index of CNF

Cellulose Sources	CrI of Starting Materials (%)	Isolation Methods	CrI of CNFs and NCC (%)	References	Notices
CrI% of NCC extracted from native cellulose sources					
Sugarcane bagasses	35.6	H_2SO_4	72.5	Kumar et al., 2014	
Cotton linter	64.4	H_2SO_4	90.5	Morais et al., 2013	
Oil palm biomass	87	H_2SO_4	88; 84	Haafiz et al., 2014	Treatment conditions
MCCs	76	HCl	84.3, 87.3, 86.5, and 88.6	Yu et al., 2013	Treatment conditions various
CrI% of NCC extracted from the waste paper					
Waste newspaper	82.0	H_2SO_4	90.2	Mohamed et al., 2015	
Office waste paper	45.5	H_2SO_4	48.7, 54.1, 61.2, 74.1, 60.1,65.7, 56.9,54.5	Lei et al., 2018	Conditions treatment effect on crystallinity index
Old corrigated container (OCC)	Not available data	H_3PO_4, Enzymes	54.1, 54.2, and 57.8	Tang et al., 2015	Treatment conditions
Waste cotton cloth	29.4, 24.6	Mixed of HCl and H_2SO_4	62.7, 55.8	Wang et al., 2017	Treatment conditions
Hardwood, cotton, and deinked paper	75.3, 82.4, 70.6	TEMPO	54.0, 62.4, and 36.9	Hai et al., 2018	Variation of cellulose sources
CrI% of NFC extraction from native cellulose					
Softwood	79.4	Supermass colloider	67.2	Hai, 2017	
CrI% of NFC extraction from the waste paper sources					
Waste corrugate paper	No available data	Grinding plus ultrasoni- cation	No available data	Wang et al., 2013	Only one paper can be found on NFC extracted from the waste paper

so on. From the above literature review, it is quite clear that CNF from both native and waste paper can be used for many different types of composites. This reflects that the CNFs regardless of the starting material would be equally valuable for the production of composite materials. Some interesting applications of CNF from waste paper sources include the bio-degradable composites (Takagi et al., 2013), green all-cellulose nanocomposites (Hai et al., 2017), PET composites (Lei et al., 2017), food packaging, bio-sensors and drug delivery (Thambiraj and Ravi Shankaran, 2017) composites, and so on. According to Takagi et al. (2013), CNFs from waste papers or native cellulose can offer higher mechanical strength to the respective biodegradable composites. However, the slightly lower mechanical improvement was noticed for the composites containing waste paper extracted CNFs than the analogous native CNF composites. Similar trends were also observed from our green all-cellulose nanocomposite where both the tensile strength and Young's modulus were increased after the addition of CNFs from waste and native too (Hai et al., 2017). Besides reinforcement, enhancement in the optical transparency and barrier properties have also been reported for waste paper extracted CNFs in PET composites (Lei et al., 2018).

18.4 Concluding Remarks

CNF gained its attention due to its attractive features like light-weight, high tensile strength, Young's modulus, and eco-friendly property. Extraction of CNFs from various sources such as wood, non-wood, and waste paper has been demonstrated. CNFs from waste paper offered many potential applications including papers, coating, lubricants, smart materials, and composites. CNFs extracted from waste cellulose sources have similar characteristics such as morphology, mechanical properties, and reinforcement as CNFs from native cellulose. Therefore, waster paper extracted CNFs would be an excellent candidate for many future applications. Hence, the extraction of CNFs from waste paper would be an exciting area to be focused on and find applications for making the more sustainable earth. We wish that more research activities, development, and applications on recycling of waste paper will be coming in front soon. We hope that this chapter may be a valuable addition in the area of waste paper management and attracts broader reader attention soon.

Acknowledgment

This work was supported by the National Research Foundation of Korea (NRF-2015R1A3A2066301).

References

Abraham E, Deepa B, Pothen LA, Cintil J, Thomas S, John MJ, Anandjiwala R, Narine SS. (2013). Environmental friendly method for the extraction of coir fibre and isolation of nanofibre. Carbohydrate Polymers. 92:1477–83.

Afra E, Mohammadnejad S, Saraeyan A. (2016). Cellulose nanofibrils as coating material and its effects on paper properties. Progress in Organic Coatings. 101:455–60.

Azizi Samir MAS, Alloin F, Sanchez JY, Dufresne A. (2004). Cellulose nanocrystals reinforced poly (oxyethylene). Polymer. 45:4149–57.

Campano C, Miranda R, Merayo N, Negro C, Blanco A. (2017). Direct production of cellulose nanocrystals from old newspapers and recycled newsprint. Carbohydrate Polymers. 110:489–96.

Charreau H, Foresti ML, Vazquez A. (2013). Nanocellulose patents trends: a comprehensive review on patents on cellulose nanocrystals, microfibrillated and bacterial cellulose. Recent Patents on Nanotechnology. 7:56–80.

Danial W, Abdul Majid Z, Mohd Muhid M, et al. (2015). The reuse of wastepaper for the extraction of cellulose nanocrystals. Carbohydrate Polymers. 118:165–9.

Dong XM, Kimura T, Revol JF, Gray DG. (1996). Effects of ionic strength on the isotropic-chiral nematic phase transition of suspensions of cellulose crystallites. Langmuir. 12:2076–82.

Dufresne A. (2012). Potential of nanocellulose as a reinforcing phase for Polymers, J-For. 2(6).

Edgar CD, Gray DG. (2001). Induce circular dichroism of chiral nematic cellulose films. Cellulose. 8:5–12.

Fahma F, Iwamoto S, Hori N, Iwata T, Takemura A. (2010). Isolation, preparation, and characterization of nanofibers from oil palm empty-fruit-bunch (OPEFB). Cellulose. 17:977–85.

Favier V, Chanzy H, Cavaille JY. (1995). Polymer nanocomposites reinforced by cellulose whiskers. Macromolecules. 28:6365–7.

Gaspar D, Fernandes SN, de Oliveira AG, Fernandes JG, Grey P, Pontes RV, Pereira L, Martin R, Godinho MH, Fortunato E. (2014).

Nanocrystalline cellulose applied simultaneously as the gate dielectric and the substrate in flexible field effect transistors. Nanotechnology. 25:094008 (11pp).

Gaspar D, Fernandes SN, Olivera AGD, Fernandes JG, Grey P, Pontes RV, Pereira L, Martins R, Godinho MH, Fortunato E. (2014). Nanocrystalline cellulose applied simultaneously as the gate dielectric and the substrate in flexible field effect transistors. Nanotechnology. 25:09008.

George J, Kumar R, Sajeevkumar VA, Ramana KV, Rajamanickam R, Abhishek V, Nadanasabapathy S, Siddaramaiah. (2014). Hybrid HPMC nanocomposites containing bacterial cellulose nanocrystals and silver nanoparticles. Carbohydrate Polymers. 105:285–92.

Gilkes NR, Jervis E, Henrissat B, Tekant B, Miller RC, Warren RAJ, Kiburn DG. (1992). The adsorption of a bacterial cellulase and its two isolated domains to crystalline cellulose. The Journal of Biological Chemistry. 267(10):6743–9.

Haafiz MKM, Hassan A, Zakaria Z, Inuwa IM. (2014). Isolation and characterization of cellulose nanowhiskers from oil palm biomass microcrystalline cellulose. Carbohydrate Polymers. 103:119–25.

Hai LV. (2017). Properties of nanofibrillated cellulose and its length-width ratio determined by a new method. Cellulose Chemistry and Technology. (7-8):549–653.

Hai LV, Chan HK, Kafy A, Zhai LD, Kim JW, Kim J. (2017). Green all-cellulose nanocomposites made with cellulose nanofibers reinforced in dissolved cellulose matrix without heat treatment. Cellulose. 8:3301–11.

Hai LV, Seo YB. (2017). Characterization of cellulose nanocrystal obtained from electron beam treated cellulose fiber. Nordic Pulp and Paper Research Journal. 32(2):170–8.

Hai LV, Son HN, Seo YB. (2015). Physical and bio-composite properties of nanocrystalline cellulose from wood, cotton linters, cattail, and red algae. Cellulose. 22:1789–98.

Hai LV, et al. (2018). Fabrication and characterization of cellulose nanofibers from recycled and native cellulose resources using TEMPO oxidation. Cellulose Chemistry and Technology. 52:215–21.

Isogai A, Kato Y. (1998). Preparation of polyuronic acid from cellulose by TEMPO-mediated oxidation. Cellulose. 5:153–64.

Kim JH, Shim BS, Kim HS, Lee YJ, Min SK, Jang D, Abas Z, Kim J. (2015). Review of nanocellulose for sustainable future materials. International Journal of Precision engineering and Manufacturing-Green Technology. 2:197–213.

Kizltas A, Kiziltas E, Boran S, Gardner D. (2013). Micro- and Nanocellulose composites for automotive applications. 13th Annual SPE Automotive Composites Conferences and Exhibition.

Klemm D, Heublein B, Fink HP, Bohn A. (2005). Cellulose: Fascinating biopolymer and sustainable raw material. Angewandte Chemie International Edition. 44(22):3358–93.

Kolakovic R, Peltonen L, Laukkanen A, Hirvonen J, Laaksonen T. (2012). Nanofibrillar cellulose films for controlled drug delivery. European Journal of Pharmaceutics and Biopharmaceutics. 82:308–15.

Kumar A, Negi YS, Choudhary V, Bhardwaj NK. (2014). Characterization of cellulose nanocrystals produced by acid-hydrolysis from sugarcane bagasse as agro-waste. Journal of Materials Physics and Chemistry. 2(1):1–8.

Lavoine N, Desloges I, Bras J. (2014). Microfibrillated cellulose coatings as new release systems for active packaging. Carbohydrate Polymer. 103:528–37.

Lavoine N, Desloges I, Dufresne A, Bras J. (2012), Microfibrillated cellulose – its barrier properties and applications in cellulosic materials: a review. Carbohydrate Polymers. 90:735–64.

Lee SY, Chun SJ, Kang IA, Park JY. (2009). Preparation of cellulose nanofibrils by high-pressure homogenizer and cellulose-based composite films. Journal of Industrial and Engineering Chemistry. 15:50–5.

Lei W, Fang C, Zhou X, Yin Q, Pan S, Yang R, Liu D, Ouyang Y. (2018). Cellulose nanocrystals obtained from office waste paper and their potential application in PET packing materials. Carbohydrate Polymers. 181:376–85.

Li W, Yue J, Liu S. (2012). Preparation of nanocrystalline cellulose via ultrasound and its reinforcement capability for poly(vinyl alcohol) composites. Ultrasonics Sonochemistry. 19:479–85.

Liu Y, Wang H, Yu G, Li Q, Mu X. (2014). A novel approach for the preparation of nanocrystalline cellulose by using phosphotungstic acid. Carbohydrate Polymers. 110:415–22.

Ma H, Burger C, Hsiao B, Chu B. (2014). Fabrication and characterization of cellulose nanofiber based thin-film nanofibrous composite membranes. Journal of Membrane Science. 454:272–82.

Mohamed MA, Salleh WNW, Jaafar J, Air SEAM, Ismail AF. (2015). Physicochemical properties of "green" nanocrystalline cellulose isolated from recycled newspaper. RSC Advance. 5:29842.

Moon RJ, Martini A, Nairn J, Simonsen J, Youngblood J. (2011). Cellulose nanomaterials review: structure, properties and nanocomposites. Chemical Society Reviews. 40:3941–94.

Morais JPS, Rosa MDF, Filho MDSMSS, Nascimento LD, Nascimento DMD, Cassales AR. (2013). Extraction an characterization of nanocellulose structures from raw cotton linter. Carbohydrate Polymers. 91:229–35.

Mukherjee and Woods. (1953). X-ray and electron microscope studies of the degradation of cellulose by sulphuric acid.

Okita Y, Saito T, Isogai A. (2009). TEMPO-mediated oxidation of softwood TMP. Holzforschung. 63:529–35.

Pullawan T, Wilkinson AN, Zhang LN, Eichhorn SJ. (2014). Deformation micromechanics of all-cellulose nanocomposites: Comparing matrix and reinforcing components. Carbohydrate Polymers. 100:31–9.

Rånby BG, Ribi E. (1950). Experientia. 6:12. Available at: https://doi.org/10.1007/BF02154044

Ranby G. 1949. Aqueous colloidal solutions of cellulose micelles. Acta Chemica Scandinavica. 3:649–50.

Revol JF, Bradford H, Giasson J, Marchessault RH, Gray DG. (1992). Helicoidal self-ordering of cellulose microfibrils in aqueous suspension. International Journal of Biological Macromolecules. 14, June.

Roberts JC. (1996). The chemistry of paper. The royal society if chemistry. Letchworth. Herts SG6 1 HN, UK.

Saito T, Kimura S, Nishiyama Y, Isogai A. (2007). Cellulose nanofibers prepared by TEMPO-mediated oxidation of native cellulose. Biomacromolecules. 8:2485–91.

Saito T, Nishiyama Y, Putaus JL, Vignon M, Isogai A. (2006). Homogeneous suspensions of individualized microfibrils from TEMPO-catalyzed oxidation of native cellulose. Bio-macromolecules. 7(6).

Santos TM, Souza Filho MDSM, Caceres CA, Rosa MF, Moraid JPS, Pinto AMB, Azeredo HMC. (2014). Fish gelatin film as affected by cellulose whiskers and sonication. Food Hydrocolloids. 41:113–8.

Shi Z, Phillips GO, Yang G. (2013). Nanocellulose electron-conductive composites. Nanoscale. 5:3194–201.

Stephen M, Catherine N, Brenda M, Andrew K, Leslie P, Corrine G. (2011). Oxolane-2,5-dione modified electrospun cellulose nanofibers for heavy metals adsorption. Journal of Hazardous Materials. 192:922–7.

Sun F, Nordli HR, Pukstad B, Gamstedt EK, Chinga-carrasco G. (2017). Mechanical characteristics of nanocellulose-PEG bionanocomposite would dressings in wets conditions. Journal of the Mechanical Behavior of Biomedical Materials. 69:377–84.

Takagi H, Nakagaito AN, Bistamam MSA. (2013). Extraction of cellulose nanofiber from waste papers and application to reinforcement in biodegradable composites. Journal of Reinforced Plastics and Composites. 32(20):1542–6.

Tang Y, Shen X, Zhang J, Gua D, Kong F, Zhang N. (2015). Extraction of cellulose nano-crystals from old corrugated container fiber using phosphoric acid and enzymatic hydrolysis followed by sonication. Carbohydrate Polymers. 125:360–6.

Thambiraj S, Ravi Shankaran D. (2017). Preparation and physicochemical characterization of cellulose nanocrystals from industrial waste cotton. Applied Surface Science. 412:405–16.

Turbak AF, Snyder FW, Sandberg KR. (1983). Microfibrillated cellulose, a new cellulose product: properties, uses, and commercial potential. Journal of Applied Polymer Science. 3:815–27.

Wang H, Li D, Zhang R. (2013). Preparation of ultralong cellulose nanofibers and optically transparent nanopapers derived from waste corrugated paper pulp. Bioresources. 8(1):1374–84.

Wang et al. (2017). Reuse of waste cotton cloth for the extraction of cellulose nanocrystals. Carbohydrate Polymers. 157:945–52.

Winter H, Carclier C, Delorme N, Bizot H, Quemener B, Cathala B. (2010). Improved colloidal stability of Bacterial cellulose nanocrystal suspensions for the elaboration of spin-coated cellulose-based model surfaces. Biomacromolecules. 11:3144–51.

Yang X, Shi K, Zhitomirsky I, Cranston ED. (2015). Cellulose nanocrystal aerogels as universal 3D lightweight substrates for supercapacitor materials. Advanced Materials. 27:6104–9.

Yu H, Qin Z, Liang B, Liu N, Zhou Z, Chen L. (2013). Facile extraction of thermally stable cellulose nanocrystals with a high yield of 93% through hydrochloric acid hydrolysis under hydrothermal conditions. Journal of Materials Chemistry A. 1:3938.

Zhou Y, Fuentes-Hernandez C, Khan TM, Liu JC, Hsu J, Shim J.W., Dindar A., Youngblood J.P., Moon R.J., Kippelen B., (2013). Recyclable organic solar cells on cellulose nanocrystal substrates. Scientific Reports. 3:1536.

19

Field Evaluation of Plastic Mulch Film for Changes in Its Mechanical Properties and Retrieval Mechanism for Its Reuse Under Onion Crop

Mintu Job[1],* and Shri Sita Ram Bhakar[2]

[1]Department of Agricultural Engineering, Birsa Agricultural University, Kanke, Ranchi
[2]Department of Soil and Water Engineering, College of Technology and Engineering, Udaipur, Rajasthan
E-mail: mintujob@rediffmail.com
*Corresponding Author

Plastic mulching is a soil moisture conservation practice that involves placing polyethylene film over raised or a flat bed to provide a more favorable environment for growth and production of onion crop. The use of plastic materials for mulching is a very common practice for horticultural crops. Black linear low density polyethylene (LLDPE) is widely used due to its excellent properties and low cost. An experiment was conducted to ascertain the effect of different types of black polyethylene mulches of varying thickness for its mechanical properties during crop growing period under onion. Changes in tensile strength, elongation at break, and tear resistance varied from 24.17 to 74.82%, 30.18 to 78.21%, and 19.45 to 60.68%, respectively, within different thicknesses of plastic mulch during the entire duration of the crop. Poly-mulch made of recycled material (50 μm) underwent early deterioration which ultimately affected yield and growth of the crop.

Tractor-operated mulch laying machine is available but its use is limited to large plots and in places where crop spacing allows it as area lost in between two rows is very large. To suit the requirement of small farmers, partial mechanization of laying of mulch film is envisaged by developing a manual multi-purpose trolley for plastic mulch laying and recollection. This machine besides being simple in design and operation also caters to a very important operation, which is otherwise overlooked, that is of retrieving the used plastic mulch by rolling it back for better storage which significantly enhances its life beside being environment friendly.

19.1 Introduction

Mulching is the process or practice of covering the soil to make more favorable conditions for plant growth, development, and efficient crop production. When compared to other mulches, plastic mulches are impermeable to water; it therefore prevents direct evaporation of moisture from the soil and thus limits the water loss and soil erosion over the surface. Wide ranges of plastic film based on different types of polymers have all been tried since the 1960s. Owing to its greater permeability to long wave radiation, which can increase the temperature around plants during the night times, polyethylene is preferred. Today, the vast majority of plastic mulch is based on LDPE. It is of paramount interest to have the right specification of mulch in terms of thickness and color. The lifetime of plastic mulching films is reduced owing to their prolonged exposure to climatic agents such as solar radiation, high air temperature, and high relative humidity and by chemical products used during crop cycle. As a consequence, plastic mulch can be used only for one or two cultivation periods and with varying degree of mechanical strength due to degradation. Agricultural films should meet a set of minimum design requirements, including adequate strength and elongation at break for mechanical installation (Briassoulis, 2004).

There are three primary non-degradable mulch types used commercially in the production of vegetable crops black, clear, and the group of white on black and silver/aluminum reflective mulch. Black, silver on black polyethylene is the most popular because of benefit of moderating the temperature according to the climatic condition, in addition to its low cost. This can be used in all the crops in general. To suit the requirement of small farmers, partial mechanization of some operations like laying of drip laterals and laying of mulch film is envisaged by developing a manual machine for plastic mulch laying and recollection. This machine besides being simple in design and operation also caters to a very important operation, which is otherwise overlooked, that is of retrieving the used plastic mulch and drip lateral by rolling it back for better storage which significantly enhances its life beside being environment friendly.

19.2 Methodology

A field experiment was conducted during 2014–15 and 2015–16 at Plasticulture farm, College of Technology and Engineering, Maharana Pratap University of Agriculture and Technology, Udaipur. Geographically, Udaipur is located at 24° 35 N latitude and 73° 44 E longitude. The altitude of the site is 582.17 m above mean sea level. The area has a sub-humid climate.

Six treatments comprising five poly-mulches (black) of different thicknesses (M_1-20 μm, M_2-25 μm, M_3-30 μm, M_4-35 μm, and M_5-50 μm-recycled) were compared with control (M_0-no mulch) in a randomized block design with four replications for understanding the changes attained in its mechanical properties with the duration in the field.

Testing of critical initial physical and mechanical properties of the films was carried out in terms of average thickness, tensile properties, elongation at break, and tear resistance. Test procedures followed were IS 2508 (1984) for tensile strength measurements and IS 13360-5-23 (1996) for tear resistance. Test specimen for tensile property determinations were of dimension 25 mm × 150 mm and 50 mm radii half circle for tear tests. Tensile strength and elongation at break along MD and TD of test specimen were conducted on a UTM. Tear resistance was tested in a Elmendorf (pendulum type) tear tester. Mulch specimen at a fixed time interval during the course of field exposure under onion crop was collected. Test specimen of mulch film from the mulch bed was collected following standard procedure for its tensile strength, elongation at break, and tear resistance at an interval of 30 days after exposure on the field till harvest of the crop. Testing was done for fresh film, 30 DAT, 60 DAT samples, 90 DAT, and 120 DAT.

19.2.1 Development of Manual Mulch Laying and Retrieving Machine

For vegetable crop like onion which is a closely spaced crop, there are some operational difficulties in using a tractor drawn mulch layer. In manual mulch laying, the cost and time requirement is very high in laying the mulch. Also in windy conditions, it becomes even more difficult. A manual mulch laying and retrieving machine was envisaged for partial mechanization of some operations involved in mulch laying like facilitating the mulch to roll and unroll while laying.

19.2.2 Constructional Details of the Machine

The main frame is made of mild steel. The frame is deliberately made rugged to provide stability and to carry the weights of its attachments like drip rolling mechanism, mulch rolling, and to work as trolley to carry agricultural produce. The overall dimension of machine is $1500 \times 750 \times 650$ mm^3 and square section pipes of 5 mm thickness is used for the main frame. Total weight of the machine is around 55 kg (without attachments). The main frame is supported by two rubberized pneumatic wheels on shaft and ball bearings for free movement. Two stands with round plates are provided for putting the trolley at rest position. Provision is made for adjustment of the transport wheel as per the raised bed width so as the machine runs on the furrow of the two adjacent beds. Rolling and unrolling provisions are made at a suitable location of the frame for plastic mulch. The detailed diagram of the machine is shown in Figure 19.1.

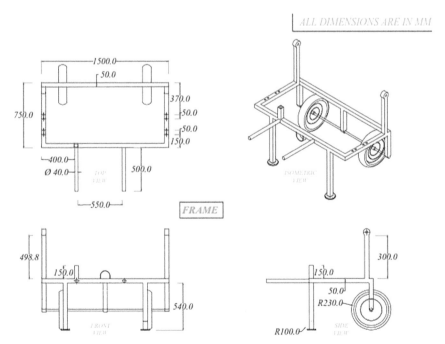

Figure 19.1   Diagrammatic details of the main frame of manual mulch cum drip laying machine.

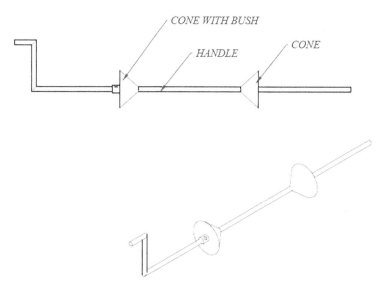

Figure 19.2 Attachments of the machine—Mulch laying and retrieving mechanism.

19.2.3 Attachments

Mulch rolling and unrolling mechanism: Two protruding square pipes near two corners are fixed with provision for putting 25 mm solid round pipe of 1750 mm. Plastic rolls are secured from both ends by adjustable solid cones with collars made of mild steel which facilitates the plastic film to roll along with the rod through a suitable handle at one end.

19.3 Results and Discussion

19.3.1 Mechanical Properties of Plastic Mulch with Duration in Field

Tensile property is one of the main parameters which decide the longevity of a plastic film. These films due to its prolonged exposure to solar radiation, heat, humidity, and other factors tend to lose its tensile strength and other mechanical properties like tear resistance, opacity, density, and thickness. Global degradation, i.e., degradation of the film irrespective of the factors, combinations, conditions, and mechanism of degradation can be roughly evaluated from loss of mechanical strength. The mechanical properties those may be measured for ascertaining it are elongation at break, tensile strength, modulus of elasticity, and tear resistance (Briassoulis and Aristopoulou,

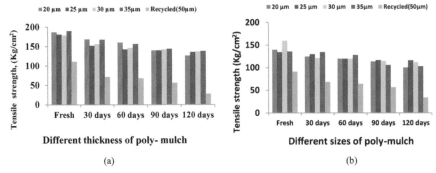

Figure 19.3   Tensile Strength of poly-mulches of different thicknesses in (a) MD and (b) TD under onion crop.

2001). The values obtained regarding mechanical properties and its changes during the crop cycle can be a measure to understand the right thickness of mulch and it can form the basis for a decision for reuse of the same mulch for using in subsequent seasons.

Tensile strength which is the force applied at any given moment to the area of original cross section (expressed in kg/cm^2) initially ranged between 178.9 and 190 kg/cm^2 for treatments M_1, M_2, M_3, and M_4. Tensile strength as prescribed by IS-2508 for LDPE should be more than 120 kg/cm^2 along the machine direction. Initial tensile strength value along the MD of the film for recycled mulch was 111.2 kg/cm^2. Tensile strength of plastic material had a decreasing trend during duration of its exposure in the field under onion crop. The decrease was seen to be more pronounced in all treatments after the first month of exposure. The reason for this can be attributed to the fact that during the first month, the film was more exposed to solar radiation as vegetative growth of onion was less.

In MD of the films, tensile strength of M_4 (35 μm mulch), M_3 (30 μm mulch), M_2 (25 μm mulch), M_1(20 μm mulch) came down to 138.5 Kg/cm^2, 137.4 kg/cm^2, 136.3 kg/cm^2, 126.60 kg/cm^2 respectively after 120 days from 190.0 kg/cm^2, 178.9 kg/cm^2, 180.8 kg/cm^2 and 186.8 kg/cm^2 respectively of the initial value. However, in M_5 (recycled mulch, 50 μm) the evolution of tensile strength was from 111.2 kg/cm^2 (initial value) to 28.0 kg/cm^2 in 120 days. Similar trend was also observed along TD of the film.

Elongation at break is the strain produced due to the force applied in the test piece expressed usually as a percentage of the original gage length of the test specimen. Reductions in elongation at break values were in the range of 30.18–78.21% after 120 days of exposure under onion crop in different

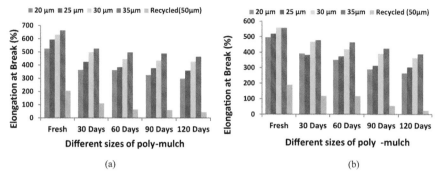

Figure 19.4 Elongation at break of poly-mulches of different thicknesses in (a) MD and (b) TD under onion crop.

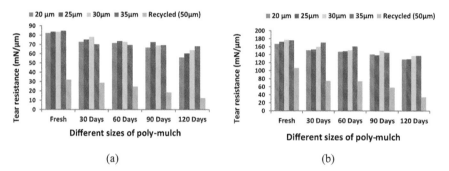

Figure 19.5 Tear resistance of poly-mulches of different thickness in (a) MD and (b) TD under onion crop.

mulches along the MD of the film. Highest reduction was seen in mulch M_5 (recycled poly mulch) where reduction was up to the tune of 70.21%. Lowest reduction of this parameter was seen in treatment M_4 (35 μm film) followed by M_3 (30 μm film) and M_2 (25 μm film). An almost similar pattern was also observed along the TD of the film where the elongation at break values reduced from 30.25 (M_4) to 80.89% (M_5). The reduction was more pronounced after 30 days of exposure across all treatments. The highest value of elongation at break was found in mulch M_4 (664.6%) followed by mulch M_3 (632.6%) and the lowest value of elongation at break was found out to be for mulch M_5 (206.6%). These values conformed to IS-2508-84 standards, which prescribes a minimum elongation at break for LDPE of thickness ranging from 12.5 to 75 μm should be more than 200 and 300% along MD and TD, respectively.

Figure 19.6   Plastic mulch laying and collection operation with manual mulch laying cum retrieving machine.

Tear resistance was more along the TD of the film compared to MD. Highest tear resistance value (176.6 mN/μm) was recorded in mulch M_4 and the lowest value (107.5 mN/μm) recorded in treatment M_5 along the TD of the film. However, along the MD, tear resistance value varied narrowly in treatments M_1, M_2, M_3, and M4 from 84.8 (M_4) to 82.4 mN/μm (M_1). Lowest value of tear resistance in MD was found in treatment M_5 (32.2). If we observe the trend during the course of exposure of the film, in treatment M_4, the drop in tear resistance was minimum (19.45%) in MD and in TD (21.97%). While in treatment M_5, a maximum drop in value was observed which was 60.86 and 67.9% in MD and TD, respectively.

Manual plastic mulch laying and recollection machine is envisaged to perform firstly to facilitate laying of drip line by unrolling the lateral from the assembly at one end of the prepared bed. The laying of mulch film can be performed by two unskilled labor one for pushing or pulling the trolley along with unrolling the film and the other to anchor the film with soil for crease free laying. Secondly, the layed film can be recollected with lesser effort and minimum tear by repeating previous process by running the trolley along the wet furrow of the bed and one person for rolling up of the retrieved mulch. The machine can also be used as push trolley by removing the attachments and can be used for transporting bags, crates, etc.

Comparative time and labor requirements for 0.1 ha area were also recorded for plastic mulch laying with machines and manual mulch laying. In manual method of plastic mulch laying, two persons were required to make raised bed of 20 cm height, two persons to press and hold the plastic roll, and one person was required to cover the edges of the plastic film with soil. In manually operated mulch laying machine, the total labor requirement

Table 19.1 Comparison of Manual Method of plastic mulching and mechanization of laying process in 0.1 ha area

Unit Operations	Manual Mulching	Manual Multi-utility Trolley	Mulching with Plastic Mulch Laying Machine
Person required to make bed of 75 cm top width	Two	Two	–
Persons required to press and hold plastic rolls	Two	One	–
No. of persons required to press the sheet and put the soil on the edges of plastic sheet	One	One	One
Total number of the persons required	Five	Four	Two (one operator and one helper)
Time required to cover 0.1 ha area,	8 h–50 min	6 h–52 min	1 h–02 min
Field capacity, ha/h	0.011	0.017	0.096
Labor requirement, man-h/ha	441.0	274.4	20.6

was reduced in the operation of pressing and holding the mulch roll. In tractor-operated mulch laying machine, the labor requirement was reduced to two (operator and helper). The time required to complete laying of plastic mulch for 0.1 ha was 8 h–50 min for manual laying, 6 h–52 min for manually operated machine, and 1 h–02 min for the tractor-operated machine. Total labor requirement for plastic mulch laying was 441, 274.4, and 20.6 man-h/ha for manual laying, manually operated mulch laying machine, and tractor-operated mulch laying machine, respectively. The other advantage of multi-utility trolley for mulch cum drip laying is its simplicity in operation and its utility for collection of drip laterals and mulch films after the end of the crop season. However, for large-scale mulching, a tractor-operated mulch laying machine is more cost effective. Mulching onion crop with black plastic has resulted in an average yield of 48.4 ton/ha as against 35.64 ton/ha for plots without mulch.

Field performance test of mulch retrieval was conducted for two consecutive years for understanding the effectiveness of the machine for mulch recollection. The machine is operated by two persons. The time required to cover 0.1 ha was 1 h 19 min with a speed of operation of 1.69 km/h. The effective field capacity was 0.10 ha/h, the field efficiency was 61.47%, and the labor requirement was 52.88 man-h/ha.

Table 19.2   Field test for field collection of mulch (total area 1000 m^2)

| S. No. | Unit Operations | Retrieval of Plastic Mulch through Multi-utility Trolley | | | |
		Test I	Test II	Test III	Average
1.	Person required to clear mulch bed of dried onion stalk and grasses before collection	2	2	2	2
2.	Persons required to push trolley	1	1	1	1
3.	Person required to operate the handle	1	1	1	1
4.	Total number of the persons required	4	4	4	4
5.	Time required to cover 0.1 ha area,	1 h–22 min	1 h–19 min	1 h–17 min	1 h–19 min
6.	Speed of operation, km/h	1.67	1.69	1.70	1.687
7.	Theoretical field capacity, ha/h	0.167	0.169	0.170	0.169
8.	Effective field capacity, ha/h	0.102	0.104	0.105	0.104
9.	Field efficiency, %	61.10	61.54	61.76	61.47
10.	Labor requirement, man-h/ha	54.66	52.66	51.33	52.88

19.4 Conclusion

The mechanical properties in all mulches except recycled black mulch did not suffer mechanical deterioration due to exposure in the field beyond permissible limit in one crop season. Furthermore, within poly mulching there is an optimum thickness which imparts more favorable condition of soil by moderating soil temperature, soil compaction for good root penetration, and proper soil moisture availability. It could also be concluded from the studies that the overall black poly-mulching has a positive impact on yield and growth parameters in onion production. Water use efficiency can be enhanced by the use of polyethylene mulching, and hence significant water savings can be achieved.

The manually operated mulch laying and retrieving machine is more suitable for closely spaced drip irrigated vegetable crop like onion in small areas. However, for large-scale cultivation and for vegetable crops with higher interbed spacing, a tractor-operated machine will be more cost effective. We can also reduce the cost of mulching by selecting the right thickness of plastic mulch which will live through the crop duration. A size increment of 5 μm in thickness of mulch film results in using an average 42 kg more plastic mulch per hectare. So care should be taken to select the right thickness of plastic mulch for laying.

Acknowledgments

The authors express gratitude to plastic mulch film manufacturer, M/s Essen Multipack Limited, Shapar, Rajkot, for providing test in its laboratory and All India Coordinated Research Project on Plasticulture Engineering and Technology, ICAR, for providing financial assistance.

References

ASAE standards X 313.2. (1998). Agricultural Engineers Yearbook of Standards. St. Joseph, Mich: ASAE.

Briassoulis D. (2004). An overview on the mechanical behavior of biodegradable agricultural films. Journal of Polymers and the Environment. 12:65–81.

Briassoulis D, Aristopoulo A. (2001). Adaptation and harmonization of standard testing methods for mechanical properties of low density polyethylene (LDPE) films. Polymer Testing. 20:615–34.

IS 13360-5-23. (1996). Determination of tear resistance of plastic films and sheeting. Elmendorf method. Bureau of Indian Standards, New Delhi.

IS 2508 (1984). Specifications for Low Density Polyethene films. Bureau of Indian Standards, New Delhi.

20

Fluoropolymer-Based Tunable Materials for Emerging Applications

Shashikant Shingdilwar, Sk Arif Mohammad, and Sanjib Banerjee*

Department of Chemistry, Indian Institute of Technology Bhilai, Raipur, Chhattisgarh, India
E-mail: sanjib.banerjee@iitbhilai.ac.in (S.B.)
*Corresponding Author

This chapter reports an overview on the synthesis, polymerizability, uses, and applications of fluorinated homo- and copolymers for emerging applications. Trifluoromethacrylic acid (MAF) and alkyl 2-trifluoromethacrylate monomers (MAF-esters, where the alkyl group stands for methyl, ethyl, t-butyl, a perfluorinated one, or a functional group) are a very important class of comonomers. Due to their easy availability and easy handling, MAF or MAF-ester based copolymers or terpolymers are used in many high-tech applications. MAF or MAF-esters do not undergo homopolymerization under radical polymerization condition. However, their radical copolymerizations with electron-donating hydrogenated monomers such as α-olefins, vinyl ethers, vinyl acetate (VAc), styrene, and norbornenes led to a wide range of copolymers bearing alternating monomer sequences. Recently, reversible deactivation radical polymerizations of VAc with *tert*-butyl-2-trifluoromethacrylate (MAF-TBE) were reported under nitroxide-mediated polymerization (NMP), reversible addition fragmentation chain transfer (RAFT), or organometallic mediated radical polymerization (OMRP) techniques leading to well-defined poly(VAc-*alt*-MAF-TBE) alternating copolymers. Thus, developed polymers might have a wide range of applications including polymer electrolyte membranes for fuel cell, solid polymer electrolytes for lithium-ion batteries, microlithography, coatings, and surfactants.

20.1 Introduction

Fluoropolymers are niche products, exhibiting unique properties such as resistance to thermal, aging, or weather aggressions and excellent inertness to a wide range of chemical environments. They have extensive uses in many high-tech applications: aeronautics and aerospace, high-performance membranes, cores and claddings of optical fibers, textile treatment, cables and wires insulation, building industries, petrochemicals (pipes and coatings as liners), automotive industries, and microelectronics (Ameduri and Boutevin, 2004; Patil and Ameduri, 2013a; Smith et al., 2014). Currently, only four commercially available homopolymers can be achieved till date, from only four monomers: trifluoroethylene (TFE), vinylidene fluoride (VDF), chlorotrifluoroethylene (CTFE) and vinyl fluoride (VF). However, these homopolymers have three major limitations: (i) high crystallinity (leading to expensive processing), (ii) poor solubility in common organic solvents (inducing to characterization issues), and (iii) difficulty in crosslinking and tuning the properties on-demand due to the lack of functionality.

One of the possible way to overcome these above-mentioned drawbacks is to introduce monomers containing pendant functional groups (Ameduri, 2009) (such as acetoxy, ethers, esters, halogens, hydroxyl, thioacetoxy, $-CO_2H$, or aryl groups) as comonomers during the radical copolymerization of the above-mentioned fluoromonomers. This may indeed improve some of the properties (such as adhesion, conductivity, crosslinking, or processability, to name a few) of the resulting polymers (Ameduri, 2009; Patil and Ameduri, 2010, 2013a). This led to investigation to explore new fluoropolymers from such above comonomers. Several aforementioned properties of the fluorinated homopolymers can also be tuned and enhanced in the presence of specific functional groups born by the backbone.

The bottleneck deals with the fact that only a few functional fluoropolymers are commercially available (Smith et al., 2014; Ameduri et al., 2016). Thus, extensive research has been carried out to synthesize new monomers that may exhibit complementary properties and also bear a few polar (ester group) functionalities to enable further adhesion or crosslinking. This led to the synthesis of fluorinated copolymers based on MAF and alkyl 2-(trifluoromethyl)acrylates (MAF-esters) (Castelvetro et al., 2000; Ito et al., 2001; Wang et al., 2006; Piletska et al., 2008; Gohy et al., 2011; Souzy et al., 2012; Zhu et al., 2012; Alaaeddine et al., 2015; Banerjee et al., 2017; Wehbi et al., 2017). Indeed, MAF or MAF-esters do not homopolymerize from perfluorinated side chain by transesterification of MAF-TBE radical polymerization conditions (Ito et al., 1984; McElroy et al., 1999).

However, they are reported to (i) homopolymerize under anionic initiation (Narita, 1999, 2010), (ii) copolymerize with various hydrogenated monomers (donating monomers such as NBs, VEs, α-olefins, VAc (Banerjee et al., 2016, 2017a,b) that led to alternating copolymers) (Ito and Miller, 2004; Ito et al., 2004a,b), (iii) even some fluoroolefins such as VDF (Souzy et al., 2004; Boyer and Ameduri, 2009; Banerjee et al., 2017; Wehbi et al., 2017), (iv) polar monomer, or (v) terpolymerize with VDF and hexafluoropropylene (HFP) (Sawada et al., 2011) or other fluorinated olefins. This has led to a significant growth in the research in the field of the co- or terpolymerization of MAF and MAF-esters. This is also boosted by the interest of several companies on the design of novel materials for various high-tech applications.

In 2013, an extensive review details the synthesis of MAF and MAF-esters, their (co)polymerizations with hydrogenated comonomers (vinyl ethers (VEs), norbornenes (NB), α-alkene), and their applications mainly as molecular imprinting materials, nanocomposites, stone protection, etc., and this present book chapter will not recall these parts. This chapter is divided into four main parts: (i) a first one will briefly review and highlight the recent progress dealing with facile synthesis of new MAF-ester-based functional monomers that bear functional groups (cyclic carbonate, dialkyl phosphonate and phosphonic acid, epoxides, cycloethers, oligo(EO) oligomers) or a perfluorinated group; (ii) then, the homopolymerization, co- and terpolymerization reactions of MAF and MAF-esters with various hydrogenated polar monomers (VAc, MMA, MAA, NB, VEs) and halogenated (VDF, HFP, etc.) monomers, as well as the properties of the resulting copolymers. This part is divided between techniques of polymerization, ranging from conventional to controlled (or RDRP) radical copolymerizations, (iii) the copolymerization kinetics (that enables to classify different reactivities of comonomers with respect to MAF or MAF-esters), and (iv) finally, the application of these copolymers in various fields including Proton-exchange membrane fuel cell (PEMFCs) for lithium-ion batteries, multi-compartmental micelles, fluoropolymers/silica composites, coatings, and fluorosurfactants.

20.2 Fluoropolymer Based Tunable Materials: Design, Synthesis and Applications

20.2.1 Fundamentals and Developments of Controlled Radical (co)Polymerization

Since the mid-1990s, extensive research has been carried out to develop controlled/living radical polymerization (Rosen and Percec, 2009;

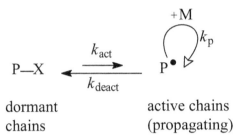

Scheme 20.1 Reversible fast exchange between active and dormant species in RDRP.

Matyjaszewski, 2012; Monteiro and Cunningham, 2012). The origin of control in RDRP process arises from the (i) nature of the (macro)radicals having very short "active" life times and (ii) fast exchange between living "active" macroradicals and dormant chains (Scheme 20.1). Both features reduce undesirable chain transfer and terminations.

In a pioneering work, Tatemoto (1979) and Tatemoto and Shimizu (1997) reported iodine transfer (co)polymerization of fluoroalkenes in 1978. Recently, NMP (Georges et al., 1993; Nicolas et al., 2013), ATRP (Matyjaszewski, 2012), reverse ITP (David et al., 2006), RAFT (Boyer et al., 2009; Gregory and Stenzel, 2012), organoheteroatom-mediated radical polymerization based on tellurium, bismuth, antimony (Yamago, 2009), and cobalt mediated radical polymerization using bis(acetylacetonato)cobalt(II) [Co(acac)$_2$] (Debuigne et al., 2007) were successfully developed as RDRP techniques. Zhang et al. (2007) reported the synthesis of chain end-functionalized fluoropolymers via RDRP using functional borane initiators. Recently, Goto et al. (2013) developed reversible complexation-mediated RDRP.

Development of RDRP methods for fluoromonomers has been hindered by the recombination of the growing macroradicals (Timmerman and Greyson, 1962). Thus, various combinations of fluorinated organic compounds have been used as initiators, monomers, CTAs, or solvents to achieve RDRP of fluoromonomers. Detailed surveys of the development of state-of-the-art strategies for RDRP of fluorinated monomers are given in two previous reviews (Ameduri, 2010, 2014). Thus, this section focuses on recent exciting discoveries in RDRP of fluoroalkenes, especially on ITP and RAFT/MADIX polymerization, followed by a second section devoted to the RDRP of fluorinated (meth)acrylates and styrenic monomers.

20.2.2 Controlled Radical Polymerization of Fluoroalkenes

20.2.2.1 ITP of fluoroalkenes

Development of ITP of fluoropolymers has been inspired by radical telom-erization of fluoroalkenes (Boutevin, 2000). In the late 1970s, Tatemoto and Shimizu (1997) were the first to introduce the concept of RDRP as ITP of fluoroalkenes. In ITP of fluoroalkenes, the terminal active bond (pseudo-living center) of the growing polymer contains a C–I bond which originates from the iodine-containing CTA and the monomer. The reactivity of this terminal group remains the same for the entire during of the polymerization and even after quenching (Ameduri, 2009).

Subsequently, Boyer et al. (2005) reported ITP of VDF utilizing various 1-iodofluoroalkanes as CTA such as $C_6F_{13}I$ and $C_6F_{13}CH_2CF_2I$. The poly-merization demonstrated a pseudo-living tendency arising out of the lability of CF_2–I bond and the rapid decomposition of the $C_6F_{13}I$ (Boyer et al., 2005). Sabine and I-u-h (2007) developed ITP of fluoromonomers in supercritical carbon dioxide ($scCO_2$) in a homogeneous-phase using $C_6F_{13}I$ as the CTA. Mladenov et al. (2006) reported the synthesis and characterization of VDF and HFP containing fluorinated copolymers using 1,6-diiodoperfluorohexane as the CTA via an ITP process.

Asandei et al. (2012) reported photo-induced RDRP of VDF and other monomers using different metal carbonyls employing $Mn_2(CO)_{10}$- as a catalyst, in the presence of iodoperfluoroalkanes (RFIs). Asandei et al. (2013) also reported hypervalent iodide carboxylate catalyzed RDRP of VDF and prepared a well-defined VDF-based block copolymer. Simpson et al. (2015) further studied the effects of metal and ligand on the ITP of VDF and block copolymerization of VDF. Although the system developed by Asandei et al. showed promise, a major drawback of this procedure is that they cannot be scaled up for commercial applications.

20.2.2.2 Reversible addition-fragmentation chain transfer/macromolecular design via the interchange of xanthates polymerization of fluoroalkenes

RAFT/MADIX polymerization has recently emerged as one of the most promising RDRP techniques because of its tolerance toward wide range of functional groups and its less-stringent experimental conditions (Hill et al., 2015). The following section will focus on recent advances toward the synthesis of specifically designed fluorinated copolymers via MADIX.

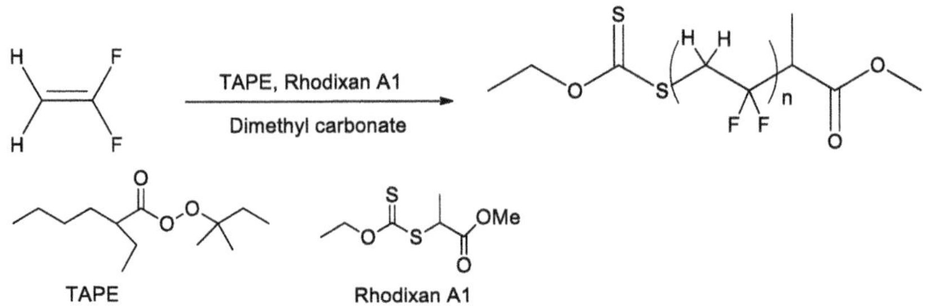

Scheme 20.2   Synthesis of PVDF by RAFT/MADIX polymerization of VDF. Reproduced with permission from American Chemical Society (Guerre et al., 2015).

RDRP of fluoromonomers, especially VDF is still very challenging. In a breakthrough research, Guerre et al. (2015) developed an efficient protocol for the first MADIX polymerization of VDF using *O*-ethyl-S-(1-methoxycarbonyl)ethyldithiocarbonate as the CTA (Scheme 20.2).

Subsequently, the same group carried out chemically modified the xanthate end-group of the above PVDF into thiols and subsequently to PVDF-methacrylate. Thus, synthesized PVDF-methacrylate macromoner was then copolymerized with MMA, to produce novel block copolymers (Guerre et al., 2016a). The as-synthesized PVDF-xanthates were also used as macroinitiators for the synthesis of PVDF-*b*-poly(VAc) block copolymers (Guerre et al., 2016b).

20.2.2.3 ATRP of fluoroalkenes

Wang and Weiss (2003) reported the unexpected ATRP of an electron-rich 4-(chloromethyl)styrene (CMS) (that acts as both monomer and initiator) and an electron-withdrawing monomer (CTFE) to prepare hyperbranched poly(CTFE-*co*-CMS) copolymers. This hyperbranched polymer was found to be soluble in common organic solvents and exhibited amorphous nature with a T_g of 88°C. This is intermediate between those of poly(CTFE) and polystyrene (PS), 57 and 100°C, respectively.

In conclusion, among all the RDRP methods employed for the synthesis of fluoropolymers from fluoroalkenes, ITP and RAFT/MADIX polymerization has emerged as the most preferred technique. Controlled polymerization of fluoroalkenes could lead to many exciting new materials with unique properties.

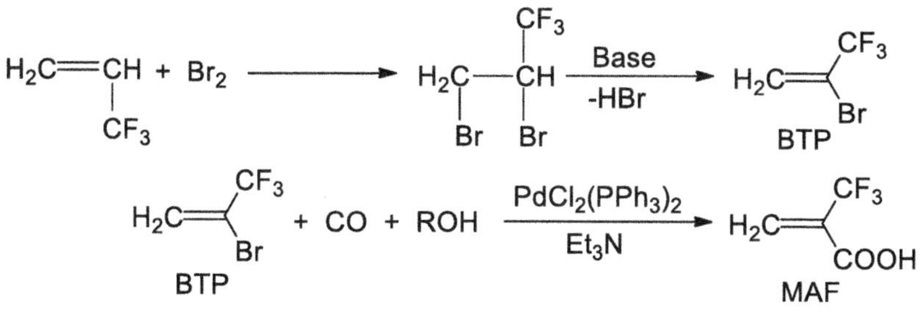

Scheme 20.3 Synthesis of MAF from TFP. Reproduced with permission from American Chemical Society (Sawada et al., 2011).

20.2.3 Synthesis of 2-(Trifluoromethyl) Acrylic Acid (MAF) and Alkyl 2-(Trifluoromethyl)Acrylates (MAF-Esters)

20.2.3.1 Commercial grade MAF and MAF-esters

Various synthesis routes have been employed for MAF synthesis (Henne and Nager, 1951; Buxton et al., 1954; Drakesmith et al., 1964; Botteghi et al., 1997). Recently, Sawada et al. (2011) reported an improved method under Heck reaction conditions using $PdCl_2(PPh_3)_2$ as the catalyst in the presence of TEA and water (Scheme 20.3). However, the yield was low, up to 67% only. Subsequently, Shaw et al. (1973) described an even more attractive method using the corresponding sodium salt with the adequate alkyl bromide or iodide.

20.2.3.2 Synthesis of non-commercial grade 2-trifluoromethacrylate monomers from MAF and MAF-esters

A series of MAF-esters containing mono 2-trifluoromethacrylate and epoxy groups and 2-trifluoromethyl styrenic), and telechelic bis(2-trifluoromethacrylate) monomers have been reviewed. Their synthesis is summarized below.

20.2.3.2.1 Mono 2-trifluoromethacrylate monomers

2-(Trifluoromethyl)acrylates containing benzoyl and ether groups

Hosoya et al. (2010) reported synthesis of synthesized 2-trifluoromethylacrylate derivatives. Later, Umino et al. (2008, 2009) reported the synthesis of phenyl 2-trifluoromethacrylate (PFA) via addition of 2-trifluoromethacryloyl chloride (MAF-COCl) to phenol and TEA at room temperature. They synthesized MAF-COCl synthesized in 80% yield via reaction of MAF with phthaloyl dichloride (Yamazaki et al., 2001; Umino et al., 2008, 2009) while recently researchers used thionyl chloride instead (Wadekar et al., 2014; Banerjee et al., 2017).

Epoxy-containing 2-trifluoromethacrylate

Shin-Etsu Chemical (Hatake yama and Takeda, 2007) reported synthesis of epoxy, alkyl and benzoyl group containing MAF-ester (Scheme 20.4). Unfortunately, no information was supplied regarding the nature and the amount of side product(s), especially when there is a possibility of oxirane ring opening.

(Trifluoromethyl)acrylates bearing a fluorinated group

Knell et al. (1968) prepared nove 1,1-dihydroperfluoroalkyl-2-(trifluoro methyl)acrylates by the reaction of MAF with a slight molar excess of phosphorus pentachloride (Scheme 20.5).

Wadekar et al. (2014) synthesized MAF bearing a perfluorinated side chain by transesterification of MAF-TBE with a fluorinated alcohol.

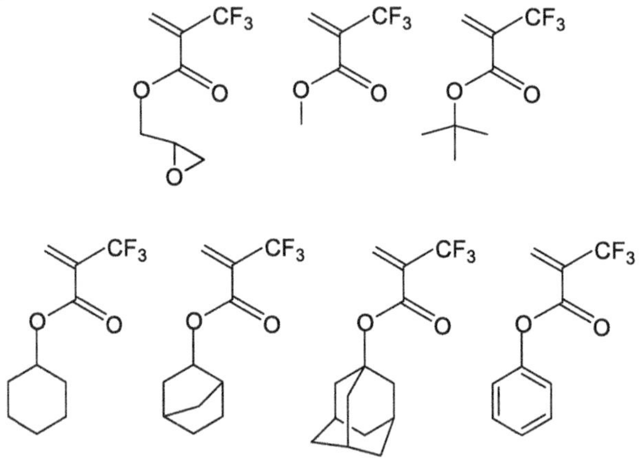

Scheme 20.4 A series of new (cyclo)alkyl 2-trifluoromethacrylate, developed claimed by Shin-Etsu company, reproduced with permission (Hatakeyama and Takeda, 2007).

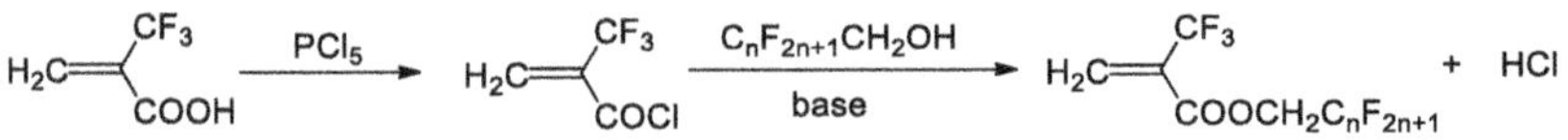

Scheme 20.5 Synthesis of 1,1-dihydroperfluoroalkyl 2-(trifluoromethyl)acrylates via esterification of 2-(trifluoromethyl) acrylic acid (MAF) with a fluorinated alcohol ($n = 1$–12), reproduced with permission from US patent office (Knell et al., 1968).

Oligo(oxyethylene) (EO)-containing 2-(trifluoromethyl)acrylates

Alaaeddine et al. (2015) reported the synthesis of original 2-(trifluoromethyl) acrylates bearing a pendant EO chain via esterification of MAF with ω-hydroxy oligo(EO). Subsequently, they produced various PVDF-*g*-oligo(EO) graft copolymers for potential application as novel polymer electrolytes for Li-ion batteries.

20.2.3.3 Synthesis of Bis(2-trifluoromethacrylate) monomers

Umino et al. (2009) and Kurakami et al. (2010) employed addition of 2-trifluoromethylacryloyl chloride onto a hydroquinone/TEA mixture to prepare telechelic monomers, bis(2-trifluoro-methacrylate)s and 1,4-phenylene bis(2-trifluoromethyl acrylate). Tosoh FineChem Corporation reported the synthesis of bis(2-trifluoromethyl) acrylic acid 1,3-adamantanediyl ester (Fuchikami et al., 2006).

20.2.4 MAF and MAF Derivatives as Precursors of Organic Chemicals

Trifluoromethyl uracyl derivative of MAF has potential application in ophthalmology. It is readily synthesized by condensation reaction of MAF with urea (Scheme 20.6) (Fuchikami et al., 1984), Hosoya et al. (2008) synthesized a series of organic MAF-esters by reaction of MAF-TBE and cyclic ethers (Scheme 20.6). Wadekar et al. (2014) developed oligo(EO), MAF-phosphonates (further modified into PA) (Banerjee et al., 2017) and MAF-cyclic carbonate (Wehbi et al., 2017), which were subsequently comonomerized with VDF to prepare VDF-functionalized copolymers. Umino et al. (2009) reported the successful double Michael addition of PFA onto ethyl cyanoacetate (ECA).

20.2.5 Polymerization of MAF and MAF-Esters

This section deals with the behaviors of MAF and MAF-esters in telomerization, radical and/or anionic homopolymerizations, and their co- and ter-polymerization with hydrogenated and/or halogenated comonomers.

20.2.5.1 Telomerization

Because of their poor reactivities, homotelomerizations of MAF or MAF-esters do not undergo homopolymerization under radical polymerization conditions (Patil and Ameduri, 2013a). Riachy et al. (2017) synthesized a new

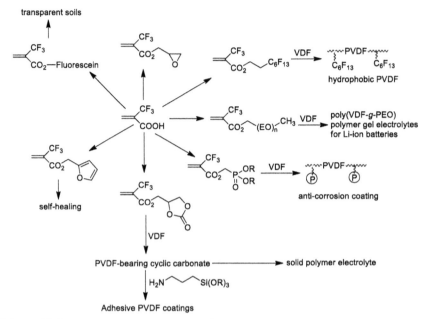

Scheme 20.6 Synthesis of a series of 2-(trifluoromethyl)acrylates (MAF-esters) and their radical copolymerizations for specific materials.

fluorinated surfactant, 2-trifluoromethyl-7,7,8,8,9,9,10,10,11,11,12,12,12-tridecafluoro-4-thia-1-dodecanoic acid (FSC) via the thiol-ene radical addition (or telomerization to produce the monoadduct) of 3,3,4,4,5,5,6,6,7,7,8,8,8-tridecafluoro-1-octanethiol onto 2-trifluoromethyl acrylic acid in 85% yield. This was subsequently used spherical wormhole-like mesostructured silica material (Scheme 20.7).

1-Iodoperfluoropropane mediated radical tertelomerization of MAF with VDF and 2H-PFP led to a tertelomer containing 52:23:25 mol% of MAF:VDF:2H-PFP (Scheme 20.8) (Ameduri et al., 2008). These original poly(MAF-*ter*-VDF-*ter*-2H-PFP) terpolymers were involved in the synthesis of original fluorosurfactants (Kostov et al., 2012).

Ito et al. (1982) and Ito and Miller (2004) also reported radical terpolymerization of two highly electron-deficient monomers (methyl 2-trifluoromethyl acrylate (MTFMA) and/or MA) with NB. Detailed kinetics study of the polymerization revealed that tBuMA was quickly converted at the initial stage of the polymerization reaction and the polymerization slowed

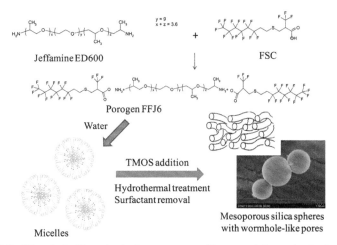

Scheme 20.7 Wormhole-like spherical mesoporous silica materials synthesized via templating approach in the presence of a fluorinated MAF-based surfactant, followed by hydrothermal treatment to remove the surfactant, reproduced with permission from Elsevier (Riachy et al., 2017).

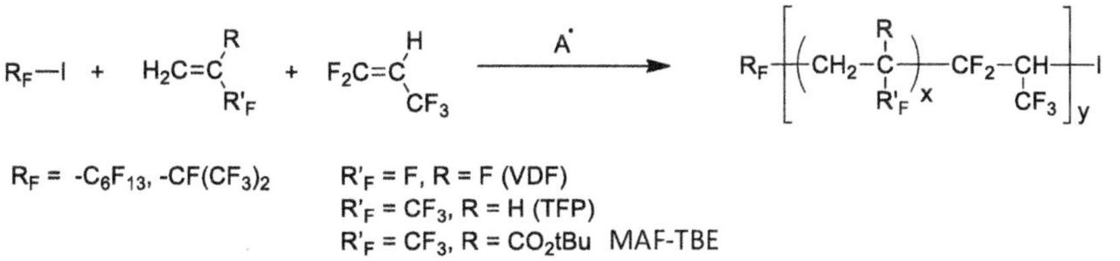

$R_F = -C_6F_{13}, -CF(CF_3)_2$

$R'_F = F, R = F$ (VDF)
$R'_F = CF_3, R = H$ (TFP)
$R'_F = CF_3, R = CO_2tBu$ MAF-TBE

Scheme 20.8 Radical co- and ter-telomerization of MAF with VDF and 2H-PFP mediated by 1-iodoperfluoroalkane as the CTA, reproduced with permission from Elsevier (Boschet et al., 2015).

down after 500 min. A similar observation was noted by the authors for MA, tBuMA, and NB (Ito et al., 2000).

20.2.5.2 Copolymerizations of MAF and MAF-esters
20.2.5.2.1 Radical copolymerization with hydrogenated monomers

Different MAF and MAF-esters have been radically copolymerized with different electron-donating monomers such as α-olefins, NB, VAc, and VEs (to lead to alternated copolymers) and also alkyl methacrylates.

20.2.5.2.2 Radical copolymerization of MAF and MAF-esters with α-olefins

Successful copolymerization of MAF with α-olefins (e.g., 1-decene, 2-methyl-1-hexene) using AIBN as the initiator was reported by Aglietto et al. (1995). Ito et al. (2004a, 2008) investigated AIBN initiated radical copolymerizations of MAF-ester with CMS and [4-(1,1,1,3,3,3-hexafluoro-2-hydroxypropyl) styrene (STHFA)]. Ito et al. (1982) also reported polymerization of methyl 2-(trifluoromethyl)acrylate and 2-trifluoromethylacrylonitrile and copolymerization of these monomers with MMA.

20.2.5.2.3 Radical copolymerization of MAF and MAF-esters with NB

Copolymers obtained via radical copolymerization of MAF and MAF-esters with NBs have potential application in 157 nm lithography (Ito et al., 2001b,c). Ito and Miller (2004) and Ito et al. (2004a) carried out an extensive investigation of the radical copolymerization of NB with MAF-TBE, initiated by AIBN (in bulk). Subsequently, the same group reported the copolymerization of methyl 2-trifluoromethyl acrylate and methyl 5-NB-2-methyl-2-carboxylate using 4.3 mol% AIBN (Ito et al., 2008). Iwatsuki et al. (1984) reported radical copolymerization of methyl 2-trifluoromethyl acrylate with MMA and styrenic monomers.

20.2.5.2.4 Radical copolymerization of MAF and MAF-esters with VEs

Aglietto et al. (1995) pioneered the radical copolymerization of MTFMA with hydrogenated VEs (such as 2-ethylhexyl vinyl ether (2EHVE), butyl vinyl ether (BuVE) and methyl vinyl ether (MVE), Scheme 20.9). Cracowski et al. (2009) revisited the same reaction with VE bearing a perfluorinated group and obtained an alternating copolymer. But the yield was very low yield (Aglietto et al., 1995). The same group reported that the obtained copolymers can be easily quantitatively transformed into the corresponding methyl ester. The nature of the sterically bulky alkyl group in alkyl VE influences the fate of the reaction. This is because the reaction goes through a charge-transfer complex increases with an increase of the bulkyness of the alkyl group (Fleš et al., 1981). Depending on the compositions and molecular weights, the obtained copolymers exhibited a wide range of glass transition temperature (T_g) ranging from 20 to 120°C.

Ito and Miller (2004) and Ito et al. (2004a) carried out a comprehensive investigation and confirmed that linear or cyclic electron-rich VEs, such

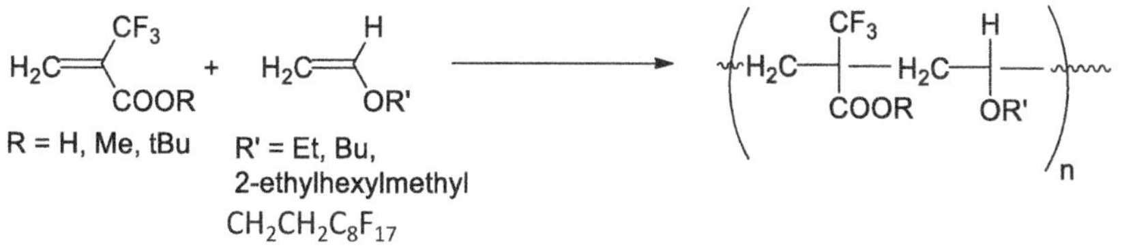

Scheme 20.9 Radical copolymerizations of VEs (R′ ranging from various hydrocarbon groups to fluorinated ones) with MAF and MAF-ester, with reproduced with permission from Wiley (Aglietto et al., 1995).

as *t*-BuVE, EVE, 3,4-dihydrofurane, and VC spontaneously undergoes radical copolymerization with MAF-TBE. Results revealed that the higher MAF-TBE content in the comonomer feed led to lowering of the lower the molecular weights of the resulting copolymer. The molecular weights of the resulting copolymer ranged from 64,000 to 51,600 g.mol^{-1}. The copolymer exhibited a T_g of ~70°C which remained constant over a wide range of comonomer feed compositions. Ito and Miller (2004) and Ito et al. (2004a) also identified a MAF-TBE–MAF-TBE–EVE triad, and the other signals attributed to EVE–MAF-TBE–EVE alternated sequences using NMR spectroscopy.

20.2.5.2.5 Radical copolymerization of MAF and MAF-esters with MMA

MAF is able to copolymerize with different acrylate monomers such as MMA. Ito et al. (1982) and Xu and Frisch (1994) reported the radical copolymerizations of MTFMA with MMA.

20.2.5.2.6 Radical copolymerization of MAF with MAA

Only very few reports are there in the literature describing the copolyerization of MAF with MAA. Sawada et al. (2010) demonstrated the radical copolymerization of MAF with MAA initiated by a perfluorodiacyl peroxide initiator. The polymerization led to almost alternating cooligomer in acceptable yield having molecular weight of 2100 g.mol^{-1}.

20.2.5.2.7 Radical copolymerization of VAc and MAF-TBE

PVAc and VAc-based copolymers have wide industrial applications in diverse areas, including functional coating and packaging. MAF-esters undergo successful copolymerization with VAc both under conventional and controlled radical conditions. The polymerization reactions were carried out both

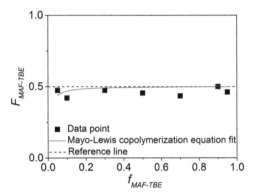

Figure 20.1   Plot showing the dependence of the instantaneous copolymer composition, $F_{MAF-TBE}$, on the initial comonomer feed composition, $f_{MAF-TBE}$, for the free radical copolymerization of VAc and MAF-TBE at 40°C. The red curve represents the Mayo–Lewis copolymerization equation fit (Banerjee et al., 2018).

under conventional radical polymerization and under RDRP conditions, as detailed below:

Under conventional radical copolymerization

Banerjee et al. (2018a) were the first to report alternating radical copolymerization of VAc and MAF-TBE initiated by V-70. Figure 20.1 depicts that the resulting copolymers exhibited a nearly perfect alternating structure over the wide range of comonomer feed compositions employed ($f_{MAF-TBE} = 0.05$ to 0.95). Monomer reactivity ratios were determined to be: $r_{MAF-TBE} = 0$ and $r_{Ac} = 0.014$ at 40°C. They also determined the Alfrey and Price parameters for MAF-TBE to be: $Q_{MAF-TBE} = 1.18$ and $e_{MAF-TBE} = 1.84$).

Under RDRP condition

Fundamental aspects of RDRP have already been described in previous section. Thus, this section focuses only on the recent important discoveries of the RDRP of MAF-TBE with other hydrocarbon monomers, especially VAc.

RAFT/MADIX polymerization

Banerjee et al. (2018b) recently reported syntheses of MAF-TBE-containing block copolymers via RAFT polymerization using a universal CTA. Thus, the prepared block copolymer was composed of one alternating copolymer and one homopolymer segment. This led to a useful and cost-effective alternative for the "on demand" synthesis and screening of fluorinated block copolymer libraries. Cyanomethyl 3,5-dimethyl-1H-pyrazole-1-carbodithioate (CDPCD) led to synthesis of well defined block

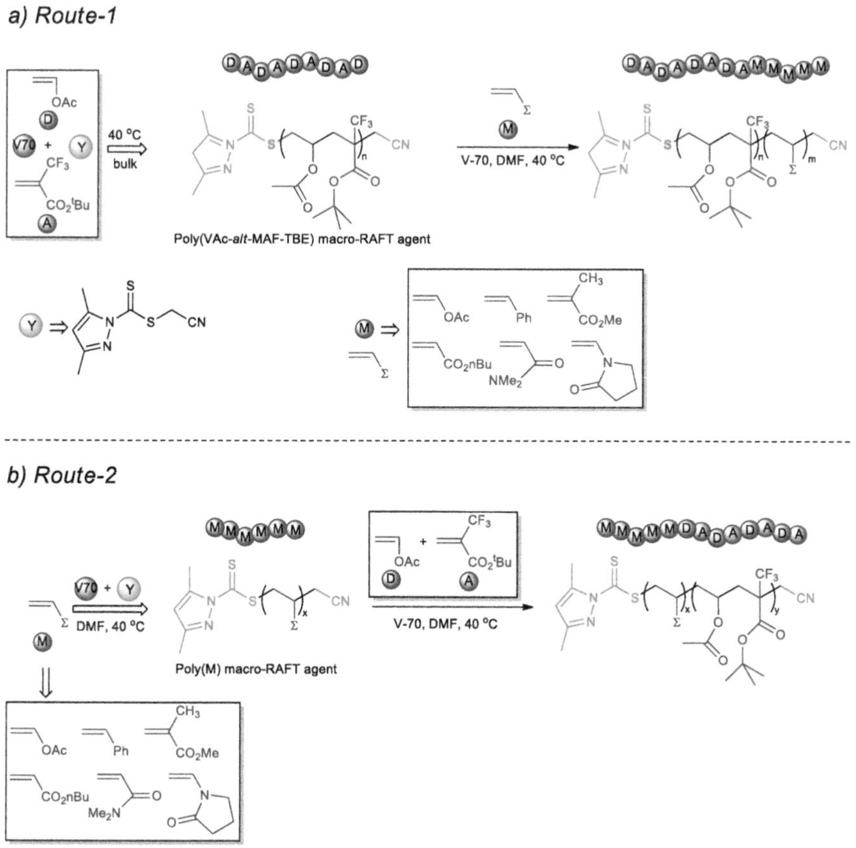

a) Route-1

b) Route-2

Scheme 20.10 RAFT copolymerization of VAc and MAF-TBE and subsequent synthesis of poly(VAc-*alt*-MAF-TBE)-*b*-poly(M) block copolymers, reproduced with permission from RSC (Banerjee et al., 2018b).

copolymer (BCP) compared to *O*-ethyl-S-(1-methoxycarbonyl)ethyldithio carbonate (CTA-XA), 2-cyano-2-propyl benzodithioate (CPDB), and 4-cyano-4-(2-phenylethanesulfanylthiocarbonyl) sulfanyl pentanoic acid (PETTC) (Scheme 20.10). SEC analysis revealed a lateral shift toward high molecular weights and a linear increase of M_n with time, maintaining low Đ values evidencing a controlled nature of the polymerization (Figure 20.2).

Results revealed that chain extension of poly(VAc-*alt*-MAF-TBE)-CDPCD with a series of monomer such as VAc, styrene (St), *n*-butyl acrylate (nBA), and DMA produced well-defined poly(VAc-*alt*-MAF-TBE)-*b*-poly(M) with acceptable dispersity values (Đ = 1.36). But, attempts to synthesize block copolymers using MMA and N-Vinylpyrrolidone (NVP)

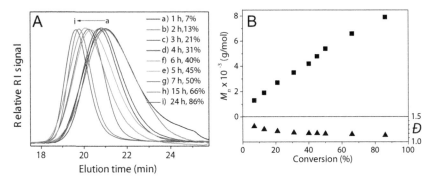

Figure 20.2 Plot showing evolutions of (A) SEC traces with reaction time and (B) Mn and Đ versus conversion for the RAFT alternating copolymerization of VAc and MAF-TBE (fMAF-TBE = 0.5) carried out at 40°C in bulk using V-70 as initiator in the presence of CDPCD (Banerjee et al., 2018b).

as the second monomer led to broadening of molecular weight distribution. Notably, attempts to synthesize the inverted poly(M)-b-poly(VAc-alt-MAF-TBE) copolymers using poly(M)-DPCD as the macroCTA led to copolymers with rather high Đ for all the cases (1.53–2.0).

Nitroxide-mediated polymerization:

Banerjee et al. (2016) also reported nitroxide-mediated alternating radical copolymerization of VAc with MAF-TBE by employing an SG1-based alkoxyamine (MAMA-SG1) at 40°C (Scheme 20.11). Linear first-order kinetics and linear evolutions of the molecular weights with time (up to 17,100 g mol^{-1}), maintaining low dispersity values (Đ = 1.33), suggested controlled nature of the copolymerization. Notable, detailed investigation revealed that none of the comonomers employed (VAc or MAF-TBE) could undergo homopolymerization under the NMP conditions, initiated by MAMA-SG1. The resulting alternating amorphous copolymer exhibited a single T_g of 59°C.

Radical copolymerization of 1,1,3,3,3-pentafluoropropylene with MAF-TBE

Organometallic mediated radical polymerization

Banerjee et al. (2017) recently developed OMRP of MAF-TBE and VAc using V-70/[Co(acac)$_2$] at 40°C leading to the synthesis of well-defined poly(VAc-alt-MAF-TBE)-b-poly(VAc) and poly(VAc-alt-MAF-TBE) copolymers. Thus, prepared copolymers had low dispersity values

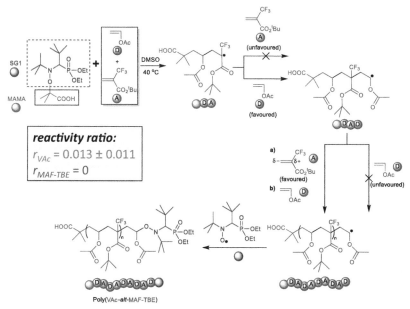

Scheme 20.11 Synthesis of poly(VAc-*alt*-MAF-TBE) alternating copolymers by NMP using MAMA-SG1. Reproduced with permission from American Chemical Society (Banerjee et al., 2016).

(≤ 1.24) and M_n up to 57,000 g/mol (Banerjee et al., 2017). The synthesized diblock copolymers exhibited two T_gs which was attributed to the alternating and homopolymer sequences.

20.2.5.2.8 Radical copolymerization of fluorinated monomers with MAF and MAF-esters

Radical copolymerization of VDF with MAF and MAF-esters

VDF-based copolymers have wide industrial applications in many areas including, but not limited to high-performance elastomers, membranes, coatings, gaskets, O-rings, and binder for lithium-ion batteries, pyro-, ferro- and piezo-electrical devices (Ameduri, 2009). However, some VDF-based copolymers suffer from three major disadvantages which hinder their performance: (i) high crystallinity which leads to increased processing cost, (ii) very poor solubility which often leads to characterization issues, and (iii) difficulty in tuning the properties which arise due to the lack of functionality (Ameduri, 2009). One of the best possible ways to overcome these limitations is to incorporate comonomers containing functional groups into the backbone (Ameduri, 2009). MAF or MAF-esters is an ideal choice because they undergo radical copolymerization with VDF both under conventional radical polymerization conditions (Souzy et al., 2004) and also under RDRP conditions (Boyer and Ameduri, 2009).

Under conventional radical copolymerization

Various functional MAF(ester) monomers have been synthesized and were successfully copolymerized with VDF targeting various potential applications. This section will provide a consolidated summary of this result. Generally, organic peroxides have been very efficient to initiate a free radical (co)polymerization. Patil et al. (2012) were the first to employ perfluoro-3-ethyl-2,4-dimethyl-3-pentyl persistent radical (PPFR) for the radical copolymerization of VDF with MAF-TBE. PPFR has a very interesting property: it is stable at room temperature, but releases $\cdot CF_3$ radicals above $80°C$ (Patil et al., 2012). The authors employed three different $[PPFR]_0/([VDF]_0+[MAF-TBE]_0)$ initial molar ratios (ca. 2.0, 1.0, and 0.5%) to produce three copolymers with molecular weights ranging from 22,000 to 105,000 g.mol^{-1}. The same team later reported a novel method to determine the molecular weights of fluoropolymers by ^{19}F NMR spectroscopy from the CF_3 end group (originating from PPFR employed as the initiator) that acts as an efficient label (Patil et al., 2013).

Banerjee et al. (2017) developed VDF-based copolymers bearing pendant PA functionality, suitable for anti-corrosion coatings for steel. For this, they first synthesized a novel (dimethoxyphosphoryl)methyl 2-(trifluoromethyl)acrylate (MAF-DMP). Then, MAF-DMP was copolymerized with VDF via conventional radical copolymerization. Simply just by varying initial comonomer feed ratio $([VDF]_0/[MAF-DMP]_0)$ a series of poly(VDF-*co*-MAF-DMP) copolymers in fair yields (47–53%) (Scheme 20.12) having different molar percentages of VDF (79–96%) and number-average molecular weights (M_n's) up to ca. 10,000 g mol^{-1} were synthesized.

Wehbi et al. (2017) recently employed cross-linking of the pendant triethoxysilane functionalized of a synthesized VDF-based copolymer to prepare a material suitable for potential applications in coatings. First, a new functional 2-trifluoromethyl acrylate cyclic carbonate monomer (MAF-cyCB) was synthesized from MAF with 70% overall yield. The target copolymer-based material was prepared via radical copolymerization of VDF with MAF-DMP, followed by the ring opening of cyCB by 3-aminopropyl triethoxysilane inducing a stable urethane function.

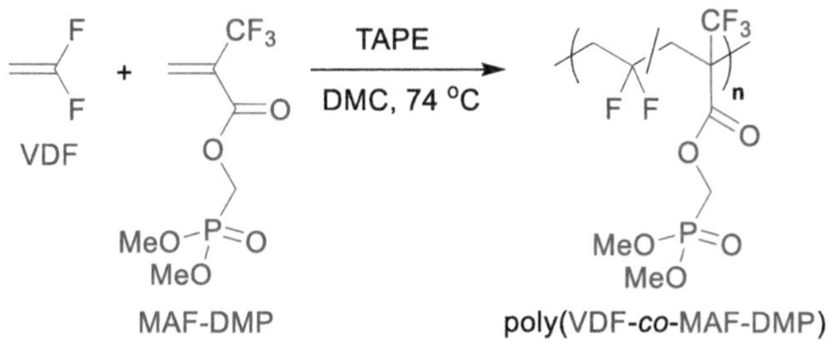

Scheme 20.12 Radical copolymerization of MAF-DMP with VDF initiated by TAPE. Reproduced with permission from American Chemical Society (Banerjee et al., 2017).

RDRP condition

The development of state-of-the-art strategies for RDRP, especially for the fluorinated monomers has already been discussed in the earlier sections (Ameduri, 2010). Thus, this part focuses only on the recent very interesting discoveries of the RDRP of fluoroalkenes with MAF or MAF-esters.

Iodine transfer polymerization

Boyer and Ameduri (2009) reported the synthesis of poly(VDF-*co*-MAF) copolymers via ITP in water in the absence of any surfactant. As per the authors, –COOH functionalities in the MAF monomer might have generated a self-sustaining emulsion system. This led to the synthesis of random poly(VDF-*co*-MAF) copolymers which is different from that obtained via solution copolymerization. The molecular weights of the resulting copolymers were in accordance with the theoretically calculated values. Recently, Banerjee et al. (2017) reported the synthesis of ω-iodo and telechelic diiodo VDF-based (co)polymers with MAF-TBE via ITP initiated by a hyperbranched radical initiator, PPFR (Scheme 20.13). Detailed kinetics and mechanistic study suggested controlled nature of the polymerization, as evidenced by the following features: (i) synthesis of iodinated polymers with different molecular weights ($M_{n,SEC}$ = 1110–5800 g mol^{-1}) and low Đ values (≤ 1.30), simply just by varying the initial comonomer feed ratio $[VDF]_0 : [CTA]_0$; (ii) M_n increased linearly with conversion, maintaining low Đs (≤ 1.28); and (iii) the presence of reactivable –CH$_2$CF$_2$–I end functionality up to 80% of monomer conversion.

RAFT/MADIX polymerization

Ameduri and Patil (2011) claimed the first RAFT copolymerization of MAF-TBE with VDF, mediated by a xanthate as the CTA. The same group also employed the as-synthesized poly(VDF-*co*-MAF-TBE)-xanthate

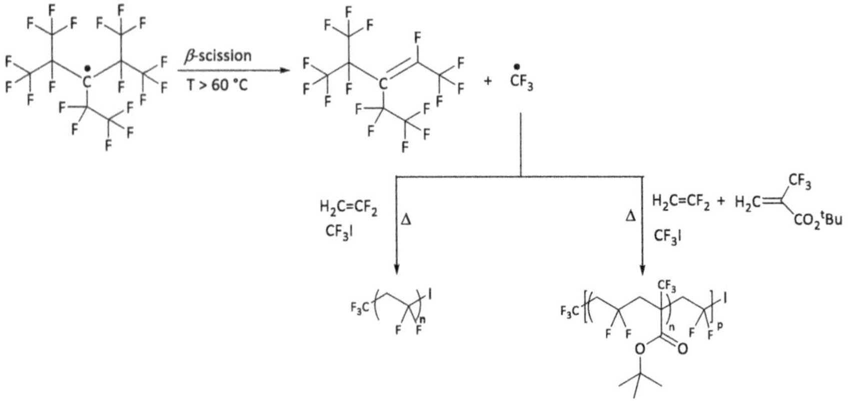

Scheme 20.13   Homopolymerization of VDF and its copolymerization with MAF-TBE initiated by ·CF$_3$ Radical released from PPFR above 75°C. Reproduced with permission from American Chemical Society (Banerjee et al., 2017).

as the macro-CTA for chain extension using VAc (Patil et al., 2013). Thus, the obtained copolymer was used as an emulsifier to produce hybrid FP/nanofillers (silicates) materials (Ameduri et al.). The mercaptan groups were subsequently oxidized into sulfonic acid and the resulting membranes displayed conductivities up to 32 mS.cm^{-1} and water uptake of 11% (Ameduri et al.). Subsequently, Banerjee et al. (2017) employed a cyclic xanthate CTA to prepare first well-defined multi-block PVDF and then multi-block VDF/MAF-TBE copolymers (Figure 20.3).

NMR and SEC analyses confirmed the synthesis of PVDF-*b*-poly(VAc-alt-MAF-TBE) copolymers. Linear increase of molecular weights with conversion, maintaining low Đ values (1.51) suggested a well-controlled polymerization.

Radical copolymerization of FAV8 with MAF-TBE

Cracowski et al. (2009) developed the radical copolymerization of MAF-TBE (an electron deficient monomer) with FAV8 (an electron-rich monomer) initiated by *tert*-butyl peroxypivalate (TBPPi). The copolymer compositions were determined using ^{19}F-NMR spectroscopic analysis. This was in good accordance with those assessed from the TGA, by calculating the isobutene released from MAF units, followed by decarboxylation. Boschet et al. (2015) were the first to report the radical copolymerization of MAF-TBE with 1,1,3,3,3-pentafluoropropylene (PFP, R1225ze). The obtained copolymer consisted of highly fluorinated random oligomers containing PFP. NMR

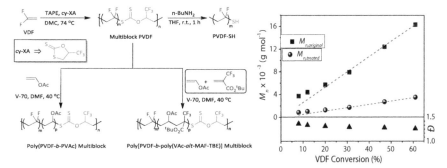

Figure 20.3 Synthesis of multi-block (PVDF-XA)n and copolymers thereof and plots of M_ns and $Ð$s of original and treated PVDF versus VDF conversion for the RAFT polymerization of VDF mediated by cy-XA, reproduced with permission from RSC (Banerjee et al., 2017).

analysis revealed that the resulting copolymer was contained mostly random architecture, containing alternating poly(VDF-alt-MAF-TBE) microblocks surrounded in both ends by one PFP unit.

Radical copolymerization of MAF-TBE with TFEMA

Boschet et al. (2010) reported radical copolymerization of two different methacrylate monomers, TFEMA and MAF-TBE, initiated by TBPPi. Detailed microstructural analysis revealed that the synthesized copolymer was a statistic poly(TFEMA-co-MAF-TBE) copolymer. Kinetics and mechanistic investigation revealed that TFEMA is more reactive than MAF-TBE. This is probably due to the presence of CF_3 substituent in MAF-TBE which is more electron-withdrawing in nature compared to CH_3, present in TFEMA (McElroy et al., 1999).

Radical copolymerization of MAF-TBE with FATRIFE

Radical copolymerization of MAF-TBE with FATRIFE was reported by Cracowski et al. (2008). They also determined the reactivity ratios of the monomer pairs using Kelen–Tüdos method: (Kelen et al., 1975) $r_{FATRIFE} = 1.71 \pm 0.01$ and $r_{MAF-TBE} = 0$ at 74°C. They further demonstrated that the refractive index of the synthesized copolymer can be successfully tuned simply just by changing copolymer compositions. Recently, Banerjee et al. (2017) reported the synthesis of poly(fluoroacrylate)s with tunable wettability and improved adhesion to steel via radical copolymerization of FATRIFE with MAF. This polymer could be useful for potential application as a functional coating. The authors noted that the thermal stability of the resulting copolymer (represented in terms of the temperature for 10% weight loss under air, $T_{d10\%}$) increased with increasing FATRIFE content in the copolymer, reaching up to 348°C (for a copolymer containing 93 mol% of FATRIFE). Notably, high MAF content in the copolymer led to good adhesion onto metal substrates.

20.2.5.2.9 Terpolymerization of MAF and MAF-esters with fluorinated monomers

Radical terpolymerization of MAF with VDF and HFP

Souzy et al. (2004) synthesized original functional poly(VDF-co-HFP) copolymers bearing –COOH pendant moieties via terpolymerization of VDF and HFP with MAF. The solution radical terpolymerization of VDF, HFP, and MAF was carried out 1 mol% peroxides as the initiator (Souzy et al., 2004). Introduction of HFP into the PVDF backbone led to increased reactivity of VDF during the copolymerization. ^{19}F NMR spectroscopy confirmed that the terpolymer is composed mainly of poly(VDF-*alt*-MAF) alternating sequences, separated by one HFP unit. Lowering of the MAF content in the terpolymer increased the thermal properties and molecular weights of the copolymers, as revealed by DSC, TGA, and SEC analyses. Thus, synthesized terpolymers could have potential applications in fuel cell membranes (Souzy et al., 2012).

Radical terpolymerization of MAF and MAF-esters with VDF and 2H-pentafluoropropylene (2H-PFP)

The knowledge obtained from the detailed investigation of the successful radical copolymerization of MAF (or MAF-esters) with VDF, described in the above section, opened the door for the radical tertelomerization of MAF with VDF and 2H-pentafluoropropylene (2H-PFP) (Ameduri et al., 2008). Notably, 2H-PFP is considered to be a difficult monomer to (co)polymerize.

Terpolymerization of MAF with VDF and HFP under RDRP conditions

Sawada et al. (2011) developed RDRP of VDF, MAF, and HFP via ITP technique in aqueous solution. The reaction was initiated by $Na_2S_2O_8$ in the absence of any surfactant and was mediated by 1,6-diiodoperfluorohexane as the CTA. ^{1}H and ^{19}F NMR spectroscopic analysis revealed three key features: (i) the molar % of each monomer in the copolymer was found to be in the range of 45–80% (VDF); 12–53% (MAF); and 1–8% (HFP), (ii) the molecular weights of the copolymer (ca. 5400–12,600 g.mol^{-1} depended on the $[monomers]_o/[IC_6F_{12}I]_o$ initial molar ratios, and (iii) the resulting copolymer mainly contained VDF-I chain end-groups.

20.2.6 Application of Copolymers Containing MAF and MAF-Esters

A few representative examples of materials for emerging applications made from copolymers containing MAF and MAF-esters are depicted below. The list includes microlithography, molecularly-imprinted polymers, PEM-FCs, polymer electrolytes for lithium-ion batteries, fluorinated copolymer/nanosilica nanocomposites, multi-compartmented micelles, self-healing materials, and functional coating materials.

20.2.6.1 Microlithography

Detailed investigation on the attempts to achieve shorter wavelengths and photolithography at 157 nm (Ito et al., 2003) (F2 excimer laser) has led to the production of new resist polymers with unique properties such as transparency. Notably chemical amplification (Ito, 2003) of these materials has remained similar to the most suitable imaging mechanism (French et al., 2009). Ito and Miller (2004) and Ito et al. (2004a,b) pioneered the application of poly(MAF-*co*-NB or VEs) copolymers as a suitable candidate for 157 nm lithography for the preparation of chemical-amplification resist polymers (Figure 20.4). Till date, all acrylate polymers suitable for 157 nm imaging applications are prepared by radical terpolymerization of 2-trifluoromethyl acrylate with another methacrylate monomers (Tegou et al., 2004; Yoshida et al., 2005). Matsushita Electric Industrial Co described that (co)polymers obtained from the radical (co)polymerization of acrylic esters

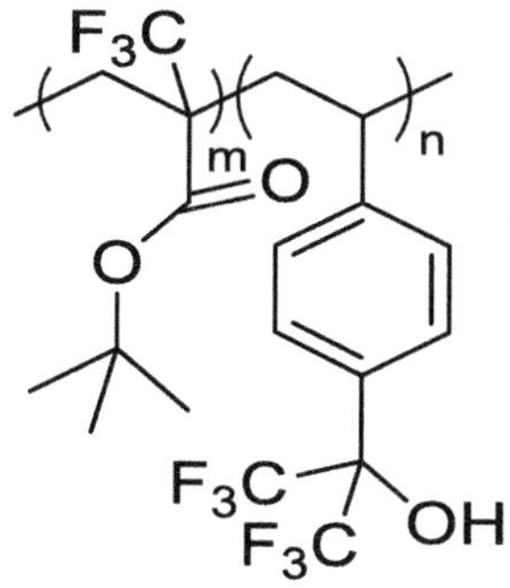

Figure 20.4 MAF-TBE-based copolymers for 157 nm photoresists (Ito et al., 2003).

could improve transparency, acid elimination, and substrate adhesion, for lithographic microfabrication (Harada et al., 2003).

20.2.6.2 Molecularity imprinted polymers

MIP techniques allow precise fabrication of recognition sites (Haginaka et al., 2008; Piletska et al., 2008), which leads to their applications as chromatographic media, sensors, artificial antibodies, and catalysts (Mayes and Mosbach, 1997). MAF-based copolymers have gained some interest in the area of MIPs (Tamayo et al., 2005a; Caro et al., 2006; Wei et al., 2006). Production of MIPs using MAF as the functional comonomer suitable for use as selective sorbents for MI solid-phase extraction procedure was reported by Tamayo et al. (2005a,b). Piletska et al. (2008) produced MIPs based on MAF-copolymers which have potential application as an efficient template recognition in water. Sambe et al. (2007) and Haginaka et al. (2008) independently reported the synthesis of uniformly sized MAF-based MIPs by a multi-step swelling and polymerization method. Yilmaz and Basan (2015) prepared MIPs using indapamide as a template molecule and MAF as a functional monomer via a non-covalent imprinting approach. Sharovi et al. (2016) successfully developed a series of MAF-based MIPs using one of the following five functional monomers: acrylamide, hydroxyethylmethacrylate, MAA, MAF, and 4-vinylpyridine. Chen and Ye (2017) prepared graphene oxide (GO)–functionalized MIP (GO–MIPs) using a non-covalent molecular imprinting approach. They applied for UPLC-PDA to sensitively detect cefadroxil in aqueous solution.

20.2.6.3 Proton exchange membranes for fuel cells

These have been a growing interest in fuel cell research for high power density clean energy conversion systems (Souzy and Ameduri, 2005; Park et al., 2011). In the 21st century, the proton exchange membrane fuel cell (PEMFC) technology has emerged as a major research activity for efficient power generation throughout the world. Extensive research has been carried out on the high-temperature PEMFC (Dodds et al., 2015). However, in order to resolve bottlenecks of materials and processing costs as well as durability, development of conventional PEM fuel cells that operate in the 50–80°C is essential. A couple of years ago, Toyota has started producing Mirail cars, equipped with fuel cell stacks. It has reached a couple of thousands of vehicles in 2017 (Tanaka). Other automobile companies such as Honda, Mercedes, and Hyundai are trying their best to commercialize Clarity, B-Class, and Tucson fuel cell electric vehicles, respectively. One of the trickiest components for

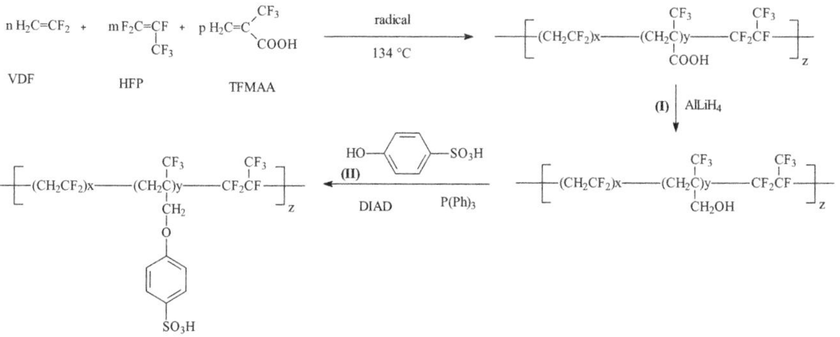

Scheme 20.14 Synthesis of poly(VDF-ter-HFP-ter-MAF) terpolymers, reduction of carboxylic acid into primary alcohols, and grafting of aryl sulfonic acid, before being processes into PEMFC. Reproduced with permission from American Chemical Society (Souzy et al., 2012).

this type of vehicles is the membrane. This has led to extensive research for the two last decades. Till date the most developed membranes for fuel cells are perfluorosulfonic acid membranes (such as Nafion®, Flemion®, BAM-3G®, Fumion®, Gore Select®, 3M®, Aciplex®, or Aquivion®). Souzy et al. (2012) employed reduction of the carboxylic acid functions followed by the Mitsunobu reaction with para-phenol sulfonic acid in order to develop original fluorinated terpolymers based on VDF, HFP, and MAF-containing aryl sulfonic acid side groups (Scheme 20.14). Unfortunately, the conductivities of these materials were low (up to 10 mS.cm^{-1}).

20.2.6.4 Polymeric electrolytes for lithium-ion batteries

SPEs for Li-ion batteries must have the following features: (i) chemical inertness, (ii) electrochemical inertness, and (iii) thermal resistance for safety concerns. This has led to slow but steady replacement of SPEs by liquid electrolytes. Typically, oligo(EO) derivatives enable solvation and complexation of Li$^+$ cation and thus during the charging and discharging cycles allow the Li$^+$ ion transport from one electrode to the other one. In order to achieve this goal, MAF was chemically modified to an original macromonomer by esterification of α-hydroxy oligo(EO) (oligo EO, MAF-TEG). This was further copolymerized with VDF to produce novel PVDF-g-oligo(EO) graft copolymers (Scheme 20.15) (Seymour and Kirshenbaum, 1987).

Ameduri et al. (2010) blended such copolymers with an ionic liquid electrolyte [1-propyl-1-methyl pyrrolidinium bis(fluorosulfonyl)imide dissolving lithium bis(trifluoromethanesulfonyl)imide (LiTFSI)] and silica

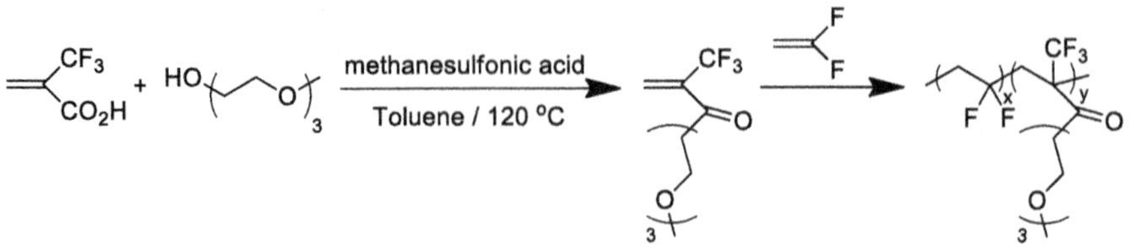

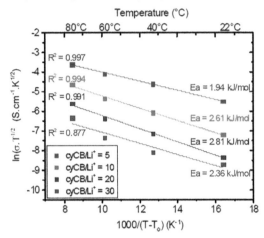

Scheme 20.15   Synthesis of 2-trifluoromethyl oligo(EO) acrylate (MAF-TEG) and its radical copolymerization with VDF initiated by tert-butyl peroxypivalate. Reproduced with permission from the Royal Society of Chemistry (Seymour and Kirshenbaum, 1987).

Figure 20.5   Temperature dependence of the ionic conductivity for the investigated SPEs using the VTF model. Reproduced with permission from Royal Society of Chemistry (Gohy et al., 2011).

nanoparticles. Results revealed that their ambient conductivities ranged from 0.2 mS.cm^{-1} and increased up to 0.5 mS.cm^{-1} (Figure 20.5), thus making such copolymers suitable for use as SPEs for lithium ion batteries (Ameduri et al., 2010; Alaaeddine et al., 2015).

Boujioui et al. (2018) recently used poly(VDF-*co*-MAF-cyCB) copolymers as SPEs. The authors mixed the copolymer with lithium perchlorate salt and the amorphous poly(VDF-*alt*-MAF-cyCB) phase of such copolymers allowed the formation of an ionic conducting phase, as evidenced by AFM. This material thus exhibited ionic conductivity values as high as 2×10^{-4} and 10^{-3} S/cm at room temperature and 80°C, respectively. Notably, these SPEs exhibited a wide electrochemical stability window (1.4–4.9 V versus Li/Li+) and relatively high values for Li^{+} ions transference numbers (0.68

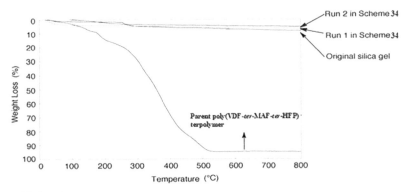

Figure 20.6 Thermograms of parent poly(VDF-ter-MAF-ter-HFP) terpolymers and of various poly(VDF-ter-TFMA-ter-HFP) terpolymers/silica nanocomposites under air. Reproduced with permission from the American Chemical Society (Sawada et al., 2011).

at 40°C), thus making such materials promising candidates for the next generation of SPEs.

20.2.6.5 Fluoropolymers/silica nanocomposites

Sawada et al. (2011) employed poly(VDF-*ter*-MAF-*ter*-HFP) terpolymers, for preparation of fluorinated polymers/silica nanocomposites by sol–gel reactions. Thus, prepared nanocomposites exhibited interesting characteristics brought by both fluorine and silica nanoparticles. They also displayed unique thermal properties as evidenced by a quasi-absence of weight loss still at 800°C (Figure 20.6). Very recently, Guo et al. (2014) reported production of fluorinated polymers/anatase titanium oxide/silica nanocomposites. This material displayed photocatalytic activity even after calcination at 1000°C.

20.2.6.6 Multi-compartment micelles

Gohy et al. (2011) synthesized telechelic diiodo-poly(VDF-*ter*-MAF-*ter*-HFP) terpolymers via ITP and then blended it in solution with poly(styrene)-*b*-poly(2-vinylpyridine)-*b*-poly(EO) triblock copolymers leading to the formation of micelles. Thus obtained micelles have the diameters ranging from 13 to 31 nm. Thus developed multi-compartment micelles acted as temperature responsive materials: they disassembled by heating above 50°C and reassembled upon cooling.

20.2.6.7 Self-healing network

Zhou et al. (2014) employed platinum nanoparticles and a unique polymer composite of poly(1-vinylimidazole) (PVI) and poly(MAF) to prepare a nature-inspired polymer nanoreactor with self-switchable catalytic ability which also exhibited "self-healing" properties. This nanoreactor demonstrated significant reactivity even at relatively high temperatures. This was probably due to the dissociation of the interpolymer interaction. This proposed strategy opens up new avenues toward the development of smart nanoreactors capable of self-switchable catalysis.

20.2.6.8 Functional coating

Ciradelli et al. (1997) reported new fluorinated acrylic polymers for improving weatherability of building stone materials. The polymer was prepared from acrylic monomers where several H-atoms in different positions were replaced with F-atoms (Ciardelli et al., 1997).

Banerjee et al. (2017) developed functional PVDF containing pendant PA functions via post-polymerization modification of the as-synthesized poly(VDF-*co*-MAF-DMP) copolymers. Thus obtained photo-cross-linkable coating material exhibited excellent corrosion resistance under simulated sea water environment (Figure 20.7).

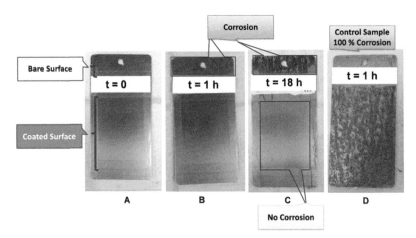

Figure 20.7   Pictures showing the status of the steel plates versus time under marine (simulated sea water) environment during corrosion experiment. Steel plates coated with PVDF-PA at the beginning of the experiment (A), after 1 h (B), and after 18 h (C). (D) Uncoated steel plate as reference sample after 1 h. Reproduced with permission from American Chemical Society (Banerjee et al., 2017).

Recently, Wehbi et al. (2017) reported the development of PVDF-based crosslinked network, generated via cross-linking of pendant triethoxysilane functionality. The material exhibited improved adhesion and could find application as functional coatings.

20.3 Conclusion

This chapter summarizes the current and future trends on the uses of MAF and MAF-esters as (i) precursors of fluorinated functional monomers and (ii) their radical and/or anionic (co)polymerization. MAF is a suitable building block to synthesize various functional (macro)onomers such as MAF-esters bearing perfluorinated alkyl, epoxide, dialkyl phosphonate, cyclocarbonate, oligo(EO), groups or organic derivatives achieved from the radical addition of various cyclic ethers or mercaptans onto MAF or MAF-esters. It can be envisaged that further modification of these monomers may lead to a wide variety of functional monomers with the desired specific property. Notably, because of the bulky alkyl substituent, MAF-esters are shown to be less reactive than MAF. In addition, MAF(esters) copolymerize readily with electron-withdrawing monomers yielding alternated copolymers and VDF under various processes such as conventional solution copolymerization and various RDRP such as ITP, RAFT/MADIX, NMP, and OMRP. The radical terpolymerization of MAF with VDF and other fluorinated alkenes also led to original fluoropolymers which bear carboxylic acid side groups. Thus, synthesized copolymers or terpolymers based on MAF and MAF-esters were involved in various applications such as fluoropolymers/silica nanocomposites, multi-compartment micelles, electrolytes for Li-ion batteries, fuel cell membranes, microlithography, photoresists, coatings, and MIPs. However, there is still lot to explore in terms of further polymeric architectures (such as graft, star, hyperbranched, or gradient copolymers) and applications or other strategies of radical copolymerization controlled by new CTAs or organometallic complexes, which should stimulate academic and industrial researchers.

The above discussion demonstrated that MAF and MAF-esters have been involved in both conventional radical copolymerization and RDRP with a wide range of hydrogenated and fluorinated monomers. A few fluoroalkenes [VDF and (VDF;HFP) and (VDF;2H-PFP) couples], α-fluoroacrylate and TFEMA were also copolymerized with some differences of reactivity. The microstructure of the copolymers can be tuned by varying the polymerization process.

Acknowledgments

SS and SAM thanks DBT and MHRD for their fellowships, respectively. SB thanks SERB, Govt. of India for the Ramanujan Fellowship Award. All coworkers (from Belgium, Bulgaria, India, Italy, Japan, Lebanon, or France) quoted in the references below are also acknowledged for their efficient contributions.

References

Aglietto M, Passaglia E, di Mirabello LM, Botteghi C, Paganelli S, Matteoli U, et al. (1995). Synthesis of new polymers containing α-(trifluoromethyl)-acrylate units. Macromolecular Chemistry and Physics. 196:2843–53.

Alaaeddine A, Vergnaud J, Rolland J, Vlad A, Gohy J-F, Ameduri B. (2015). Synthesis of an original fluorinated triethylene glycol methacrylate monomer and its radical copolymerization with vinylidene fluoride. Its application as a gel polymer electrolyte for Li-ion batteries. Polymer Chemistry. 6:6021–8.

Ameduri B. (2009). From vinylidene fluoride (VDF) to the applications of VDF-containing polymers and copolymers: recent developments and future trends. Chemical Reviews.109:6632–86.

Ameduri B. (2010). Controlled radical (co)polymerization of fluoromonomers. Macromolecules. 43:10163–84.

Ameduri B. (2014). Journal of Taiwan Institute of Chemical Engineers. 45:3124.

Ameduri B, Boutevin B (Eds). (2004). Well-Architectured Fluoropolymers: Synthesis, Properties and Applications. Amsterdam: Elsevier.

Ameduri B, Brandstradter S, Kostov G. (2008). Preparation of pentafluoropropene copolymers with different fluorinated and non-fluorinated monomers. In: PCT Int. Appl. W, editor.: assigned to Great Lakes Ltd Co.

Ameduri B, Gerard J-F, Bounor LV, Seck S, Buvat P, Bigarre J. Process for preparing an ion-exchange composite material comprising a polymer matrix and a filler consisting of ion-exchange particles.

Ameduri B, Kostov G, Brandstrater S, Edwards EB. (2008). Telomer compositions and production processes. In: PCT Int. Appl. W, editor.: assigned to Great Lakes Ltd Co.

Ameduri B, Patil Y. (2011). Copolymerization radicalaire controlee de monomers fluores par un xanthate ou trithiocarbonate. In: 1161017 FP, editor.: assigned to Arkema.

Ameduri B, Sawada H. (2016). Fluoropolymers: from fundamentals to applications. Oxford: RSC.

Ameduri B, Vergnaud J, Galiano H. (2010). Nouveaux copolymères greffés à squelette PVDF et à greffons poly(oxyethylene) pour électrolytes polymères pour batteries Lithium Ion ou purification d'eau. In: EP2014/2596032B1 Wa, editor.: assigned to CEA and CNRS.

Asandei AD, Adebolu OI, Simpson CP. (2012). Mild-temperature Mn2(CO)10-photomediated controlled radical polymerization of vinylidene fluoride and synthesis of well-defined poly(vinylidene fluoride) block copolymers. Journal of the American Chemical Society. 134:6080–3.

Asandei AD, Adebolu OI, Simpson CP, Kim J-S. (2013). Visible-light hypervalent iodide carboxylate photo(trifluoro)methylations and controlled radical polymerization of fluorinated alkenes. Angewandte Chemie International Edition. 52:10027–30.

Banerjee S, Bellan E, Gayet F, Debuigne A, Detrembleur C, Poli R, et al. (2017) Bis(formylphenolato)cobalt(II)-mediated alternating radical copolymerization of tert-butyl 2-trifluoromethylacrylate with vinyl acetate. Polymers. 9:702.

Banerjee S, Domenichelli I, Ameduri B. (2016). Nitroxide-mediated alternating copolymerization of vinyl acetate with tert-butyl-2-trifluoromethacrylate using a SG1-based alkoxyamine. ACS Macro Letters. 5:1232–6.

Banerjee S, Guerre M, Améduri B, Ladmiral V. (2018). Syntheses of 2-(trifluoromethyl)acrylate-containing block copolymers via RAFT polymerization using a universal chain transfer agent. Polymer Chemistry. 9:3511–21.

Banerjee S, Ladmiral V, Debuigne A, Detrembleur C, Rahaman SMW, Poli R, et al. (2017). Organometallic-mediated alternating radical copolymerization of tert-butyl-2-trifluoromethacrylate with vinyl acetate and synthesis of block copolymers thereof. Macromolecular Rapid Communication. 38: DOI: 10.1002/marc.201700203.

Banerjee S, Ladmiral V, Totee C, Ameduri B. (2018). Alternating radical copolymerization of vinyl acetate and tert-butyl-2-trifluoromethacrylate. European Polymer Journal. DOI; 10.1016/j.eurpolymj.2018.04.037.

Banerjee S, Patil Y, Gimello O, Ameduri B. (2017). Well-defined multiblock poly(vinylidene fluoride) and block copolymers thereof: a missing piece of the architecture puzzle. Chemical Communication. 53:10910–3.

Banerjee S, Patil Y, Ono T, Ameduri B. (2017). Synthesis of w-iodo and telechelic diiodo vinylidene fluoride-based (co)polymers by iodine transfer polymerization initiated by an innovative persistent radical. Macromolecules. 50:203–14.

Banerjee S, Tawade BV, Ladmiral V, Dupuy LX, MacDonald MP, Ameduri B. (2017). Poly(fluoroacrylate)s with tunable surface hydrophobicity via radical copolymerization of 2,2,2-trifluoroethyl [small alpha]-fluoroacrylate and 2-(trifluoromethyl)acrylic acid. Polymer Chemistry. 8:1978–88.

Banerjee S, Wehbi M, Manseri A, Mehdi A, Alaaeddine A, Hachem A, et al. (2017). Poly(vinylidene fluoride) containing phosphonic acid as anticorrosion coating for steel. ACS Applied Materials Interface. 9:6433–43.

Boschet F, Kostov G, Ameduri B, Yoshida T, Kawada K. (2010). Kinetics of the radical copolymerization of 2,2,2-trifluoroethyl methacrylate with tert-butyl α-trifluoromethacrylate. Journal of Polymer Science Part A: Polymer Chemistry. 48:1029–37.

Boschet F, Kostov G, Raynova H, Ameduri B. (2015). Telomers of 1,1,3,3,3-pentafluoropropylene. European Polymer Journal. 73:487–99.

Botteghi C, Lando C, Matteoli U, Paganelli S, Menchi G. (1997). Studies on the preparation of 2-(trifluoromethyl)acrylic acid and its esters from 3,3,3-trifluoropropene via hydrocarbonylation reactions. Journal of Fluorine Chemistry. 83:67–71.

Boujioui F, Zhuge F, Damerow H, Wehbi M, Améduri B, Gohy J-F. (2018). Solid polymer electrolytes from a fluorinated copolymer bearing cyclic carbonate pendant groups. Journal of Materials Chemistry A. 6: 8514–22.

Boutevin B. (2000). From telomerization to living radical polymerization. Journal of Polymer Science A: Polymer Chemistry. 38:3235–43.

Boyer C, Ameduri B. (2009). Iodine transfer copolymerization of vinylidene fluoride and α-trifluoromethacrylic acid in emulsion process without any surfactants. Journal of Polymer Science, Part A: Polymer Chemistry. 47:4710–22.

Boyer C, Bulmus V, Davis TP, Ladmiral V, Liu J, Perrier S. (2009). Bioapplications of RAFT polymerization. Chemical Reviews. 109:5402–36.

Boyer C, Valade D, Sauguet L, Ameduri B, Boutevin B. (2005). Iodine transfer polymerization (ITP) of vinylidene fluoride (VDF). Influence of the defect of VDF chaining on the control of ITP. Macromolecules. 38:10353–62.

Buxton MW, Stacey M, Tatlow JC. (1954). Studies upon [small alpha]-trifluoromethylacrylic acid, [small alpha]-trifluoromethylpropionic acid, and some derived compounds. Journal of Chemical Society. 366–74.

Caro E, Marce RM, Borrull F, Cormack PAG, Sherrington DC. (2006). Application of molecularly imprinted polymers to solid-phase extraction of compounds from environmental and biological samples. Trends in Analytical Chemistry. 25:143–54.

Castelvetro V, Ciardelli F, Francini G, Baglioni P. (2000). On the surface properties of waterborne fluorinated coating polymers. Macromolecular Materials and Engineering. 278:6–16.

Chen X, Ye N. (2017). A graphene oxide surface-molecularly imprinted polymer as a dispersive solid-phase extraction adsorbent for the determination of cefadroxil in water samples. RSC Advance. 7:34077–85.

Ciardelli F, Aglietto M, Montagnini di Mirabello L, Passaglia E, Giancristoforo S, Castelvetro V, et al. (1997). New fluorinated acrylic polymers for improving weatherability of building stone materials. Progress in Organic Coatings. 32:43–50.

Cracowski J-M, Montembault V, Hardy I, Bosc D, Améduri B, Fontaine L. (2008). Free radical copolymerization of 2,2,2-trifluoroethyl α-fluoroacrylate and tert-butyl α-trifluoromethylacrylate: thermal and optical properties of the copolymers. Journal of Polymer Science Part A: Polymer Chemistry. 46:4383–91.

Cracowski J-M, Montembault V, Odobel F, Ameduri B, Fontaine L. (2009). Synthesis and characterization of poly(fluorinated vinyl ether-alt-tert-butyl α-trifluoromethacrylate) copolymers. Journal of Polymer Science Part A: Polymer Chemistry. 47:6116–23.

David G, Boyer C, Tonnar J, Ameduri B, Lacroix-Desmazes P, Boutevin B. (2006). Use of iodocompounds in radical polymerization. Chemical Reviews. 106:3936–62.

Debuigne A, Poli R, Jerome C, Jerome R, Detrembleur C. (2009). Overview of cobalt-mediated radical polymerization: roots, state of the art and future prospects. Progress in Polymer Science. 34:211–39.

Dodds PE, Staffell I, Hawkes AD, Li F, Grünewald P, McDowall W, et al. (2015). Hydrogen and fuel cell technologies for heating: A review. International Journal of Hydrogen Energy. 40:2065–83.

Drakesmith FG, Stewart OJ, Tarrant P. (1968). Preparation and reactions of lithium derivatives of trifluoropropene and trifluoropropyne. Journal of Organic Chemistry. 33:280–5.

Fleš D, Vuković R, KureševićV, Radičevic R. (1981). The influence of steric factors on the mechanism of copolymerization of phenylvinyl alkyl ethers and maleic anhydride. Journal of Polymer Science Part A: Polymer Chemistry. 19:35–43.

French RH, Wheland RC, Qiu W, Lemon ML, Zhang E, Gordon J, et al. (2009). Novel hydrofluorocarbon polymers for use as pellicles in 157 nm semiconductor photolithography: fundamentals of transparency. Journal of Fluorine Chemistry. 122.

Fuchikami T, Wakasa N, Tokuhisa K, Mimura H, Arai S. (2006). Method for producing fluorine-containing acrylate. Assigned to Tosoh F-Tech, Inc.

Fuchikami T, Yamanouchi A, Ojima I. (1984). An effective and convenient route to 5-trifluoromethyl-5,6-dihydrouracils and their thio derivatives. Synthesis. 9:766–8.

Georges MK, Veregin RPN, Kazmaier PM, Hamer GK. (1993). Narrow molecular weight resins by a free-radical polymerization process. Macromolecules. 26:2987–8.

Gohy J-F, Lefevre N, D'Haese C, Hoeppener S, Schubert US, Kostov G, et al. (2011). Multicompartment micelles from blends of terpolymers. Polymer Chemistry. 2:328–32.

Goto A, Ohtsuki A, Ohfuji H, Tanishima M, Kaji H. (2013). Reversible generation of a carbon-centered radical from alkyl iodide using organic salts and their application as organic catalysts in living radical polymerization. Journal of the American Chemical Society. 135:11131–9.

Gregory A, Stenzel MH. (2012). Complex polymer architectures via RAFT polymerization: from fundamental process to extending the scope using click chemistry and nature's building blocks. Progress in Polymer Science. 37:38–105.

Guerre M, Ameduri B, Ladmiral V. (2016). One-pot synthesis of poly(vinylidene fluoride) methacrylate macromonomers via thia-Michael addition. Polymer Chemistry. 7:441–50.

Guerre M, Campagne B, Gimello O, Parra K, Ameduri B, Ladmiral V. (2015). Deeper insight into the MADIX polymerization of vinylidene fluoride. Macromolecules. 48:7810–22.

Guerre M, Wahidur Rahaman SM, Ameduri B, Poli R, Ladmiral V. (2016). RAFT synthesis of well-defined PVDF-b-PVAc block copolymers. Polymer Chemistry. 7:6918–33.

Guo S, Yoshika H, Kato Y, Kakehi H, Miura M, Isu N, et al. (2014). Preparation of fluorinated polymers/anatase titanium oxide/silica nanocomposites possessing photocatalytic activity even after calcination at 1000°C. European Polymer Journal. 58:79–89.

Haginaka J, Tabo HK, C. (2008). Uniformly sized molecularly imprinted polymers for d-chlorpheniramine: influence of a porogen on their morphology and enantioselectivity. Journal of Pharmaceutical and Biomedical Analysis. 46:877–81.

Harada Y, Hatakeyama J, Kawai Y, Sasago M, Endo M, Kishimura S, et al. (2003). Alkoxymethyl acrylic esters containing fluorine atoms at α-position. In: 0100791 UP, editor.: assigned to Shin-Etsu Chemical Co., Ltd., Japan; Matsushita Electric Industrial Co., Ltd.; Central Glass Co., Ltd.

Hatakeyama J, Takeda T. (2007). JP patent 2007/140461A (assigned to Shin-Etsu Chemical Industry Co., Ltd., Japan).

Henne AL, Nager M. (1951). Trifluoropropyne. Journal of the American Chemical Society. 73:1042–3.

Hill MR, Carmean RN, Sumerlin BS. (2015). Expanding the scope of RAFT polymerization: recent advances and new horizons. Macromolecules. 48:5459–69.

Hosoya A, Hamana H, Narita T. (2010). Synthesis and polymerization of novel fluoroalkyl 2-trifluoromethylacrylate possessing tetrahydrofuran moiety. Journal of Polymer Science Part A: Polymer Chemistry. 48:3497–500.

Hosoya A, Umino Y, Narita T, Hamana H. (2008). Carbon–carbon bond formation by radical addition of α-trifluoromethylacrylate with cyclic ethers. Journal of Fluorine Chemistry. 129:91–6.

Ito H. (2003). Chemical amplification resists: inception, implementation in device manufacture, and new developments. Journal of Polymer Science Part A: Polymer Chemistry. 41:3863–70.

Ito H, Giese B, Engelbrecht R. (1984). Radical reactivity and Q-e values of methyl alpha-(trifluoromethyl)acrylate. Macromolecules. 17:2204–5.

Ito H, Miller DC. (2004). Radical copolymerization of 2-trifluoromethylacrylic monomers. i. kinetics of their copolymerization with norbornenes and vinyl ethers as studied by in situ 1H NMR analysis. Journal of Polymer Science Part A: Polymer Chemistry. 42:1468–77.

Ito H, Miller DC, Willson CG. (1982). Polymerization of methyl α-(trifluoromethyl)acrylate and .alpha.-trifluoromethylacrylonitrile and copolymerization of these monomers with methyl methacrylate. Macro-molecules. 15:915–20.

Ito H, Miller D, Sveum N, Sherwood M. (2000). Investigation of the radical copolymerization and terpolymerization of maleic anhydride and nor-bornenes by an in situ1H NMR analysis of kinetics and by the mercury method: evidence for the lack of charge-transfer-complex propagation. Journal of Polymer Science Part A: Polymer Chemistry.38:3521–42.

Ito H, Okazaki M, Miller DC. (2004a). Radical copolymerization of 2-trifluoromethylacrylic monomers. II. Kinetics, monomer reactivities, and penultimate effect in their copolymerization with norbornenes and vinyl ethers. Journal of Polymer Science, Part A: Polymer Chemistry. 42:1478–505.

Ito H, Okazaki M, Miller DC. (2004b). Radical copolymerization of 2-trifluoromethylacrylic monomers. III. Kinetics and monomer reactivities in the copolymerization of t-butyl 2-trifluoromethylacrylate and methacrylate with styrene bearing hexafluoroisopropanol. Journal of Polymer Science Part A: Polymer Chemistry. 42:1506–27.

Ito H, Truong HD, Okazaki M, DiPietro RA. (2003). Fluoropolymer resists: pregress and Properties. Journal of Photopolymer Science and Technology. 4:523–36.

Ito H, Trinque BC, Kasai P, Willson CG. (2008). Penultimate effect in radi-cal copolymerization of 2-trifluoromethylacrylates. Journal of Polymer Science Part A: Polymer Chemistry. 46:1559–65.

Ito H, Wallraff GM, Brock P, Fender N, Truong H, Breyta G, et al. (2001a). Polymer design for 157-nm chemically amplified resists. Proc SPIE. 4345:273–84.

Ito H, Wallraff GM, Fender N, Brock PJ, Hinsberg WD, Mahorowala A, et al. (2001b). Development of 157nm positive resists. Journal of Vacuum Science and Technology. B19:2678–84.

Ito H, Wallraff GM, Fender N, Brock P, Larson CE, Truong HD, et al. (2001c). Novel fluoropolymers for use in 157nm lithography. Journal of Photopolymer Science and Technology. 14:583–93.

Iwatsuki S, Kondo A, Harashina H. (1984). Free radical copolymer-ization behavior of methyl Î±-(trifluoromethyl)acrylate and Î±-(trifluoromethyl)acrylonitrile: penultimate monomer unit effect and monomer reactivity parameters. Macromolecules. 17:2473–9.

Kelen T, Tuds F. (1975). Analysis of the linear methods for determining copolymerization reactivity ratios. I. A new improved linear graphic method. Journal of Macromolecular Science: Part A: Chemistry. 9:1–27.

Knell M, Dexter M. (1968). 1,1-Dihydroperfluoroalkyl α-trifluoromethacry lates. USA: Greenburgh, N.Y., a corporation of Delaware.

Kostov G, Holan M, Ameduri B, Hung MH. (2012). Synthesis and characterizations of photo-cross-linkable telechelic diacrylate poly(vinylidene fluoride-co-perfluoromethyl vinyl ether) copolymers. Macromolecules. 45:7375–87.

Kurakami G, Hosoya A, Hamana H, Narita T. (2010). Synthesis and polymerization of novel bis(2-trifluoromethylacrylate). Journal of Polymer Science Part A: Polymer Chemistry. 48:2722–4.

Matyjaszewski K. (2012). Atom transfer radical polymerization (ATRP): current status and future perspectives. Macromolecules. 45:4015-39.

Mayes AG, Mosbach K. (1997). Molecularly imprinted polymers: useful materials for analytical chemistry. Trends in Analytical Chemistry. 16:321–32.

McElroy KT, Purrington ST, Bumgardner CL, Burgess JP. (1999). Lack of polymerization of fluorinated acrylates. Journal of Fluorine Chemistry. 95:117–20.

Mladenov G, Ameduri B, Kostov G, Mateva R. (2006). Synthesis and characterization of fluorinated telomers containing vinylidene fluoride and hexafluoropropene from 1,6-diiodoperfluorohexane. Journal of Polymer Science A: Polymer Chemistry. 44:1470–85.

Monteiro MJ, Cunningham MF. (2012). Polymer nanoparticles via living radical polymerization in aqueous dispersions: design and applications. Macromolecules. 45:4939–57.

Narita T. (1999). Anionic polymerization of fluorinated vinyl monomers. Progress in Polymer Science. 24:1095–148.

Narita T. (2010). Stimulation on the addition reactivity of fluorinated vinyl monomers—Facile carbon–carbon bond formation by the aid of fluorine substituents. Journal of Fluorine Chemistry. 131:812–28.

Nicolas J, Guillaneuf Y, Lefay C, Bertin D, Gigmes D, Charleux B. (2013). Nitroxide-mediated polymerization. Progress in Polymer Science. 38:63–235.

Park CH, Lee CH, Guiver MD, Lee YM. (2011). Sulfonated hydrocarbon membranes for medium-temperature and low-humidity proton exchange membranes fuel cells (PEMFCs). Progress in Polymer Science. 36:1443–98.

Patil Y, Alaaeddine A, Ono T, Ameduri B. (2013). Novel method to assess the molecular weights of fluoropolymers by radical copolymerization of vinylidene fluoride with various fluorinated comonomers initiated by a persistent radical. Macromolecules. 46:3092–106.

Patil Y, Ameduri B. (2013a). Advances in the (co)polymerization of alkyl 2-trifluoromethacrylates and 2-(trifluoromethyl)acrylic acid. Progress in Polymer Science 38:703–39.

Patil Y, Ameduri B. (2013b). First RAFT/MADIX radical copolymerization of tert-butyl 2-trifluoromethacrylate with vinylidene fluoride controlled by xanthate. Polymer Chemistry. 4:2783–99.

Patil Y, Ono T, Ameduri B. (2012). Innovative trifluoromethyl radical from persistent radical as efficient initiator for the radical copolymerization of vinylidene fluoride with tert-butyl î±-trifluoromethacrylate. ACS Macro Letters. 1:315–20.

Piletska EV, Guerreiro AR, Romero-Guerra M, Chianella I, Turner APF, Piletsky SA. (2008). Design of molecular imprinted polymers compatible with aqueous environment. Analytica Chimica Acta. 607:54–60.

Riachy P, Lopez G, Emo M, Stébé M-J, Blin J-L, Ameduri B. (2017). Investigation of a novel fluorinated surfactant-based system for the design of spherical wormhole-like mesoporous silica. Journal of Colloid and Interface Science. 487:310–9.

Rosen BM, Percec V. (2009). Single-electron transfer and single-electron transfer degenerative chain transfer living radical polymerization. Chemical Reviews. 109:5069–119.

Sabine B, I-u-h M. (2007). Homogeneous phase polymerization of vinylidene fluoride in supercritical CO_2: surfactant free synthesis and kinetics. Macromolecular Symposia. 259:210–7.

Sambe H, Hoshina K, Haginaka J. (2007). Molecularly imprinted polymers for triazine herbicides prepared by multi-step swelling and polymerization method. Their application to the determination of methylthiotriazine herbicides in river water. Journal of Chromatography A 1152:130–7.

Sawada H, Tashima T, Kikuchi M. (2010). Fluoroalkyl end-capped oligomers possessing non-flammable and flammable characteristics in silica gel matrices after calcination at 800° under atmospheric conditions. Polymer Journal. 42:167–71.

Sawada H, Tashima T, Nishiyama Y, Kikuchi M, Goto Y, Kostov G, et al. (2011). Iodine transfer terpolymerization of vinylidene fluoride, α-trifluoromethacrylic acid and hexafluoropropylene for exceptional thermostable fluoropolymers/silica nanocomposites. Macromolecules. 44:1114–24.

Seymour RB, Kirshenbaum GS. (1987). History of High Performance Polymers. New York, USA: Elsevier.

Shaw JE, Kunerth DC, Sherry JJ. (1973). A simple quantitative method for the esterification of carboxylic acids. Tetrahedron Letters. 14:689–92.

Shoravi S, Olsson GD, Karlsson BCG, Bexborn F, Abghoui Y, Hussain J, et al. (2016). In silico screening of molecular imprinting pre-polymerization systems: oseltamivir selective polymers through full-system molecular dynamics-based studies. Organic Biomolecular Chemistry.14:4210–9.

Simpson CP, Adebolu OI, Kim J-S, Vasu V, Asandei AD. (2015). Metal and ligand effects of photoactive transition metal carbonyls in the iodine degenerative transfer controlled radical polymerization and block copolymerization of vinylidene fluoride. Macromolecules. 48:6404–20.

Smith DW, Iacono ST, Iyer SS. (2014). Handbook of Fluoropolymer Science and Technology. New York: Wiley.

Souzy R, Ameduri B. (2005). Functional fluoropolymers for fuel cell membranes. Progress in Polymer Science. 30:644–87.

Souzy R, Ameduri B, Boutevin B. (2004). Radical copolymerization of α-trifluoromethylacrylic acid with vinylidene fluoride and vinylidene fluoride/hexafluoropropene. Macromolecular Chemistry and Physics. 205:476–85.

Souzy R, Boutevin B, Ameduri B. (2012). Synthesis and characterizations of novel proton-conducting fluoropolymer electrolyte membranes based on poly(vinylidene fluoride-ter-hexafluoropropylene-ter-α-trifluoromethacrylic acid) terpolymers grafted by aryl sulfonic acids. Macromolecules. 45:3145–60.

Tatemoto M. (1979). The first regular meeting of Soviet-Japanese fluorine chemists. Tokyo.

Tatemoto M, Shimizu T. (1997). In: Scheirs J (Ed). Modern Fluoropolymers. New-York: Wiley. pp. 565–76.

Tamayo FG, Casillas JL, A M-E. (2005a). Clean up of phenylurea herbicides in plant sample extracts using molecularly imprinted polymers. Analytical and Bioanalytical Chemistry. 381:1234–40.

Tamayo FG, Casillas JL, A M-E. (2005b). Evaluation of new selective molecularly imprinted polymers prepared by precipitation polymerization for the extraction of phenylurea herbicides. Journal of Chromatography A. 1069:173–81.

Tanaka Y. Development of the Mirai Fuel Cell vehicle. In: Sasaki K, Li HW, Hayashi A, Yamabe J, Ogura T, Lyth SM (Eds). Hydrogen Energy Engineering, A Japanese Perspective. Tokyo: Spinger. pp. 461–76.

Tegou E, Bellas V, Gogolides E, Argitis P, Evon D, Cartry G, et al. (2004). Polyhedral oligomeric silsesquioxane (POSS) based resists: material design challenges and lithographic evaluation at 157 nm. Chemical Materials. 16:2567–77.

Timmerman R, Greyson W. (1962). The predominant reaction of some fluorinated polymers to ionizing radiation. Journal of Applied Polymer Science. 6:456–60.

Umino Y, Narita T, Hamana H. (2008). Initiation reactivity of anionic polymerization of fluorinated acrylates and methacrylates with diethyl(ethyl cyanoacetato)aluminum. Journal of Polymer Science Part A: Polymer Chemistry. 46:7011–21.

Umino Y, Nozaki H, Hamana H, Narita T. (2009). Novel anionic polyaddition of 2-trifluoromethylacrylate by double Michael reaction with ethyl cyanoacetate. Journal of Polymer Science Part A: Polymer Chemistry. 47:5698–708.

Wadekar MN, Patil YR, Ameduri B. (2014). Superior thermostability and hydrophobicity of poly(vinylidene fluoride-co-fluoroalkyl 2-trifluoromethacrylate). Macromolecules. 47:13–25.

Wang C, Weiss RG. (2003). Thermal cis → trans isomerization of covalently attached azobenzene groups in undrawn and drawn polyethylene films. Characterization and comparisons of occupied sites. Macromolecules. 36:3833–40.

Wang Z-Y, Fan H-Q, Su K-H, Wen Z-Y. (2006). Structure and piezoelectric properties of poly(vinylidene fluoride) studied by density functional theory. Polymer. 47:7988–96.

Wei S, Jakusch M, Mizaikoff B. (2006). Capturing molecules with templated materials-Analysis and rational design of molecularly imprinted polymers. Analytica Chimica Acta. 578:50–8.

Wehbi M, Banerjee S, Mehdi A, Alaaeddine A, Hachem A, Ameduri B. (2017). Vinylidene fluoride-based polymer network via cross-linking of pendant triethoxysilane functionality for potential applications in coatings. Macromolecules. 50:9329–39.

Xu Q, Frisch HL. (1994). Structural study of methyl methacrylate–α-trifluoromethacrylic acid copolymer by 1H-NMR. Journal of Polymer Science Part A: Polymer Chemistry. 32:2803–7.

Yamago S. (2009). Precision polymer synthesis by degenerative transfer controlled/living radical polymerization using organotellurium, organostibine, and organobismuthine chain-transfer agents. Chemical Reviews. 109:5051–68.

Yamazaki T, Ichige T, Takei S, Kawashita S, Kitazume T, Kubota T. (2001). Effect of allylic CH3-nFn groups (n = 1−3) on π-facial diastereo selection. Organic Letters. 3:2915–8.

Yılmaz H, Basan H. (2015). Preconcentration of indapamide from human urine using molecularly imprinted solid-phase extraction. Journal of Separation Science. 38:3090–5.

Yoshida T, Kanaoka S, Aoshima S. (2005). Photosensitive copolymers with various types of azobenzene side groups synthesized by living cationic polymerization. Journal of Polymer Science Part A: Polymer Chemistry. 43:4292–7.

Zhang Z-C, Wang Z, Chung TCM. (2007). Synthesis of chain end functionalized fluoropolymers by functional borane initiators and application in the exfoliated fluoropolymer/clay nanocomposites. Macromolecules. 40:5235–40.

Zhou Y, Zhu M, Li S. (2014). Self-switchable catalysis by a nature-inspired polymer nanoreactor containing Pt nanoparticles. Journal of Materials Chemistry A. 2:6834–9.

Zhu L, Wang Q. (2012). Novel ferroelectric polymers for high energy density and low loss dielectrics. Macromolecules. 45:2937–54.

21

Recycling and Reuse of Metal Catalyst: Silica Immobilized Palladium Complex for C–C Coupling Reaction

Tahshina Begum[1]*, Sanjib Gogoi[1], and Pradip K. Gogoi[2]

[1]Applied Organic Chemistry Group, CSTD, CSIR-NEIST, Jorhat
[2]Department of Chemistry, Dibrugarh University, Dibrugarh, Assam, India
E-mail: tahshi.du@gmail.com
*Corresponding Author

We have designed an efficient Pd-catalyst combining palladium, hydroxyl-functionalized propyl-bridge N-heterocyclic carbene (NHC) ligand, and silica. The NHC-based heterogeneous catalyst exhibited excellent activity in the Suzuki–Miyaura coupling of aryl bromides with arylboronic acids in aqueous *i*PrOH at room temperature and was reusable up to six cycles without significant loss of its catalytic activity. The prepared catalyst was characterized by Fourier transform infrared (FT-IR), X-ray diffraction (XRD), scanning electron microscope (SEM)-energy-dispersive X-ray (EDX), and inductively coupled plasma-atomic emission spectrometry (ICP-AES) analysis.

21.1 Introduction

The palladium catalyzed Suzuki–Miyaura cross-coupling reactions between aryl halides (or pseudohalides) (Badone et al., 1997; Nan et al., 2011; Yang et al., 2012; Zarei et al., 2012; Li et al., 2014) and arylboronic acids are an efficient and clean method for the preparation of biaryl organics (Mondal and Bora, 2012; Ishiyama et al., 1992; Miyaura and Suzuki, 1995; Kotha et al., 2002; Littke and Fu, 2002; Frisch and Beller, 2005; Balanta et al., 2011; Fihri et al., 2011), which are important building blocks for pharmaceutical and agricultural chemicals (Miyaura and Suzuki, 1995).

This methodology has been widely utilized in various applications ranging from academia to industry due to the ease of preparation of organoboron compounds, their relative stability to heat, air, and water, combined with the relatively mild reaction conditions, as well as the formation of non-toxic by-products. An essential part of practicing Suzuki–Miyaura coupling is the homogeneous catalysis involving electron-rich phosphane-based palladium catalyst. Since the early development of cross-couplings, numerous novel homogeneous catalytic approaches have been reported with improved efficiency relative to a wide range of electronically diverse substrates. This includes simple tertiary phosphines, hemilabile-type phosphines, sterically crowded biphenyl-type phosphine-based ligands, etc. (Bellina et al., 2004; Fleckenstein and Plenio, 2010; Suzuki, 2011). Although phosphines are excellent in stabilizing palladium species, they are often associated with drawbacks, viz., high cost, low stability, toxicity, and side-products generation, which hamper the isolation and purification process of desired products. To overcome this, various phosphine-free protocols involving N-based ligands were designed, developed, and executed during the last decade, e.g., NHC, oxime, imine, amine, Schiff-base, acetanilide, etc. Among these, NHCs are one of the most explored ligands (Herrmann et al., 1998; Li and Liu, 2004; Navarro et al., 2004; Tao and Boykin, 2004; Xiong et al., 2004; Marion and Nolan, 2008; Zhou et al., 2009; Banik et al., 2012; Amadio et al., 2013; Costa and Nobre, 2013; Dewan et al., 2014 a, b).

NHCs act as neutral, two-electron, strong σ-donor ligand with very little or no metal to ligand π-back-bonding ability; there is concomitant increase of electron density on the metal center. This enables to form a strong bond with palladium facilitating the oxidative addition step, and simultaneously helps in preventing the decomposition of the active catalyst, which provides longer catalyst lifetime and a constant reactivity throughout the course of the reaction. Thus, compared to phosphines, NHCs are considerably a better alternative and are widely used. Consequently, numerous palladium-based complexes bearing NHC-ligands have been prepared and successfully applied in a number of cross-coupling processes.

However, major problems with these homogeneous catalytic systems are efficient separation and removal of the metal catalyst, which other-wise contaminates the final products and poses a serious issue for large-scale synthesis. For example, industries, in the synthesis of pharmaceutical molecules, must meet the government directives of <5 ppm residual metal in medicinal products. To overcome this problem, heterogeneous catalysis seems particularly well suited, as the metal immobilized on solid supports remains insoluble in water and can be removed by common filtration or decantation with minimum contamination of the products. Moreover, in most cases, the heterogeneous catalysts are recyclable and retain their reactivity even after consecutive cycles, without any loss in reactivity. Palladium can be immobilized on several solid supports such as activated carbon, organic polymers, zeolites, sepiolite, triazole, hydrogel, supported on natural source, etc. (Biffis et al., 2001; Papp et al., 2005; Seki, 2006; Wu and Yin, 2013; Nehlig et al., 2015; Paul et al., 2015). However, the most common heterogeneous catalysts are based on silica supports, due to their low cost, good accessibility, excellent stability, large surface area, and excellent porosity, which allow organic groups to bind effectively on the surface to provide improved catalytic centers. Thus, numerous silica-supported palladium catalysts were developed in recent years to execute efficient Suzuki–Miyaura reaction. However, many of these methods still rely on the use of toxic organic or biphasic solvents, elevated temperature, and/or high catalyst loading. Moreover, many of these catalysts suffer from the drawbacks of poor recyclability due to aggregation and growth of less reactive large palladium particles.

21.2 General Information

All chemicals were obtained commercially and used without further drying or purification. FT-IR measurements were recorded in KBr or $CHCl_3$ on a Shimadzu Prestige-21 FT-IR spectrophotometer. N_2 adsorption/desorption isotherms were measured using Quantachrome Instrument, BOYNTON BEACH, FL 33426 at liquid N_2 temperature. The surface area of the samples was calculated according to the BET equation and pore size distribution was evaluated using BJH algorithm. ^1H spectra were recorded in $CDCl_3$ on a JEOL, JNM ECS NMR spectrometer operating at 400 MHz. GC-MS was carried out on an Agilent Technologies GC system 7820A coupled with a mass detector 5975 and SHRXI-5MS column (15 m length, 0.25 mm inner diameter, and 0.25 micron film thickness). BUCHI B450 melting point apparatus was used for the determination of melting points of the products. Palladium loading in the catalyst and its possible leaching after catalytic reaction was determined using ICP-AES on a Thermo Electron IRIS Intrepid II XSP DUO. The SEM-EDX data were obtained from "JEOL, JSM Model 6390 LV" model. The course of the reaction was monitored by TLC using n-hexane-ethyl acetate as eluent. Coupling products were confirmed by comparing the ^{1}H spectra with those reported in the literature. Silica gel (60–120 mesh) was purchased from SRL Chemicals, India. CPTES was purchased from Sigma–Aldrich. Pd(OAc)$_2$ was purchased from Spectrochem Pvt. Ltd., India.

Due to insolubility of the silica-supported NHC–Pd complex in all common organic solvents, its structural investigations were limited only to its physicochemical properties, like FTIR, XRD, SEM-EDX, ICP-AES, and N$_2$ adsorption–desorption spectral data.

21.3 Results and Discussion

The preparation of CPTES modified silica gel (60–120 mesh) was carried out as follows (Scheme 21.1). Due to insolubility of the silica-supported NHC–Pd complex in all common organic solvents, its structural investigations were limited only to its physicochemical properties, like FT-IR, XRD, SEM-EDX, and ICP-AES analysis.

The FTIR spectra of the samples prepared are presented in Figure 21.1. For the samples, the band around 1638 and 3445 cm^{-1} can be assigned for the ν_{O-H} stretching and bending vibrations of the adsorbed water. The absorption peaks around 810 and 1087 cm^{-1} were found due to the Si–O–Si structure of the silica framework.

A characteristics band at 955 cm^{-1} was seen due to the silanol group of silica, but after functionalization with CPTES and methyl imidazole, the intensity of this band for all the samples decreases or even disappears, indicating the formation of –Si–O– bond, which is formed by the reaction between silanol groups of silica with the (C$_2$H$_5$O)$_3$Si– group of CPTES. In comparison with the simple silica, the bands at 2985 and 2935 cm $^{-1}$

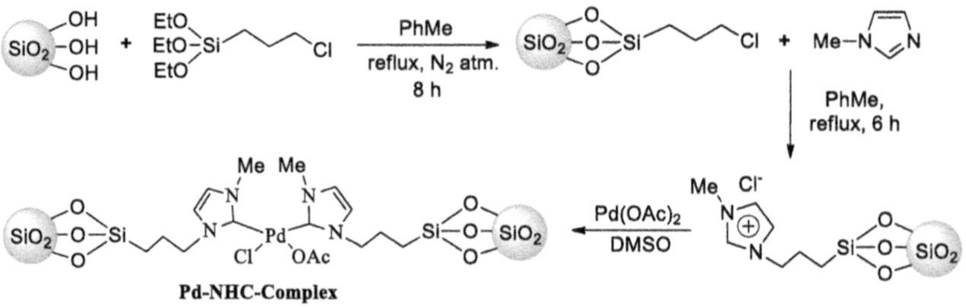

Scheme 21.1   Preparation of silica-supported Pd-NHC complex.

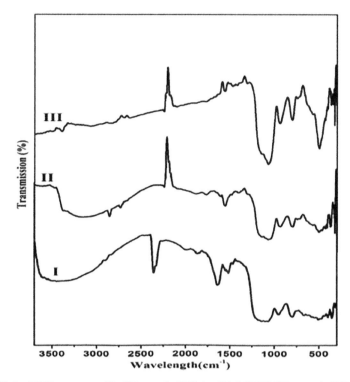

Figure 21.1 FTIR spectra; (I) Silica gel (SiO$_2$), (II) NHC@SiO$_2$, and (III) Pd-NHC complex.

for the sample silica functionalized NHC was due to C–H stretching of CPTES. Compared with the unfunctionalized silica, the significant features of the spectrum of the Pd catalyst were the appearance of the peak at 3185 (sp^2 C–H stretching vibration of the imidazole moiety), 2825 (N–CH$_2$ stretching vibration) and 1580 cm^{-1} (C–N and C=C vibrations of the imida-zole ring), which indicates the definite attachment of the all materials. In the low-frequency region, the band detected around 475 cm^{-1} was assigned to Pd–C stretching frequency.

The X-ray powder diffraction patterns of the silica and Pd(II) complex are shown in Figure 21.2, which are recorded over $2\theta = 20$–$60°$. Both silica and Pd(II) complex shows a strong diffraction peak at low angles corresponding to $2\theta = 21.99°$ (100) and corroborate that the hexagonal mesoporous structure of silica remained roughly intact during the course of preparing the catalyst.

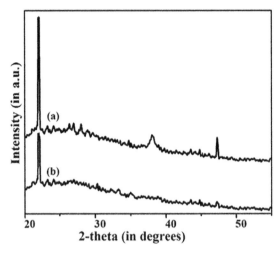

Figure 21.2   XRD pattern of (a) free silica and (b) Pd-NHC complex.

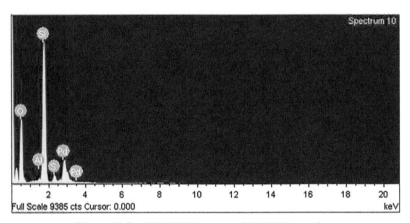

Figure 21.3   SEM-EDX spectra of Pd-NHC complex.

An XRD spectrum of silica is highly ordered, showing two weak diffractions for the 110 and 200 planes. No wide-angle XRD peak was observed in the spectra (Figure 21.2b) suggesting the absence of metallic palladium in the complex.

Considering the results of the ICP-AES and SEM-EDX interpretation, we were able to find out the metal attachment on the surface of solid supports (Figure 21.3).

EDX analysis of the metal complex shows the metal content along with the O and Si and suggests the formation of the metal with the anchored ligand at the various sites of the silica support. The palladium content of the supported catalyst measured by ICP-AES analysis suggested that 1 mg solid catalyst contains 0.00030 mmol palladium.

21.3.1 Pd-NHC Complex Catalyzed Suzuki–Miyaura Reaction

21.3.1.1 Catalyst screening and base–solvent optimization

To optimize the reaction conditions, a series of experiments under varied conditions, in terms of solvents, base, and catalyst, for a model cross-coupling reaction of 4-bromonitrobenzene with phenylboronic acid were carried out. We have performed the reaction in a different solvent system like acetonitrile, *i*PrOH, EtOH, and PEG$_{400}$ and found that the use of *i*PrOH as solvent and K$_2$CO$_3$ as base at room temperature gives the highest yield of 75% (Table 21.1, entries 1–4). The reaction gave a poor yield of the product, when H$_2$O was used as a solvent only (Table 21.1, entry 5). However, when H$_2$O was used as a co-solvent, quantitative yield of the product was achieved (Table 21.1, entry 6).

Table 21.1 Optimization of the Suzuki–Miyaura reaction of 4-bromonitrobenzene with phenylboronic acid in the presence of Pd-NHC complex[a]

Entry	Solvent	Base	Time (h)	Yield [b](%)
1	CH$_3$CN	K$_2$CO$_3$	5	50
2	*i*PrOH	K$_2$CO$_3$	5	65
3	Ethanol	K$_2$CO$_3$	7	68
4	PEG$_{400}$	K$_2$CO$_3$	7	60
5	H$_2$O	K$_2$CO$_3$	7	35
6	*i*PrOH/H$_2$O (1:1)	K$_2$CO$_3$	3	98
7	*i*PrOH/H$_2$O (1:1)	Na$_2$CO$_3$	3.5	94
8	*i*PrOH/H$_2$O (1:1)	KOH	7	57
9	*i*PrOH/H$_2$O (1:1)	Et$_3$N	7	52
10	*i*PrOH/H$_2$O (1:1)		7	
11	*i*PrOH/H$_2$O (1:1)	K$_2$CO$_3$	24	82[c]

[a] Reaction conditions: 4-bromonitrobenzene (0.5 mmol), phenylboronic acid (0.6 mmol), base (1.0 mmol), solvent (4 mL), Pd complex (2.5 mg), base (1.5 mmol).
[b] Isolated yield.
[c] Pd-NHC complex (1 mg).

Among the solvents examined, *i*PrOH-H$_2$O (1:1) is found to be the best reaction medium. As we know that, a well-chosen base is essential for the success of the coupling reaction. Several different types of inorganic and organic bases such as Na$_2$CO$_3$, KOH, and NEt$_3$ were examined for the model cross-coupling reaction (Table 21.1, entries 7–9). Among them, K$_2$CO$_3$ is found to be the most effective one and gave the desired product in a 98% yield in 3 h (Table 21.1, entry 6). However, the coupling reactions did not proceed in the absence of base (Table 21.1, entry 10).

We have also studied the amount of catalyst loading in the reaction. Decreasing the catalyst loading to 1 mg resulted in poor conversion even after increasing the reaction time up to 24 h (Table 21.1, entry 11). Therefore, the most appropriate reaction condition where the Pd catalyst proceeds efficiently is the presence of K$_2$CO$_3$ and *i*PrOH-H$_2$O (1:1) at room temperature.

We next explored the scope and limitations of substrates for the Pd-NHC complex catalyzed Suzuki–Miyaura coupling reaction under the optimized conditions using K$_2$CO$_3$ as a base in *i*PrOH-H$_2$O (1:1) at room temperature. The results are illustrated in Table 21.2. A wide range of electronically and structurally diverse aryl bromides gave the cross-coupling products with excellent isolated yields. It is often observed that, compared to the aryl halides bearing electron donating substituent, the reactivity of Suzuki–Miyaura coupling reaction is low when aryl halides having electron withdrawing

Table 21.2   Suzuki–Miyaura reaction of aryl bromides with different arylboronic acids[a]

R^1⟨ ⟩–Br + (HO)$_2$B–⟨ ⟩$_{R^2}$ →[Pd-NHC Complex / K$_2$CO$_3$, *i*PrOH-H$_2$O, r.t.] R^1⟨ ⟩–⟨ ⟩$_{R^2}$

Entry	R^1	R^2	Time (h)	Yield[b] (%)
1	OMe	H	2.5	98
2	OMe	4-F	2	97
3	NO$_2$	H	3	98
4	NO$_2$	4-F	3.5	96
5	CHO	H	4	96
6	CHO	4-OMe	3.5	97
7	CHO	4-F	4	95
8	COMe	H	5	96
9	COMe	4-OMe	4	97
10	H	H	2	100
11	H	4-Me	3.5	95

[a]Reaction conditions: aryl halide (0.5 mmol), arylboronic acid (0.6 mmol), Pd-NHC complex (2.5 mg), K$_2$CO$_3$(1.5 mmol), *i*PrOH/H$_2$O (4 mL), *ca.* in air unless otherwise noted.
[b]Isolated yield.

Table 21.3 Catalyst recycle of the Suzuki–Miyaura coupling reaction[a]

$$O_2N-\langle\rangle-Br + (HO)_2B-\langle\rangle \xrightarrow[\text{K}_2\text{CO}_3, \text{iPrOH-H}_2\text{O}]{\text{Pd-NHC Complex}} O_2N-\langle\rangle-\langle\rangle$$

Entry	Run	Time (min)	Yield [%][b]
1	1st	180	98
2	2nd	185	95
3	3rd	195	94
4	4th	200	91
5	5th	210	89
6	6th	220	85

[a] Reaction conditions: aryl halide (2 mmol), arylboronic acid (2.4 mmol), Pd-NHC complex (10 mg) K_2CO_3 (6 mmol), iPrOH/H_2O (15 mL), *ca.* in air unless otherwise noted.
[b] Isolated yield.

substituent are used. However, no such variation was observed under the present reaction condition as both varieties of substrates were successfully coupled with excellent isolated yield (Table 21.2).

To verify the heterogeneity, we carried out the reusability test for the Pd-NHC complex catalyzed Suzuki–Miyaura reaction. On the basis of our recyclability test, we found that the Pd-NHC complex catalyst could be continually used up to many consecutive cycles in the coupling reactions of phenylboronic acid with 4-bromonitrobenzene (Table 21.3).

Being a solid catalyst, it could be easily recovered by centrifugation after each cycle and washed with iPrOH/H_2O followed by EtOAc then dried in oven for 5 h. The recovered catalyst was used in the next run with further addition of substrates in appropriate amount under optimum reaction conditions. The catalyst shows unchanged catalytic activity up to six reaction cycles. No catalyst decline was observed, confirming the high stability and enhanced reactivity of the heterogeneous catalyst under the present reaction conditions.

21.4 Conclusion

In summary, we have synthesized a symmetrically substituted bis-NHC palladium complex functionalized with CPTES anchored on silica. The resultant palladium complex showed good excellent catalytic activity in the Suzuki–Miyaura coupling reactions under aqueous conditions. In terms of green chemistry approach, the present catalyst is easily recoverable and can be reused up to many cycles without loss of its activity.

Acknowledgments

T. Begum thanks the DST-SERB, New Delhi, for a National Post-Doctoral Fellowship (PDF/2017/002066).

References

Amadio E, Scrivanti A, Beghetto V, Bertoldini M, Alam MM, Matteoli U. (2013). RSC Advance. 3:21636–40.

Badone D, Baroni M, Cardamone R, Ielmini A, Guzzi U. (1997). Journal of Organic Chemistry. 62:7170.

Balanta A, Godard C, Claver C. (2011). Chemical Society Reviews. 40:4973.

Banik B, Tairai A, Shahnaz N, Das P. (2012). Tetrahedron Letters. 53:5627–30.

Bellina F, Carpita A, Rossi R. (2004). Synthesis. 15:2419–40.

Biffis A, Zecca M, Basato M. (2001). Journal of Molecular Catalysis A: Chemistry. 173:249.

Costa DP, Nobre SM. (2013). Tetrahedron Letters. 54:4582–4.

Dewan A, Bora U, and Borah G. (2014). Tetrahedron Letters. 55:1689–92.

Dewan A, Buragohain Z, Mondal M, Sarmah G, Borah G, Bora U. (2014). Applied Organometallic Chemistry. 28:230–3.

Fihri A, Bouhrara M, Nekoueishahraki B, Basset J-M, Polshettiwar V. (2011). Chemical Society Reviews. 40:5181.

Fleckenstein CA, Plenio H. (2010). Chemical Society Reviews. 39:694–711.

Frisch AC, Beller M. (2005). Angewandte Chemie International Edition. 44:674.

Herrmann WA, Reisinger CP, Spiegler MJ. (1998). Organometallic Chemistry. 557:93–6.

Ishiyama T, Abe S, Miyaura N, Suzuki A. (1992). Chemistry Letters. 691.

Kotha S, Lahiri K, Kashinath D. (2002). Tetrahedron. 58:9633.

Littke AF, Fu GC. (2002). Angewandte Chemie International Edition. 41:4176.

Li JH, Liu WJ. (2004). Organic Letters. 6:2809–11.

Li Y, Liu W, Tian Q, Yang Q, Kuang C. (2014). European Journal of Organic Chemistry. 3307.

Marion N, Nolan SP. (2008). Accounts of Chemical Research. 41:1440–9.

Miyaura N, Suzuki A. (1995). Chemical Reviews. 95:2457.

Mondal M, Bora U. (2012). Green Chemistry. 14:1873.

Nan G, Zhu F, Wei Z. (2011). Chinese Journal of Chemistry. 29:72.

Navarro O, Kaur H, Mahjoor P, Nolan SP. (2004). Journal of Organic Chemistry. 69:3173–80.

Nehlig E, Waggeh B, Millot N, Lalatonne Y, Motte L, Guénin E. (2015). Dalton Transactions. 44:501.

Papp A, Galbacs G, Molnar A. (2005). Tetrahedron Letters. 46:7725.

Paul S, Islam MM, Islam SM. (2015). RSC Advance. 5:42193.

Seki M. (2006). Synthesis. 2975.

Suzuki A. (1999). Journal of Organometallic Chemistry. 576:147–68.

Suzuki A. (2011). Angewandte Chemie International Edition. 50:6722–737.

Tao B, Boykin DW. (2004). Journal of Organic Chemistry. 69:4330–5.

Wu L, Yin Z. (2013). European Journal of Inorganic Chemistry. 6156.

Xiong Z, Wang N, Dai M, Li A, Chen J, Yang Z. (2004). Organic Letters. 6:3337–40.

Yang J, Liu S, Zheng J-F, Zhou J. (2012). European Journal of Organic Chemistry. 6248.

Zarei A, Khazdooz L, Hajipour AR, Rafiee F, Azizi G, Abrishami F. (2012). Tetrahedron Letters. 53:406.

Zhou J, Guo X, Tu C, Li X, Sun HJ. (2009). Organometallic Chemistry. 694:697–702.

MODULE 4

RRR Strategy and Atmosphere

22

Investigation on Sound Absorber Performance, Insulation Property, and Dielectric Constant of Sugarcane Bagasse

Sakti Prasad Mishra[1], **Punyatoya Mishra**[2], **and Ganeswar Nath**[1,*]

[1]Department of Physics, V S S University of Technology, Sambalpur, Odisha, India
[2]Department of Physics, Parala Maharaja Engineering College, Berhampur, Odisha, India
E-mail: ganesh_nath99@yahoo.co.in; saktimishra27@gmail.com; punyatoya.phy@gmail.com
*Corresponding Author

Non-biodegradability in environment, sophisticated method of fabrication, and economical soundness of synthetic fibers make its alternative source for the preparation of other composites from environmentally eco-friendly materials. Many bio-based agricultural waste materials having a very low environmental impact can be used in different scientific applications than manmade synthetic materials. Sugarcane is a renewable, natural agricultural resource which can be used in many products for major scientific applications after extraction of juice. Ultrasonically determined blended aqueous solution of NaOH is used for the surface modification of the dried fibers. The scanning electron microscope (SEM)characterization of treated fiber and the bio-composite prepared from suitable composition of epoxy resin as matrix and sugarcane as filler confirms the suitability of the material for thermal, electrical, and sound absorption property. Measurement of electrical and thermal conductivity of the sample indicates the insulating behavior of the sample. The dielectric constant of the material decreases with increase of log frequency indicating its energy storage performance. The presence of more porosity in network structure improves the sound absorption coefficient computed from the data obtained experimentally.

22.1 Introduction

The issue of environmental pollution and its relation with global warming have encouraged the scientists and engineers to develop new technologies which are more suitable for the nature. Recently different composite materials developed from various synthetic fibers like foams, rubber, glasswool, polyester are hazardous to human health, disruptive at workplaces, and harmful for the environment. Thus, instead of these manmade fibers, natural fibers which are cheaper, easily available, biodegradable, porous in nature, and pose lower health risk should be used for various scientific applications (Putra et al., 2013). Over the last three decades, composite materials which have multifunctional system properties have been dominant emerging materials. The composites made from jute, coir, sisal, ramie, bamboo, banana, and sugarcane are the current challenge to develop several innovative materials. Sugarcane is abundantly available in India and after extraction of its juice, the bagasse can be utilized for the development of various composite materials (Verma et al., 2012). It is chemically composed of cellulose (35%), hemicelluloses (25%), lignin (22%), and ash (22%). Due to lignin content, it can be considered as more durable compared to other natural fibers. After cultivation and extraction of juice, these wastes even if unable to use as cattle food and burned to clear the field for the next crop. Thus, some efforts have also been done for utilization of the fibers that were extracted from the bagasse. The extracted soft fibers may be very weak, but when it is coupled with epoxy resin, then the formation of bio composite material is physically very tough. This is because of the development of close chain between epoxy resin and sugar cane fibers (El-Tayeb, 2008; Jeefferie et al., 2011; Rezende et al., 2011; Cesar et al., 2015; Othmani et al., 2017; Zunaidi et al., 2017). Here both physical and acoustic properties of the prepared sample were investigated.

22.2 Materials and Methods

By using the conventional mixture juice extractor, all the juice was separated from the sugarcane. After extraction of juice, the left residue known as bagasse was ready for the extraction of fibers. The soft core part was removed

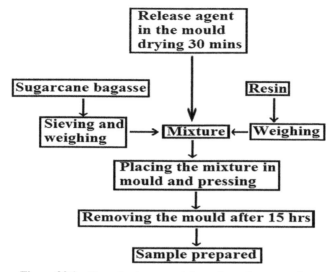

Figure 22.1 Flow chart representation of sample preparation.

to get the rind. These rinds were cut across the length and the cut portions should be free from the nodes. The extracted fibers were now kept in hot water at 90°C for 1 h to remove the sugar traces. Now the samples were dried under the sunlight for 1 week. Then all the fibers were chemically treated with 1N solution of NaOH under atmospheric pressure in order to remove the extra impurities. Then the sample fibers were kept in an oven at 600°C for 1 day to remove the moisture. The small pieces of fiber were now grinded and by the help of test sieve method, the particles were separated in to 150 µm size. A metallic mold was taken whose bottom was spread by plastic releasing agent and applied silicon spray for the easily removal of the composite (Asagekar and Joshi, 2014). By using the following formula in Equation (22.1), the fiber dust and epoxy were mixed together and stirred for 30 min to form a matrix for the formation of different weight percentages of the sample. To avoid air bubbles by pressing the matrix poured over the fiber and curing for 24h. The method of preparation of sample is represented in Figure 22.1.

$$\text{Fibre wt \%} = \left(\frac{\text{Weight of the fibre}}{\text{Weight of the polymer} + \text{Weight of the fibre}} \right) \times 100 \tag{22.1}$$

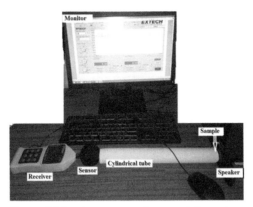

Figure 22.2   Arrangement for sound absorption.

22.3 Experimental Arrangements

22.3.1 Sound Absorption Measurement

For the measurement of sound absorption coefficient, the sample was placed at one end of a cylindrical tube and a microphone was placed at the other end which was connected to the computer. The instrument measured the sound pressure in decibels and indicated the sound absorption coefficient by software SC8100. Figure 22.2 indicates the experimental arrangement of sound absorption coefficient.

22.3.2 Measurement of Insulation Property

Both thermal and electrical conductivities were performed to investigate insulation property of the sample. By using Lee's apparatus, thermal conductivity was investigated as shown in Figure 22.3. Figure 22.4 indicates the experimental arrangement of electrical conductivity in which the sample was placed between two copper plates whose one end was fitted to the negative terminal of the 12V battery, which was fitted to a digital multi-meter. By short circuiting, the fixed resistance was calculated (Nath and Mishra, 2016).

22.3.3 Dielectric Measurement

HP Impedance Analyzer E4980A was used for the measurement of dielectric properties of the composite material. The sugar cane bagasse fiber composites were analyzed by using two contacting metal electrode method in the frequency range 1 kHz–1 MHz with ASTM D150-11standard.

Figure 22.3 Lee's apparatus.

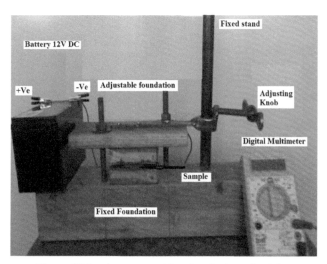

Figure 22.4 Arrangement for electrical conductivity.

22.4 Results and Discussion

The morphological behavior of the sample was analyzed by SEM. Figure 22.5(a) shows the SEM of the untreated fiber which determines the presence of bundles of fibrils and wax which can reduce the bonding of the fiber and the polymer. Figure 22.5(b) indicates 1N NaOH chemical treatment of fiber in which the distribution of fibrils in more surface area and provides more space for the bonding with superior mechanical strength. Figure 22.5(c) indicates 15 wt% of the sample under SEM analysis and showing the porous nature of the sample.

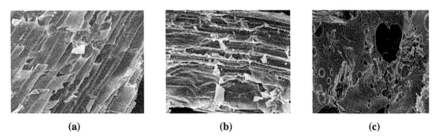

(a)   (b)   (c)

Figure 22.5   SEM (a) Untreated bagasse fiber. (b) Treated bagasse fiber. (c) 15 wt % of bagasse.

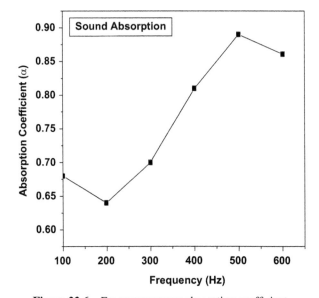

Figure 22.6   Frequency versus absorption coefficient.

In Figure 22.6, sound absorption coefficient increases with the increase in frequency and the optimum value of sound absorption is 0.89 at 500 Hz. This is due to the random orientation of fiber particles with the polymer. Again due to high porosity in the sample as determined by the SEM analysis when sound wave enters into the material, due to multiple internal reflections, it loses its energy and being absorbed. As the density of the material increases, there is a decrease in thermal conductivity as shown in Figure 22.7. Again in Figure 22.8 increasing of electrical conductivity in 10^{-5} order with the

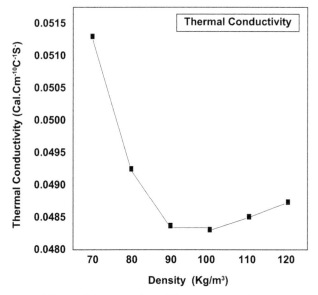

Figure 22.7   Variation of thermal conductivity.

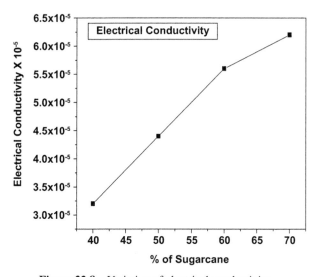

Figure 22.8   Variation of electrical conductivity.

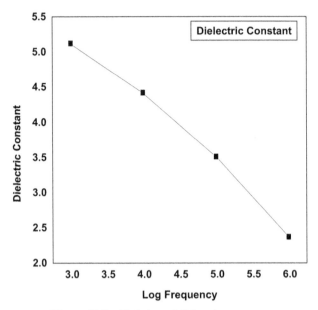

Figure 22.9   Variation of dielectric constant.

increase percentage of sugarcane fiber. So it has a very low electrical conductivity. Thus, it behaves like a thermally and electrically insulator as the air is entrapped inside the porous regions of the sample. Hence, it can be utilized as a building composite for the reduction of heat transfer in air-conditioned buildings (Phanindra et al., 2014).

The study of dielectric constant shows that the dielectric constant of sugarcane bagasse decreases with the increase in log frequency due to the interfacial polarization between the epoxy and bagasse molecules as shown in Figure 22.9. Again at higher frequencies, there is a reduction of ability to complete dipole orientation. Hence, it has less ability to store electrical energy in an electric field [12].

22.5 Conclusion

The result of this study indicates that the NaOH-treated sugar cane bagasse reinforced composites are very light in weight, tough, and exhibit good insulating, dielectric, and noise reduction properties. Hence, it can be utilized in the interior parts of the automobiles, electronic instruments, construction of air condition buildings, defense, and sports tools.

References

Asagekar S D, Joshi V K. (2014). Characterstics of sugarcane bagasse. Indian Journal of Fiber and Textile Research. 39:180–4.

Cesar A A S, Flórez J B, Mori F A, Rabelo G F. (2015). Acoustic characterization of sugarcane bagasse particle board panels. Materials Research. 18(4):821–7.

El-Tayeb N S M.(2008). A study on the potential of sugarcane fibers/polyester composite for tribological applications. Wear. 3:39–45.

Jeefferie A R, Fariha O N, Warikh A R M, Yuhazri M Y, Sihombing H, Ramli J. (2011). Preliminary study on the physical and mechanical properties of tapioca starch/sugarcane fiber cellulose composite.Journal of Engineering and Applied Science. 6:7–15.

Nath G, Mishra S P.(2016). Study of physical and sound absorbing property of epoxy blended coir dust biocomposite. Materials Science and Engineering.149.DOI:10.1088/1757-899X/149/1/012101.

Othmani C, Taktak M, Zain A, Hantati T, Dauchez N, Elnady T, Fakhfakh T, Haddar M. (2017).Acoustic characterization of a porous absorber based on recycled sugarcane wastes. Applied Acoustics. 120:90–7.

Phanindra A, Vidyasagar T, Kumar N P, Teja D S. (2014). Thermal characteration of sugarcane fibrereinfoeced composite. International Journal of Engineering Research and Technology. 3(4):2393–8.

Putra A, Abdullah Y, Efendy H, Farid W M. (2013).Utilizing sugarcane wasted fibers as a sustainable acoustic absorber. Professional Engineering.53:632–8.

Rezende C A, de Lima M A, Maziero P, deAzevedo E R, Garcia W, Polikarpov I. (2011).Chemical and morphological characterization of sugarcane bagasse submitted to a delignification process for enhanced enzymatic digestibility. Bio-tech for Bio Fuels. 4(54):1–18.

Verma D, Gope P C, Maheshwari M K, Sharma R K. (2012).Bagasse fiber composites: a review. Journal of Materials and Environmental Science. 3(6):1079–92.

Zunaidi N H, Tan W H, Majid M S A, Lim E A. (2017). Effect of physical properties of natural fibre on the sound absorption coefficient. Journal of Physics: Conference Series.DOI:10.1088/1742-6596/908/1/012023

23

Synthesis and Characterization of Microwave Absorbed Material from Agricultural Wastes

Sakti Prasad Mishra[1], Ganeswar Nath[1]* and Punyatoya Mishra[2]

[1]Department of Physics, V S S University of Technology, Sambalpur, Odisha, India
[2]Department of Physics, Parala Maharaja Engineering College, Berhampur, Odisha, India
E-mail: ganesh_nath99@yahoo.co.in; saktimishra27@gmail.com; punyatoya.phy@gmail.com
*Corresponding Author

Recent technology has paying much attention on recycling and reuse of different types of agricultural wastes for the development of technologically active materials. The carbon-rich agricultural waste not only serves the basic needs of human life but also ergs its application in many important fields like concentrating the electromagnetic waves, communications, radar, and satellite systems for both civil and military systems. This paper encloses the synthesis of coconut coir powder-based microwave absorbing material. The reflectivity and absorption characteristics of the synthesized samples show the circular shape of the material having a good absorbing capacity of microwave in the frequency range 8.2–12.4 GHz. The epoxy blended composite of coconut coir powder having thickness 0.14 mm possesses the maximum reflection loss −20.0 dB at 10 GHz.

23.1 Introduction

The digital life system of human society is now searching for new functional green materials for communication purposes rather than a fully chemical synthesized material. In recent years, though the carbon-based nanomaterials (Liu et al., 2007; Kakirde et al., 2009), which are synthesized by different chemicals have good EM shielding and absorption properties and high reliability, these materials produces various issues in environment and the synthesis of these materials is very costly. All these technical issues can be easily handled with the eco-friendly materials such as agricultural waste like coconut coir which is abundant in nature and rich in carbon and can also be used as an alternative to such high cost materials (Tyagi et al., 2011). Coconut coir dusts are the residue or the waste of coconut which are abundantly available in most of the parts of India and play a significant role in the agrarian economy of India.

23.2 Theory for Computational Parameter

As the agricultural wastes are carbon-rich materials and have no metallic composition, the material has only electrical permittivity. Thus, for a dielectric material, the complex permittivity is given as:

$$\varepsilon^* = \varepsilon' - j\varepsilon'' \tag{23.1}$$

where ε' is the dielectric constant that corresponds to storage of EM energy and ε'' is the loss factor EM energy to heat. The effectiveness of a material as an absorber is determined by loss tangent which results in attenuation of EM wave as:

$$\tan \delta = \frac{\varepsilon''}{\varepsilon'} \tag{23.2}$$

$$\alpha = \frac{\sqrt{2}\pi f}{C} \left(\sqrt{(\mu''\varepsilon'' - \mu'\varepsilon') + \sqrt{\left((\mu''\varepsilon'' - \mu'\varepsilon')^2 + (\mu''\varepsilon' + \mu'\varepsilon'')^2\right)}} \right) \tag{23.3}$$

where "f" is the frequency of EM wave on the surface of the medium determined by "S" parameter properties which include reflected signal S_{11}

and transmitted signal S_{21} from which the absorption rate of the material can be determined (Baharudin et al., 2017) as:

$$AR(\omega) = 1 - |S_{11}|^2 - |S_{21}|^2 \tag{23.4}$$

23.3 Materials

The raw coir fibers are cut into small pieces of size 10 mm and are bleached with the ultrasonically determined blend of acetone and ethanol (50–50). The dried fibers are grinded to make fine size up to 150 μm by test sieve. The epoxy resin and methyl ethyl ketone peroxide are used for fabrication process in a circular mold of diameter 29 mm and thickness 0.14 mm.

23.3.1 Characterization of the Coconut Fiber and its Composite

The SEM and EDS of bleached coconut fiber and its composites are shown in Figures 23.1(a–c), respectively. Figure 23.1 shows that the fiber surface has small voids that may facilitate the epoxy impregnation on to the fiber. The EDS analysis performed at a micro region of the fiber surface reveals that the coir fiber is mostly rich with carbon as organic matter. Surface morphology of the composite indicates the presence of more networking structure of coir dust with that of the epoxy resin making the composites with high mechanical strength.

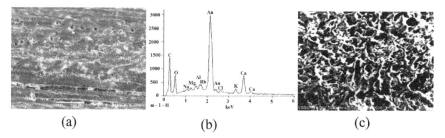

(a)   (b)   (c)

Figure 23.1   (a) SEM and (b) EDS of a coir fiber and (c) its composite.

23.4 Results and Discussion

In the present study, since the material is dielectric in nature, we have kept $\mu_r' = 1$ and $\mu_r'' = 0$ .The dielectric properties of the coconut composite are measured over the frequency range of 8.2–12.4 GHz (X-band) using a commercial dielectric probe by means of network analyzer with Agilent technologies 85070 software.

The frequency dependence of both ε', ε'' and tanδ of composite at RT are shown in Figures 23.2(a–c), respectively. The ε' values of composite are found to decrease with increase in frequency. The decrease in ε' values with increase in frequency can be explained on the basis of decrease in polarization. Polarization of a material is the sum of dipolar, electronics, ionic, and interfacial polarizations (Shukla et al., 2016). At low frequencies, all the polarizations can respond easily to the time-varying electric field, but as the frequency increases, the contribution of different polarizations filters out. Thus, the net polarization of the material decreases. As a result of which the ε' values of coir samples decrease (Shukla et al., 2016). Figures 23.2(b) and (c) show RT variation of ε'' and tangent loss of coir fiber with frequency. The increase of both ε'' and tan δ with increase in frequency attains a maximum and then decreases at a higher frequency. The variation of frequency shows one relaxation process. The decrease in their values at low frequency is due to migration of ions. The variation of ε'' and tan δ at higher frequency can be attributed to vibration of ions only (Fares, 2011; Shukla et al., 2016). The absorption coefficient increase with increase in frequency which may be due to random migration of carbon particles from one point to other which

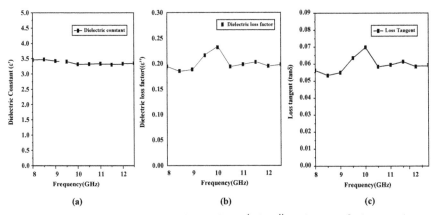

Figure 23.2 RT frequency dependence of (a) ε', (b) ε'', and (c) tan δ of composite.

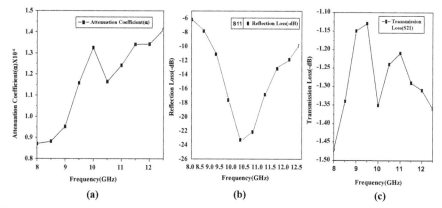

Figure 23.3 Variation of (a) absorption coefficient, (b) reflection loss (S_{11}), and (c) transmission loss (S_{21}) with frequency of composite.

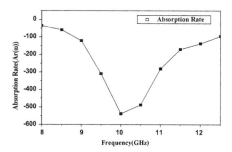

Figure 23.4 Variation of absorption rate with frequency.

causes loss of energy within the material as shown in Figure 23.3(a). The S_{11} parameters show -23 dB reflection loss at 10 GHz which corresponds to 90% power attenuation of EM wave and S_{21} parameters show a transmission loss of -1.35 at 10 GHz frequency indicating that the material has a good absorbing property as shown in Figures 23.3(b) and (c), respectively.

Figure 23.4 shows the variation absorption rate with frequency which has maximum reflection loss at 10 GHz indicating that the material will act as a good microwave absorber at X-band.

23.5 Conclusion

It is concluded that the dielectric constant decreases slightly with the increase in frequency. The absorption effect is pronounced especially at 10 GHz frequency of the EM wave. The tendency is the same in the case of dielectric loss angle tangent, like dielectric loss. The variation of S_{11} parameter and absorption rate shows the same trend which is found to be 90% power attenuated at 10 GHz. Thus, the coir dust composite has a good microwave absorbing property and finds its application for synthesis of RAM and stealth material for application in defense.

Acknowledgment

The authors are thankful to DRDO, New Delhi, for financial support under the project ERIP/ER/1203150/M/01/1559 and DMSRDE Lab, Kanpur, for microwave characterization.

References

Baharudin E, Ismail A, Alhawari ARH, Zainudin ES, Majid DLAA, Seman FC. (2017). Investigate of wave absorption performance for oil palm frond and empty fruit bunch at 5.8 GHz. Journal of Fundamental and Applied Science. 9(3S):335–48.

Fares S. (2011). Frequency dependence of the electrical conductivity and dielectric constants of polycarbonate (Makrofol-E) film under the effects of γ-radiation. Nature Science. 3(12):1034–9.

Kakirde A, Sinha B, Sinha SN. (2009). Development and characterization of nickel-zinc spinel ferrite for microwave absorption at 2ů4 GHz. Bulletin of Material Science. 31:767–70.

Liu Z, Bai G, Huang Y, Li F, Ma Y, Guo T, He X, Lin X, Gao H, Chen Y. (2007). Microwave absorption of single-walled carbon nanotubes/soluble cross-linked polyurethane composites. Journal of Physical Chemistry C. 111:13696–700.

Shukla N, Kumar V, Dwivedi DK. (2016). Dependence of dielectric parameters and a.c. conductivity on frequency and temperature in bulk $Se_{90}Cd_8In_2$ glassy alloy. Journal of Non-Oxide Glasses. 8(2):47–57.

Tyagi S, Baskey HB, Agrwalla RC, Agarwalaa V, Shami TC. (2011). Core–shell-type organogel–alginate hybrid microparticles: a controlled delivery vehicle. Ceramics International. 37:2631–41.

24

Recycling/Purification of Atmospheric Air—CFD Analysis of Flow Through Filters

Sowjanya Makarla

CVR College of Engineering, Hyderabad, Telangana
E-mail: madireddisowjanya@gmail.com

Pollution of air is of major concern in any part of the world, chiefly in developing countries. The reasons include emissions from automobiles, industries, burning of agriculture waste, and recycling of various materials which include burning or melting. Harmful gases in the atmosphere like NO_x, SO_2, and CO lead to severe health hazards including cancers. Dust particles released during various processes lead to respiratory infections. Dust particles in the air may accumulate in the respiratory system causing damage of lungs. In view of this, a device is designed to continuously filter the atmospheric air removing the dust particles of various sizes. However, flow through filters of different sizes leads to drop in pressure. This drop in pressure if negative may stagnate the air inside the device. It is very crucial to maintain the required pressure inside the chamber after every filter for out flow of air. The present work develops a numerical model of filter chamber with single and multi-filters. The filter chamber selected is of rectangular, divergent, and convergent outlets. Computational fluid dynamics is employed for the simulations. Turbulent flow equation with standard k-ε is solved along with mass and momentum equations. The flow simulations are performed for various porosities of the filter material at different intake air velocities. A three-filter chamber with 0.5 filter porosity, fan at the diverging end, and outflow of air from the converging end shows better results in the form of positive pressures at the outlet chamber for an intake air velocity of 5 m/s.

24.1 Introduction

The demand for pure atmospheric air is growing all over the world as there is an increase in pollution due to industrialization, vehicular, and burning of waste. The fuel as an energy source generates harmful gases after combustion leading to pollution of the atmospheric air. The major sources of this pollution are the automobiles, even though measures are taken by automobile manufacturers using catalytic converters before leaving the exhaust into the atmosphere. Particulate matter in the air due to pollution from industry and by burning of waste in open areas and fields causes respiratory problems if inhaled leading to pulmonary diseases. National AQI published by the Central Pollution Control Board (CPCB) states that the AQI should be less than 100. The report on February 2018 shows that the AQI of Hyderabad city is under the moderate zone of 101–200 AQI. This means that the possible health impact can be breathing discomfort in children and older adults and also in people with lung and heart diseases. Prominent parameters in the air are PM_{10} and $PM_{2.5}$.

Zhang et al. (2016) studied the air pollution in Beijing to propose control strategies and improve the air quality. Zhang and Batterman (2013) studied the effect of air pollution on the health of population who reside on and near road. They observed that the traffic congestion leads to increase in health hazards during rush hour. Liu et al. (2018) focused their research on the air pollution inside the house. The concentration of finer size particulate matter was observed to be higher indoors than outdoors. Miao et al. (2017) observed that the air pollution ($PM_{2.5}$) not only effects the population health but also staples the yields of the food crops. Cheng et al. (2017) studied the air pollution during winter and summer seasons. It was observed that the burning of straw during summer is a strong reason for the increase in air pollution. It was also stated that transportation and atmospheric reactions also contributing the raise in pollution levels. However, pollution during winter is typically seasonal. Xie et al. (2017) studied the usage of vehicles based on odd and even number license plate to reduce air pollution. The rule was observed to give short-term positive result only. Lyu et al. (2017) studied the effect of various pollution control methods between 2005 and 2013 in Hong Kong. They compared "solvent" program, "source appointment," "DCV," and emission reduction plan. All the plans focused on control of volatile organic compounds only. PPM pollution was ignored. Duo et al. (2018) investigated the characteristics of all the pollutants for a year at Lhasa of Tibetan Plateau. It was observed that the burning of biomass and dust suspension and factors due to meteorology were responsible for the air pollution. Brugha et al. (2018) observed that air pollution contributes to respiratory and cardiovascular diseases. By decreasing exposure to air pollution, the problem of cystic fibrosis can be reduced. Zhang et al. (2018) studied the particulate matter (PM_{10}, $PM_{2.5}$, and trace elements) using online pollution monitoring system, source terms retrieval system, and forecast warning system. It was observed that dust and combustion are the highest contributors for the air pollution. Pollution near concrete plants was observed to be three times that of pollution due to vehicular traffic. Song et al. (2015) studied the PM content in the office indoors. It was observed that PAE were 72.64% in the $PM_{2.5}$ matter. However, in a study room after construction, PAE was decreased by 50% after 2 years.

The present work simulates air flow in an air filter designed for the continuous recycling of air at traffic signals. Simulations are performed by modeling the device to estimate the pressure drop and recirculation zones inside the chamber. This helps to redesign the testing chamber to improve the air flow rate for better air purification. The test rig is made of a wooden chamber with divergence at the end. The sides are closed with glass for visibility of the chamber inside. Bottom is closed with a wooden sheet. Top is covered with a grooved wooden sheet. Three filters are placed at different locations. The position of the filters is fixed. Each filter is filled with different materials. In the present simulation process, this sample design is used as a geometry and domain for simulation. Figure 24.1 shows the test rig.

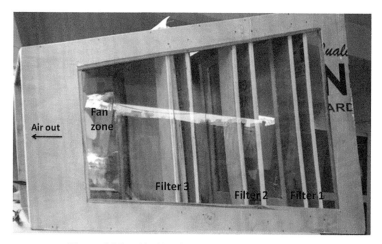

Figure 24.1 Air filtration test rig with three filters.

In the above test rig, the filters are laminated between two wooden frames to firmly hold in position. The filter frames are replaceable for testing with various conditions and types of filters. The major problem encountered during this design is the selection of position of the filter frames for improved air flow. As the groves are fixed in the top wooden sheet, change of position of filters is difficult. Hence, numerical simulations are of great use for testing filters at different locations and of different permeabilities. Simulations will also be useful to predict the pressure drop and velocity profile of air after every filter; thereby, the number of filters can be decided. The following section describes the modeling of the filter chamber.

Various materials show specific permeability values. The most common filter materials which are used for air filtration are activated carbon, polyurethane foam, filter papers, saw dust, natural fibers, etc.

24.2 Numerical Model

Figure 24.2(a) shows the geometry with single filter considered as the initial case and Figure 24.2(b) shows the high-quality quad mesh generated for the simulations. The grid shows the computational domain discretized into control volumes to apply the numerical methods. Viscous model with turbulence K-epsilon (two equations) is considered for the formulations.

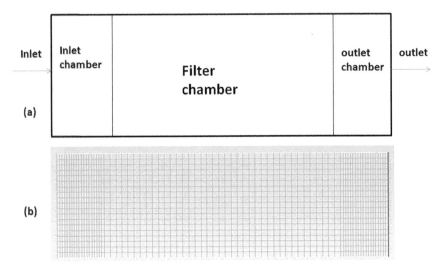

Figure 24.2   (a) Air filtration domain using single filter, (b) Quad mesh generated for the domain.

Standard k-epsilon model with near wall treatment of standard wall functions is applied. Air is the fluid entering and leaving the domain. The air inlet and outlet chambers are modeled as two separate domains and filter chamber is modeled as a separate chamber with inertial and viscous resistances and porosity. The numerical experiments are conducted at various inlet air velocities for various porosities of the material. This helps to understand the pressure and velocity distributions in the filter chambers and useful to identify any stagnation zones or negative pressure zones. The following are the assumptions and boundary conditions considered for the simulations.

Assumptions:

- The filter is similar to that of a catalytic converter in automobiles.
- Fan is replaced with inlet air boundary with velocity.
- Outside air is at atmospheric pressure.
- Free space in inlet and outlet chambers.
- Filter chamber is filled with the filter material.
- Flow resistances are constants.
- Air sent into the chamber with a velocity means using a blower.
- If air is drawn through the filter means using an exhaust fan.

Boundary conditions:

- Inlet boundary is velocity inlet with 10% turbulence intensity.
- Outlet is pressure outlet boundary with 5% backflow turbulent intensity.
- Filter chamber inlet and outlet are interior boundaries to the zone.

Interior boundaries allow the flow to pass from one domain into the other domain. The porosity of the material may alter the flow dynamics within the filter chamber. Each air chamber is 20 mm and filter chamber is 10 mm wide. The height of the total domain is 50 mm.

Two more models are also generated one with three filters of 10 mm wide and with divergence at the end. Figure 24.3(a) shows the model with three filters. There are two more air chambers between the filters. Filters 1–3 are placed equidistantly within the region of filter chamber. The width of the inlet and outlet chambers and filter chamber is not changed. Figure 24.3(b) shows the corresponding mesh generated for simulations. Assumptions and boundary conditions are similar to the first model.

Another geometry created is as shown in Figure 24.4 with divergence at the end. The angle of divergence is taken as 10°. This model is developed to reduce the negative pressures developed after the third filter. As negative pressures trap the air in the closed chamber, it is required to modify the

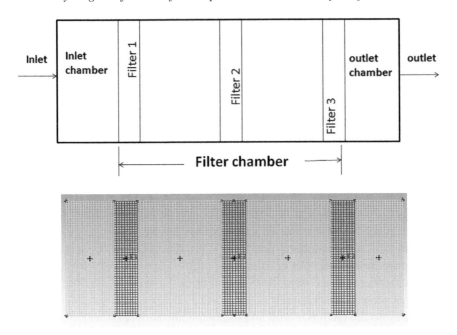

Figure 24.3   (a) Air filtration domain using three filters. (b) Quad mesh generated for the domain.

end chamber for directing the air flow into the environment. Conservation equations with turbulence model considering porous zone are solved using commercial CFD software Ansys 15.

Equations:

For the porous media, standard fluid flow equation is added with a momentum source term.

$$\frac{\partial(\gamma\rho\emptyset)}{\partial t} + \nabla\cdot(\gamma\rho\bar{v}\emptyset) = (\gamma\tau\nabla\emptyset) + \gamma S_\emptyset \qquad (24.1)$$

Equations representing continuity (Equation 24.2) and momentum (Equation 24.3) equations are given below:

$$\frac{\partial(\gamma\rho)}{\partial t} + \nabla\cdot(\gamma\rho\bar{v}) = 0 \qquad (24.2)$$

$$\frac{\partial(\gamma\rho\bar{v})}{\partial t} + \nabla\cdot(\gamma\rho\overline{vv}) = -\gamma\nabla p + \nabla\cdot(\gamma\overline{\tau}) + \gamma\overline{B} + f \quad - \left|\frac{\gamma^2}{K}\mu\bar{v} + \frac{\gamma^2 C_2}{2}\rho\overline{v\,v}\right| \qquad (24.3)$$

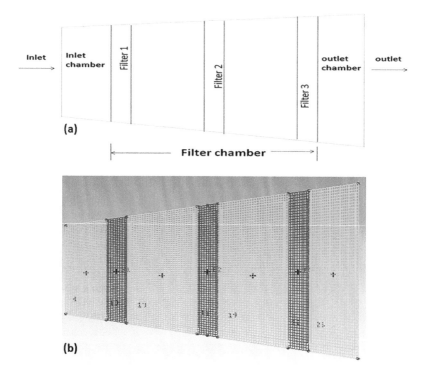

Figure 24.4 (a) Air filtration using three filters with divergence at the end. (b) Quad mesh generated for the domain.

24.3 Results and Discussion

24.3.1 Single Filter—Effect of Porosity and Flow Velocity

Numerical experiments are conducted for various air velocities using the first model with the entire filter chamber filled with only one filter material. The velocities and porosity changed for each experiment are as given in Table 24.1. Figure 24.5 shows the contours of pressure in the chamber across the filter zone of 0.5 porosity at 20 s. The pressure near the inlet of the filter zone is 718 Pa and that after the filter zone and at the entry of outlet zone is −46 Pa. Total drop in pressure is 764 Pa. As the porosity is small, there is a high drop in pressure. Figure 24.6 shows the pressure contours in the chamber for velocity 10 m/s with porosity 0.25–0.9. The flow is smooth for filters with high porosity because of less resistance. For very low inlet air velocity,

Table 24.1   Pressure drop across each chamber for single-filter air filtration system

Numerical Experiment	Velocity (m/s)	Porosity	Inlet Chamber (Pa)	Filter Chamber (Pa)	Outlet Chamber (Pa)	Any Negative Pressures
				Pressure Drop in Each Chamber		
1		0.25	23	80.8	24	Majority of filter chamber and outlet chamber
2	5	0.5	14	366	11	In the last portion of filter chamber and in outlet chamber
3		0.75	5	372	2	Near the inlet to outlet chamber and in the outlet chamber
4		0.9	2	373	0.4	No negative pressures
5		0.25	301	848	289	Majority of filter chamber and outlet chamber
6	10	0.5	46	764	46	In the last portion of filter chamber and in outlet chamber
7		0.75	17	805	10	At the outlet of filter chamber, no negative pressures in the outlet chamber
8		0.9	8	808	0.97	No negative pressures
9		0.25	3740	27842	3840	In the last portion of filter chamber and in outlet chamber, HIGH TURBULENCE
10	50	0.5	972	3955	877	Majority of filter chamber and outlet chamber
11		0.75	356	6421	310	In the last portion of filter chamber and in outlet chamber
12		0.9	129	6510	76.98	No negative pressures
13	100	0.25	16622	120855	16710	Majority of filter chamber and outlet chamber

Table 24.1 (Continued)

Numerical Experiment	Velocity (m/s)	Porosity	Inlet Chamber (Pa)	Filter Chamber (Pa)	Outlet Chamber (Pa)	Any Negative Pressures
14		0.5	3832	8585	3591	Whole filter chamber and outlet chamber
15	100	0.75	1376	18675	1219	At inlet to outlet chamber
16		0.9	1138	1.9X10	308	No negative pressures

718 Pa

-46 Pa

Figure 24.5 Pressure contours across the filter chamber with 0.5 porosity (Pa).

the pressure drop is observed to increase with increase in porosity. As the porous material causes obstruction to air flow, the momentum of low-speed air eventually drops by the time the flow reaches the end of the filter. Hence, it is required to either increase the porosity or increase the flow velocity. Pressure drop in the filters with different porosities is presented in Table 24.1.

High inlet velocities 50 and 100 m/s are observed to create more turbulence in the chamber. Hence, it is preferable to have air velocity lower than 50 m/s. The low air velocity will also provide sufficient time for the filter material to obstruct the flow of particulate matter. For filters with 0.25 and 0.5 porosity, pressure drop is negative in the filters and in the outlet chambers. Positive pressures assist the air to flow out of the filter chamber. Negative pressures in flow chambers lead to air recirculation/accumulation inside the chamber or backflow from atmospheric air from rear end into the chamber. Filter with 0.75 porosity results in positive pressures in the outlet chamber but slightly negative at the filter outlet. Filter with 0.9 porosity always leads

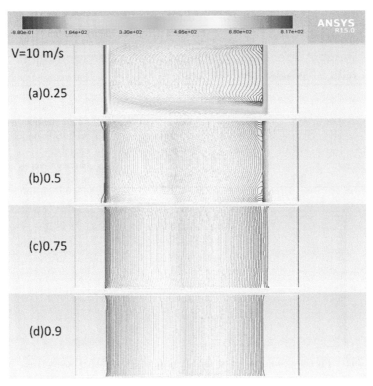

Figure 24.6   Pressure (Pa) contours across the filter chamber with different porosities at 10 m/s inlet air velocity.

to positive pressures in the outlet chamber. However, to filter the air with particulate matter, it is preferred to have a material with low porosity as filter. Hence, filters with porosity 0.5 are to be considered.

24.3.2 Three-filter Chamber

Using the second model, single filter is replaced with three filters of 10 mm thick. The simulations are performed for inlet air velocity of 5 m/s. Figure 24.7(a) shows the pressure contours in a three-filter chamber and Figure 24.7(b) shows the drop in pressure across each filter and air chamber. It is interesting to note that there is a decreasing trend in each air chamber succeeding the previous chamber. Similarly, pressure also decreases for every filter compared to the previous filter. However, the outlet chamber is maintained at positive pressures. Hence, the chamber with three filters

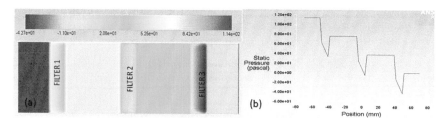

Figure 24.7 (a) Pressure contours across the three-filter chamber with porosity 0.5 and inlet air velocity at 5 m/s. (b) Pressure drop along the center line.

is preferable than a single filter. It is also interesting to note that the fluid pressure drops in each filter and slightly increases at each filter exit. This is due to change in the zone type from porous to normal fluid zone at the contact line. Pressure drop in filters 1–3 is around 35 Pa. The pressure at the outlet is at atmospheric pressure. However, for the air to flow out of the filtration chamber, the pressure at the outlet is to be higher than the atmospheric pressure.

24.3.3 Three-filter Chamber with Divergence

Now the third model with divergence at the outlet is simulated for the similar conditions of 0.5 porosity and 5 m/s inlet air velocity. Figure 24.8 shows the pressure contours in the divergent chamber with three filters. Even for this model the outlet from the chamber is at atmospheric pressure. Hence, in the next simulations, the outlet chamber boundary condition is changed from atmospheric pressure outlet to velocity with 5 m/s, i.e., placing a fan at the outlet chamber, air is drawn from the converging end. The inlet of the chamber is at atmospheric pressures. This change has shown negative pressures in the entire filtration chamber. Now repeating the same with rectangular chamber with three filters, the pressures are found to be negative in the second and third filters and outlet chamber.

Figure 24.9 shows the corresponding pressure contours in the rectangular chamber. Even for higher outlet velocity of 10 m/s, the negative pressures exist at the outlet for the rectangular chamber. Hence (filter with 0.5 porosity), the three-filter chamber with divergence can be of help to create at least the atmospheric pressures in the outlet chamber. However, the air at higher pressures than atmospheric can only be able to flow out of the chamber. Hence, the flow simulations are extended for the chamber with convergence at the outlet to verify that weather the new design can create pressures higher than atmospheric.

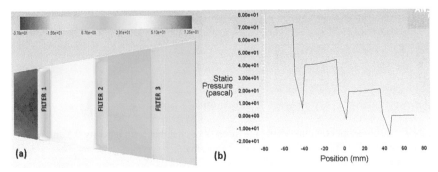

Figure 24.8   (a) Pressure contours across the three filter divergent chamber with porosity 0.5 and inlet air velocity at 5 m/s. (b) Pressure drop along the center line (positive pressures).

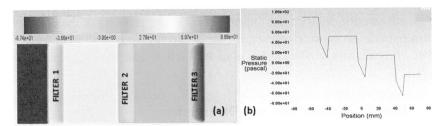

Figure 24.9   (a) Pressure contours across the three filter rectangular chamber with porosity 0.5 and inlet and outlet air velocities at 5 m/s. (b) Pressure drop along the center line.

24.3.4 Chamber with Fan on Both Sides

Inlet velocity in the single-/three-filter chambers is created with the help of a fan. The above models have employed single fan at the inlet or at the outlet. However, these models are creating either negative pressures or pressures just equal to atmospheric at the outlet/exit chamber. To generate effective flow of air, the next model is employed with two fans at either side of the chamber. In this case, the velocity of the air is set at 5 m/s for the simulations.

Figure 24.10 shows the velocity and pressure contours in the chamber. It is interesting to observe that the pressure in the outlet chamber is completely positive and above atmospheric pressure (9.34 Pa). This clearly indicates the use of fans on either side of the chamber to generate the required positive pressures which are above the atmospheric values. The small flow velocity selected helps to give sufficient time for the air to get filtered in the three chambers. But the two fans on both sides of the chamber make the unit bulky. If weight is not a concern factor, this setup can be used even for much lower fan speeds.

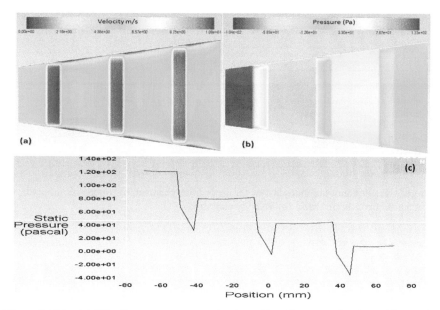

Figure 24.10 (a) Velocity contours across the three-filter divergent chamber with porosity 0.5 and inlet and outlet air velocities at 5 m/s. (b) Pressure contours. (c) Pressure distribution along the center line.

24.3.5 Three-filter Chamber with Convergence

As per the continuity equation with decrease in the flow area, velocity increases and vice versa. This can be applied for the present system by making a converging flow chamber. In this case, the three-filter chamber with divergence is used with boundary conditions of negative velocity (-5 m/s) at inlet and pressure inlet at the outlet boundary shown in Figure 24.4. This is similar to a fan at the converging zone dragging the atmospheric air through the diverging area. By this, the flow direction is changed in the negative x-axis direction. Hence, the same model can be considered for the case of chamber with convergence. By rotating the computational domain, we can present the case of converging chamber results. Figure 24.11(a) shows the pressure contours in the chamber and Figure 24.11(b) shows the velocity contours. It has been observed that the entire chamber is at negative pressures. To further investigate the use of convergent chamber, the fan is now located at the diverging area. This in fact has shown very interesting results. Figure 24.12(a) shows the velocity vectors in when the fan is at diverging end and air is pushed out of the converging chamber. Figure 24.12(b) shows the earlier case of fan

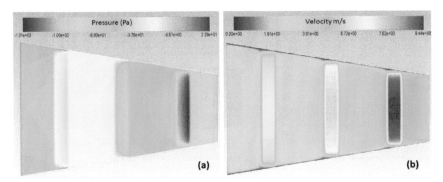

Figure 24.11   (a) Pressure contours across the three-filter chamber with porosity 0.5 and velocity at 5 m/s at converging area. (b) Velocity contours.

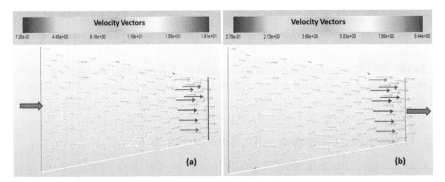

Figure 24.12   Velocity vectors with fan at the (a) converging end and (b) diverging end.

at the converging end and the air is dragged from atmosphere through the diverging end into the chamber.

It has been observed that the velocity from the chamber is increased when the fan is at the diverging end when compared to the placing of fan at the converging end. By comparing the values, continuity equation is applied in both the cases. The equation is as given below:

$$A_1V_1 = A_2V_2$$

By solving the above equation for both cases, for placing the fan at the diverging end, the velocity is increased to 10 m/s at the exit of the chamber at the converging end. For the case of placing the fan at the converging end, the velocity of air at the diverging end is 2.5 m/s. The simulation results are the same as calculated. This validates the model for predicting the required results. Hence, to avoid the weight of the chamber and make use of the continuity principle, it is useful to place the fan at the divergent end of the chamber.

24.4 Conclusion

A numerical model is developed for a filter chamber useful for the filtration of dust or soot present in the atmospheric air. Three models are developed one with single filter and two with three filter materials. The filter chamber is once again of rectangular and converging/diverging design. The flow simulations of rectangular chamber with single filter are compared for various inlet velocities and porosities of the filter material. The flow simulations for three-filter chamber rectangular/convergent/divergent are compared for positive pressure development for a porosity of 0.5 of the filter material. The flow simulations are also compared for single fan and two fans, location of single fan. The inferences are as given below:

- Single filter simulated for various porosities and intake velocities; high turbulence for high velocities and high pressure drop for highly porous filter. Hence, low velocity or fan speeds are preferred to provide sufficient time for the filter material to obstruct the flow of particulate matter. Filters with very low porosity lead to high negative pressures due to resistance to flow. However, to filter the air with particulate matter, it is preferred to have a material with a low porosity of 0.5 as filter.
- Three-filter rectangular chamber; negative pressures in the third filter and atmospheric pressure at the outlet chamber
- Three filter with diverging end—exhaust fan at the diverging end; negative pressures in the entire chamber, similar to the case of the three-filter rectangular chamber
- Three filter with diverging end—fans on both sides; positive pressures at the outlet chamber, can be employed if weight is not a concern
- Three filter with diverging end—exhaust fan at the converging end; negative pressures in the entire chamber
- Three filter with diverging end—blower at the diverging end; resulted in the required positive pressures and higher air velocity from the converging end.

This model and simulation can be repeated for any design of the chambers to filter the particulate matter in the air. The devices fabricated using this information after testing can be employed for filtering the air coming out of the industries as well to keep the environment safe for the living beings.

References

Brugha R, Edmondson C, Davies JC. (2018). Outdoor air pollution and cystic fibrosis. Paediatric Respiratory Reviews. Available at: https://doi.org/10.1016/j.prrv.2018.03.005

Cheng N, Li Y, Cheng B, Wang X, Meng F, Wang Q, Qiu Q. (2017). Comparisons of two serious air pollution episodes in winter and summer in Beijing. Journal of Environmental Sciences. Available at: https://doi.org/10.1016/j.jes.2017.10.002

Liu W, Shen G, Chen Y, Shen H, Huang Y, Li T, Wang Y, Fu X, Tao S, Liu W, Huang-Fu Y, Zhang W, Xue C, Liu G, Wu F, Wong M. (2018). Air pollution and inhalation exposure to particulate matter of different sizes in rural households using improved stoves in central China. Journal of Environmental Sciences. 63:87–95.

Lyu XP, Zeng LW, Guo H, Simpson IJ, Ling ZH, Wang Y, Murray F, Louie PKK, Saunders SM, Lam SHM, Blake DR. (2017). Evaluation of the effectiveness of air pollution control measures in Hong Kong. Environmental Pollution. 220(Part A):87–94.

Miao W, Huang X, Song Y. (2017). An economic assessment of the health effects and crop yield losses caused by air pollution in mainland China. Journal of Environmental Sciences. 56:102–13.

Song M, Chi C, Guo M, Wang X, Cheng L, Shen X. (2015). Pollution levels and characteristics of phthalate esters in indoor air of offices. Journal of Environmental Sciences. 28:157–62.

Xie X, Tou X, Zhang L. (2017). Effect analysis of air pollution control in Beijing based on an odd-and-even license plate model. Journal of Cleaner Production. 142(2):936–45.

Zhang H, Wang S, Hao J, Wang X, Wang S, Chai F, Li M. (2016). Air pollution and control action in Beijing. Journal of Cleaner Production. 112(2):1519–27.

Zhang K, Batterman S. (2013). Air pollution and health risks due to vehicle traffic. Science of the Total Environment 450-451:307–16.

Zhang S, Zhang Q, Albergel A, Buty D, Yu L, Wang H, Bi W, Cheng P, Chen F, Fang J, Hou R, Luan X, Shu C, Su J. (2018). Particulate matter pollution in Kunshan High-Tech zone: source apportionment with trace elements, plume evolution and its monitoring. Journal of Environmental Sciences. https://doi.org/10.1016/j.jes.2018.03.022

Index

About the Authors

Adam Cenian, Senior research assistant, Polish Academy Science, 1983–1984; assistant professor, Polish Academy Science, since 1985; visiting researcher, Heat and Mass Transfer Institute, Minsk, 1984–1986; postdoctoral, Freie University, Berlin, since 1987; professor, Technology U., Gdansk, 1993–1995. Program committee, organizer various conferences.

Ahmad K. Jassim has expertise in sustainable manufacturing process, nonconventional forming process, waste management, production and metallurgical engineering, and welding technology. He has three patents in the field of refractory materials. He got his B.Sc. degree in production and metallurgical engineering from university of technology in Baghdad (Iraq), M.Sc. degree in global production engineering from technical university Berlin (Germany), and his Ph.D. degree in mechanical engineering from university of Basrah (Iraq). He worked as a research and development manager, and planning and studies manager in the state company for iron and steel. He is author for two books that published in Lambert publisher and InTech open publisher. He published more than 25 articles and work as reviewer for different international journals. He won seven Golden Medals and one silver medal for his patents. He is teaching at university of Basra, University of Basra for gas and oil as well as at southern technical university.

Amrita Kaurwar is a PhD scholar at the department of Mechanical Engineering at IIT Jodhpur. Her research revolves around preparing low cost and potable clay-ceramic devices for household drinking water treatment by keeping rural traditions and workmanship alive. She uses material characterization, solid waste reutilization and mechanical behaviour as core tools to carry out her research studies. In future, she would like to knit threads between materials education, natural wastes and mechanical properties for the improvement of rural life. She has a Masters in Materials and Metallurgical Engineering from IIT Roorkee.

Anand Plappally, is a faculty at the department of Mechanical Engineering at IIT Jodhpur. His areas of expertise are in materials for water management, flow mechanics, hydrology and energy for water. He has worked in the area of point-of-use water treatment devices over the last decade. Further he designs large agriculture irrigation pumping systems based on micro-hydro on rivulets. He has a doctorate in Food, Agricultural and Biological Engineering from the Ohio State University.

Anubha Kaushik is Professor & former Dean at University School of Environment Management, and Director, International Affairs, Guru Gobind Singh Indraprastha University (GGSIPU), New Delhi, India. She has more than 34 years of research, teaching and administrative experience in various universities. She has 08 gold medals to her credit along with Nehru Memorial Foundation Prize, National B. R. Ambedkar Fellowship Award, and Distinguished Author Award. She has published more than 125 research articles in international and national scientific journals of repute and 05 books. Her current research interests are focused on bioremediation, wastewater treatment using constructed wetland technology, Biohydrogen production, Microbial Fuel Cell technology, Ecosystem studies, Ecosystem services, Ecorestoration, Solid Waste Management and Environmental Impact Assessment (EIA) Studies.

Aparna Bhardwaj holds a doctorate in Environmental Science from Guru Jambheshwar University of Science & Technology, Hisar. She was awarded M.Sc. and M.Tech degrees from same University in 2007 and 2009 respectively. Her areas of research interest include wastewater treatment, soil quality analysis and constructed wetland technology. She has published a couple of research papers and has participated in several national and international conferences.

Ashok N. Bhaskarwar currently works at Department of Chemical Engineering, Indian Institute of Technology Delhi, India. His areas of research are Petroleum Engineering, Chemical Engineering, Materials Engineering, Interfacial Engineering, Bioengineering, and Nanotechnology. The current projects range from new reactors and materials through renewable energy, environment, and healthcare including targeted drug delivery for cancer treatment. He has been awarded the BOYSCAST Fellowship of Department of Science and Technology, Government of India, and INSA Medal for Young Scientist. He is a member of the advisory boards of several journals. He has published 75 papers in international journals, 88 in conferences, and has to his credit 28 patents (applied or granted), 4 books, and 6 book-chapter contributions. He has also delivered 12 invited lectures.

Bartosz Hrycak graduated from the Gdynia Maritime University and obtained a master science degree. Currently, he is continuing his education at doctoral studies; at the Institute of Fluid-Flow Machinery.

Biswajit Debnath, is a PhD scholar in Chemical Engineering Department, Jadavpur University, India. He has a B.Tech and M.E. in Chemical Engineering. He is an executive member of International Society of Waste Management, Air and Water (ISWMAW) & Life associate member of Indian Institute of Chemical Engineers (IIChE). He is a Commonwealth Split-Site Scholar visiting Aston University, Birmingham, UK. His **area of Specialisation is** E-waste Valorisation and sustainability. His **Research Interests** include E-waste management, SDG, Supply Chain, Circular Economy etc. He has worked in UKIERI projects and published nearly 55 articles including conference proceedings, peer-reviewed journals and contributed to eight book chapters. He is a CPCB certified trainer for the six waste management rules and provided training on e-waste & plastic waste rules on invitation. He has won best paper award several times in International Conferences which fetched him four invited lectures. He has completed two collaborative (unfunded) projects with colleagues in USA and India. He was the secretariat of International Conference on Solid Waste Management (6th and 7th issue). He is reviewer of reputed journals published by Springer; ACS and Elsevier. Additionally, he is a self-taught singer-songwriter active in spreading awareness by making songs on e-waste or general waste.

Britika Mazumdar is a student of Master in Technology from National Institute of Technology, Durgapur (2016–2018) with specialization in Environmental Science and Technology. She has done her B.Tech in Civil Engineering. Environment has always fascinated her and she feels that when technology meets environment, the objective must always remain to reduce load in terms of pollution from our Mother Nature. As Eric Knight quoted "Waste not the smallest thing created, for grains of sand make mountains, and atomies infinity", she firmly believes that even wastes when treated properly, can turn to some other beneficial by products. She has done her M.Tech dissertation project on the treatment of whey water from food processing units using hybrid methods. Currently she is working in the field of Environmental Impact Assessment and wishes to pursue higher degree in the near future.

Chander Prakash Kaushik is currently working as a Professor in Amity School of Earth and Environmental Sciences, Amity University Haryana, Gurgaon, India. He is a Member of National Accreditation Board for Education and Training – Accreditation Committee (2016–2018; 2018–2020) of Quality Council of India. He discharged duties of different administrative posts in GJ University of Science and Technology, Hisar, Haryana, and other Indian universities. An alumnus of the University of Delhi Prof. Kaushik has a total teaching experience of four decades in various universities. He has been actively involved in research and consultancy. His research interests pertain to Environmental Pollution, Toxicology, Solid waste management, Biohydrogen production, Bioremediation, EIA etc. He has completed 12 international and national research and consultancy projects (of about Rs.110 million), guided 17 Ph.D., 83 M.Tech., and 03 M.Phil. students and has 123 publications including more than 100 research papers in international and national journals, 05 books (co-authored), and conference proceedings.

Dariusz Czylkowski, graduated from the Faculty of Applied Physics and Mathematics at the Gdansk University of Technology. Currently, he works as specialist at Institute of Fluid Machinary.

Deniz Genc Tokgoz is a research fellow, and conducted the current experiments at the Center for Advanced Materials, Qatar University. Her research interests are in the area of concrete technology and Civil Engineering materials. Currently she studies on developing new ways of manufacturing concretes using routine waste materials in suitable forms as a partial replacement to sand.

Ganeswar Nath is working as Associate Professor, Department of Physics, Veer Surendra Sai University of Technology, Sambalpur, Odisha. He has completed his Ph.D (2010) from Ravenshaw University in Physics. Currently, he is actively engaged in the field of Ultrasonic, Acoustics, Material Science, Nano materials, Plasma Science and Laser Science. He has published fifty papers in different international and national journals and more than fifty papers in national and international conferences. He is executive member and life member of different academic bodies like USI, ASI, PSSI, IAPT, ISCA, OPS, ISTE.

Gargi Biswas is now working as Research Associate in Central Pollution Control Board, Delhi, India. She is presently working on inventory and risk-assessment study of contaminated sites in India. She is also associated with several remediation projects in the field of wastewater treatment, solid-waste management and water microbiology. Previously, her research exposures were on valorization of waste materials and abatement of pollutants through biological and chemical channel. The partial focus of her research was also in the field of Bio-fuel production from algal lipid. She had published 09 research articles in renowned national and international journals and 01 book chapter during her academic research work. She did her PhD from National Institute of Technology Durgapur, India under the department of Chemical Engineering. Before this, she had completed her B.Tech in Biotechnology and M.Tech in Environmental Science & Technology.

Gaweł Sołowski has completed his Master Degree at Silesian University of Technology at Gliwice in 2014. He is the PhD student at Institute of Fluid Machinary of Polish Academy Science from 2014. He has been serving assistant editor in Open Chemistry in de Gruyter Open and published 3 papers in reputed journal.

Gyanashree Bora completed her B.Sc from Debraj Roy College, Golaghat, Assam, India and Post-graduation from Dibrugarh University, Dibrugarh, Assam in 2014. Currently she is perusing her PhD under the guidance of Prof. Jyotirekha G. Handique, Faculty, Department of Chemistry, Dibrugarh University as a Senior Research Fellow. The area of interest of her research work are – isolation of lignin and oxidation of lignin to vanillin from woods/isolation of vanillin from various natural resources synthesis of phenolic Schiff bases from vanillin and other hydroxybenzaldehydes and aromatic diamines using different catalysts and greener protocolos, Study of the antioxidant, antimicrobial and cation as well as anion sensing property of the Schiff bases, Computational study of their antioxidant behaviour and sensing mechanism by using Density functional theory (DFT).

Hyun-Chan Kim, He received B.E (Mechanical Engineering) from INHA University, Korea in 2014. He is now pusuing M.S (Mechanical Engineering) at INHA University. His research interest includes Sensor, haptic actuator, MEMs.

Izabela Konkol graduated from the Faculty of Chemistry at the Gdansk University of Technology and obtained a master of science degree. Currently, she is continuing her education at doctoral studies; at the same time she is an employee at the Institute of Fluid-Flow Machinery.

Jaehwan Kim is director of Creative Research Center for Nanocellulose Future Composites which is supported by National Research Foundation of Korea (NRF). He is an associate editor of Smart Materials and Structure, and Smart Nanosystems in Engineering and Medicine; an editor of International Journal of Precision engineering and Manufacturing, Frontier Materials – Smart Materials; editorial board member of International Journal of Precision Engineering and Manufacturing-Green Technology, Journal of Materials Science and Engineering, Actuators. He is a pioneer of cellulose smart materials and his research interests are smart materials, cellulose, electroactive polymers, smart sensors, polymer based MEMS and biomimetic devices.

Juma Haydary is an associate professor of Chemical Engineering at the Department of Chemical and Environmental Engineering of the Slovak University of Technology in Bratislava. He holds a PhD. in Chemical Engineering and Process Control from the same university. His main research activities include pyrolysis, cracking and gasification of waste, biomass and petroleum residues. He is involved in research and teaching computer aided chemical process design, unit operations in chemical engineering and separation of multicomponent mixtures. He has provided wide-ranging contributions in these fields over the past 15 years. Up to 2018, he has been the principle investigator of 10 research and educational projects.

Jung Ho Park, He received B.E (Mechanical Engineering) from Inha University, South Korea in 2017. He is now pusuing M.S (Mechanical Engineering) at INHA University. His research interest includes Nanocellulose.

Dr. Jyotirekha G. Handique is a Professor and former Head of the Department of Chemistry, Dibrugarh University and also the Director (Honorary) of the Educational Multimedia Research Centre (EMRC), Dibrugarh University. She completed her M.Sc from Gauhati University with Specialization in Organic Chemistry and received her Ph.D. from Indian Institute of Technology – Guwahati. Her research areas of interest are Synthetic Organic Chemistry and Natural Products Chemistry. Currently she is working on Synthesis of dendritic polyphenols and studies of their antioxidant properties, antioxidant properties of herbal food and beverages of North East India, Chemical investigation of medicinal plants of North-East India etc. She has worked as a reviewer of various national and international journals, viz., RSC Advances, European Journal of Medicinal Chemistry, National Academy Science Letters, Natural Product Research, Journal of Food and Nutritional Disorders, Advances in Pharmacological Sciences etc. She is actively engaged in popularisation of science through writing popular science articles in newspapers and magazines, delivering talks and participating in interactive sessions.

Keshaw Ram Aadil, M.Sc. Ph.D. (Biotechnology) is Assistant Professor (contract) in the Centre for Basic Science, Pt. Ravishankar Shukla University Raipur, Chhattisgarh, India. He did his Post-Doctoral work from National Institute of Technology Raipur, Chhattisgarh, India. His research focuses on biomaterials and nanomaterials based value-added products for food packaging and biomedical applications. He has five years of research experience. He has published 15 research papers in peer-reviewed journals and one book in his credit.

Krzysztof Pastuszak, received the M.Sc. in computer science in 2016. Since 2016 he is a Ph.D. student in computer science. He works as assistant in Department of Algorithms and Modelling of Systems. Main areas of his research are the design of efficient algorithms and the analysis of mathematical properties (including computational complexity) of discrete optimization problems and bioinformatics.

Kulasekaran Jaidev is working as a research fellow in LARPM-CIPET Bhubaneswar, after his M.Tech in Nanotechnology. Presently he is pursuing PhD degree in Biju Patnaik University of Technology (BPUT)-Odisha. His major interests of researches are mechanical recycling and value addition of E-waste plastics, ageing studies of recycled plastics, development of recycled polymer based nanocomposites.

Le Van Hai is a Postdoctoral Research Fellowship at Creative Research Center for Nanocellulose Future Composites, Inha University, South Korea. He received Ph.D. degree from Chungnam National University, South Korea in 2015. He is a lecturer at Pulp and Paper Technology Dept., Phutho College of Industry and Trade, Vietnam. His research interest includes Pulp and

Paper Technology, waste paper recycling, Nanocellulose and Nanocellulose composite, Green composites, hydro-gel, chitin, chitosan and so on.

Lovelesh Dave is a Research Fellow on the SERB-DST project "Local composite geotextile mats for Soil and Water Conservation in Western Rajasthan" at IIT Jodhpur. The areas of specialization are into manufacturing geo-textiles for use in soil and water conservation, hydrological/dam site construction, construction technology and water treatment. He has Bachelors in Civil Engineering from Sir Padampat Singhania University, Udaipur.

Manish Kumar Mishra, Associate Professor at Department of Chemistry, Sardar Patel University, Vallabh Vidyanagar, Gujarat. He obtained his Ph.D. in Heterogeneous Catalysis from Discipline of Inorganic Materials and Catalysis from CSIR – Central Salt & Marine Chemicals Research Institute (CSMCRI) Bhavnagar, Gujarat (2008). His current research interests are synthesis of mesoporous catalytic materials, photocatalysis using Titania and modified Titania semiconductors and Micellar Catalysis. Also, he works on the synthesis of Metal-Organic Frameworks (MOFs) for the catalytic activities.

Mintu Job is working as Assistant Professor, Department of Agricultural Engineering, Birsa Agricultural University, Ranchi, Jharkhand, India. He has vast experience in the field of Dryland Agriculture, Protected cultivation, Micro Irrigation and Watershed management. He has published more than 20 research papers in National and International Journals of repute. He has two books written in his credit. Apart from this he has also guided 2 post graduate students as major advisor of PG thesis.

Mona Sharma is an Assistant Professor at Department of Environmental Science & Engineering, Guru Jambheshwar University of Science & Technology, Hisar, Haryana, India. She was the recipient of various research fellowships from UGC and CSIR during her doctorate. She has bagged many publications in International journals with high impact factor and also in proceedings of International and National conferences and books.

Nesibe Gozde Ozerkan is a consultant in Civil Engineering. In the past she worked as an Assistant Professor at the Center for Advanced Materials, Qatar University. She specialises in the area of autogenous self-healing concretes and their durability characteristics. Her research involved experimental characterisation of concrete materials and their correlations to the functional properties of concretes at the bulk scale.

Omdeo K. Gohatre earned his M.Tech in polymer technology and presently continuing his PhD in BPUT-Odisha. His major research interests are recovery and recycling of E-waste plastics, recycling of vinyl and flame retardant polymers, nanocomposites and minimization of hazardous gas evolution during recycled polymer processing.

Parsanta Verma is a research scholar at Department of Chemical Engineering, Indian Institute of Technology Delhi, India. Before this, she worked as a project assistant at National Physical Laboratory (NPL), New Delhi, India. She has an M.Tech in Nanotechnology, and an M.Sc. in Biotechnology. She has been awarded the best oral presentation at an international conference in 2018.

Prof. Pradip K. Gogoi did M.Sc. in Chemistry in 1975 from Dibrugarh University and Ph.D. from Indian Institute of Technology, Bombay in 1981. In 1985, he joined the Department of Chemistry, Dibrugarh University and became Professor in 2001. In 1992–93 he was a Marie Curie Post-doctoral Fellow at Imperial College of Science, Technology & Medicine, London. He was also a Visiting Scientist at LPV Institute of Physical Chemistry, Ukrainian Academy of Sciences, Kiev during 2008. His fields of interest are co-ordination chemistry, catalysis, cross-coupling reactions, synthesis of novel materials, petroleum technology and coal chemistry. During his career, he guided 17 Ph.D. and 10 M.Phil. students. He has published over 111 scientific papers and is a life member of the National Academy of Sciences, India.

Pratima Gupta, M.Sc. Ph.D. (Microbiology) is a Head and Associate Professor, in the Department of Biotechnology, National Institute of Technology Raipur, Chhattisgarh, India. She has more than 15 years of teaching and research experience. Her research area includes biopolymers based scaffolds development for biomedical applications, Microbial products, and process technology mainly in Biofuel production. She authored about 25 peer-reviewed papers and 2 book chapters. She also worked in two centrally funded research projects as Principle investigator.

Pravin Kumar is working as an Associate Professor in the Department of Mechanical Engineering, Delhi Technological University, Bawana Road, Delhi-110042. He has 18 years of teaching and research experience. He obtained his Ph.D. degree in Supply Chain Management from IIT Delhi. He received his M.Tech. in Industrial Management from IIT, Banaras Hindu University, Varanasi. His research area is supply chain and operations management. He has authored a book on engineering economics, Wiley India Pvt. Ltd. and a book on Industrial Engineering and Management from Pearson Learning, India. He has also published more than 50 research papers in international journals and conferences. He has published papers in the journals such as Applied Soft Computing; Annals of Operations Research; Clean Technology and Environmental Policy; Environment, Development and Sustainability; International Journal of System Assurance Engineering and Management, Journal of Modelling in Management, International Journal of Mobile Communications, International Journal of Business Excellence etc.

Punyatoya Mishra is working as Assistant Professor, Department of Physics, Parala Maharaja Engineering College, Berhampur, Odisha. She has completed her Ph.D (2014) from NIT Rourkela in Physics. Currently, she is actively involved in research in the field of Ultrasonic, Material Science, Nano materials. She has published ten papers in international journals and life member of different societies like USI, OPS, etc.

Pushpa Jha, the author of a paper entitled "Application of agro-residues based activated carbon as adsorbents for phenol sequestration from aqueous streams: A Review."

Prof. Pushpa Jha has got her PhD degree from Department of Chemical Engineering, Indian Institute of Technology, Delhi in 1997. She did her B.E. (Chem. Engg.) in 1987 and M.E. (Chem.Engg.) in 1989 from Dr S. S. Bhatnagar University Institute of Chemical Engg. & Tech., Punjab University, Chandigarh. She has teaching and research experience of more than 20 years at various Engineering colleges of repute. Presently, she is a Professor at Sant Longowal Institute of Engineering & Technology (Established by MHRD, Govt. of India), Sangrur, Punjab, India. She has completed sponsored projects under TAPTEC & MODROB scheme of AICTE in Chemical Engineering discipline. Her research interests include Bio-Resource Technology, Alternate Sources of Energy, Waste Management & Recycling and Nanotechnology.

Rajesh Kumar Singh is Professor at Management Development Institute, Gurgaon, India. He has twenty-five years teaching and research experience. He has also worked as Associate Professor in Operations and Supply Chain Management area at Indian Institute of Foreign Trade (IIFT), Delhi and Delhi Technological University, Delhi, India. His areas of research are Total Quality Management, IT applications and Cyber Security issues in supply chains, Warehousing and Logistic management, Quantitative Techniques, Operations and Supply Chain Management. He has handled research projects sponsored by All Indian Council for Technical Education (AICTE), Department of Science and Technology (DST), Ministry of External Affairs, Ministry of Micro, Small and Medium enterprises, Govt of India, Telanagana State Govt etc. He has about 170 research papers published in reputed international/national journals and conferences. His papers have figured in International journals published by Tayler and Francis, Elsevier, Emerald, Springer, Inderscience, IGI Global (USA) and Sage publications such as International Journal of Production Research, Annals of Operations Research, Production Planning and Control, Journal of Cleaner Production, Resource, Conservation and Recycling, Journal of Industrial and Production Engineering, Industrial Management Data Systems etc.

Dr. Rajnarayan Saha, Professor, Dept. of Chemistry, NIT, Durgapur. Dr. Saha is a Chemist as well as Environmental Engineer with M.Sc. (Chemistry from Burdwan University), M.Tech and Ph.D (from Env. Sci. & Engg from IIT, Bombay). Earlier, he was a key faculty member to developed and started Department of Environmental Science, Burdwan University. Dr. Saha is not only a good teacher but also a very good researcher. Eleven (11) Ph.D. students has already been awarded under his guidance and several are working under him with various problems. Several M.Sc. and M.Tech students completed their Dissertation work under his supervision. Dr. Saha working in the areas on Physico-Chemical and biological treatment of wastewater, Food chain contamination of Arsenic, Municipal solid waste management and environmental application of nano materials. Regularly handling research projects sponsored from DST, UGC by Dr. Saha and published several research papers in national and international journals. Dr. Saha visited several countries like USA, Germany, Netharlands, Italy, Oman, Spain for his collaborative research and academic activities. Dr. Saha is a very good organiser to organised several workshops, winter and summer schools for college and university research scholar and faculty members.

Ruth M. Muthoka is a candidate of The integrated PhD in Mechanical Engineering at Inha University, Korea. She is currently a full time researcher at the Creative Research Center for Nanocellulose Future Composites, Inha University. Her research interests center around Mechanical Engineering and Material Science with an interplay between experimental techniques (fabrication & characterization) and computational approaches (Molecular Dynamic simulations) to explore the applications of Nanocellulose and it's composites in various areas.

Sakti Prasad Mishra received his M.Phil. Degree in Physics from Veer Surendra Sai University of Technology, Sambalpur, Odisha in the year 2015. Currently he is perusing his Ph.D in material science especially on biocomposites for electro – acoustic stealth material using bio waste material.

Sandeep Gupta is a PhD Scholar at the department of Mechanical Engineering at IIT Jodhpur. His research focuses on computational approaches to simulate fluid flows, tomography, porous media flow analysis, separation techniques for removal of contaminants from water, interactive thermo-chemical kinetics, engineering design and water–energy relationships. He holds a Masters from the Center of Excellence in Energy at IIT Jodhpur.

Sanjay K. Nayak holds the position of Professor and Chair of R&D wings of CIPET with 30 years of experience in teaching and research. He is an active researcher in the field of polymer science and technology, with several publications in peer-reviewed international journals and patented technologies in the area of advanced polymeric materials and characterization.

Sanjib Banerjee is an Assistant Professor at the Department of Chemistry, Indian Institute of Technology Bhilai, India. He obtained his Ph.D. in chemistry from Indian Association for the Cultivation of Science, working under the mentorship of Prof. Tarun K. Mandal on Synthesis of Polymer Nanostructures via Controlled Polymerizations. Following his doctoral work, he carried out his postdoctoral work at the University of Massachusetts Lowell with Prof. Rudolf Faust, working on polyisobutylene-based materials for application as motor oil additives and self-healing sealants. Subsequently, he was an ANR postdoctoral fellow at ENSCM, France working with Dr. Bruno Ameduri. During this time, he worked on designing fluoropolymers for application in functional coatings. He had a brief stay as a Research

Scientist at Solvay Research & Innovation Centre, India. He is a recipient of the SERB Ramanujan Fellowship.

Dr. Sanjib Gogoi obtained his M.Sc. degree from Dibrugarh University in 2001. He received his Ph.D. degree in 2008 under the supervision of Dr. N. P. Argade at the Pune University (CSIR-National Chemical Laboratory). He has three years of postdoctoral experience at the University of Texas at San Antonio (UTSA, San Antonio, USA) and Wayne State University, Detroit, USA. He is currently a Senior Scientist at the Applied Organic Chemistry Group in CSIR-North East Institute of Science and Technology, Jorhat, Assam. His research interests include metal-catalyzed reactions and asymmetric catalysis.

Shanmugasundaram O. Lakshmanan is a Professor in the Department of Textile Technology at K.S.Rangasamy College of Technology, India. He graduated in B.Tech. Textile Technology from Bannari Amman Institute of Technology, India. He aced the GATE Examination with 83.21 percentile score [All India Rank: 93]. He received his M.Tech. Degree in Textile Technology from Alagappa College of Technology, Anna University Campus, Chennai and completed his Ph.D. degree in Medical Textiles from Anna University.

His tenure of service for 16 years in an Engineering college has honed his skills in teaching, research and administration. He serves as an editorial board member and reviewer in many peer-reviewed journals and is also acting as organizing committee member and scientific committee member in 100 International and 60 National Conferences in India and abroad.

He disseminated his research results through regular publications in scholarly and peer reviewed Journals. He has authored 40 research papers in reputed journals. He has to his credit 30 papers presented in International Conferences and 7 others in National Conferences. His current research activities focus on Healthcare and Hygiene Textiles, Bio-Medical Textiles, Tissue Engineering and Regenerative Medicine, Synthesis of nanoparticles/polymers and its applications in wound dressings.

Shashikant Shingdilwar is a Ph.D. student at the Department of Chemistry, Indian Institute of Technology Bhilai, India. He obtained his B.Tech. and M.Tech. in Biotechnology from National Institute of Technology, Raipur and West Bengal University of Technology, Kolkata, respectively. Subsequently, he qualified for GATE and DBT-JRF to start his research career. Currently, he is pursuing his doctoral research on the "Design, Synthesis and Applications of Functional Mesoporous Molecular Materials".

Shri Sita Ram Bhakar is working as Professor in the Department of Soil and Water Engineering, College of Technology and Engineering, Maharana Pratap University of Agriculture and Technology, Udaipur, Rajasthan. He has worked for more than 30 years in the field of Irrigation management, Protected Cultivation and Precision Agriculture. He is also the Principal Investigator of All India Coordinated project on Plastic Engineering Technology of Udaipur Centre. He has published more than 30 research papers in National and International Journal of repute. He has also many books, practical manuals published to his credit and has guided 6 PhD students.

Simon Joseph Antony, FRSC (London) is an Associate Professor at the School of Chemical and Process Engineering, University of Leeds. His primary research interest is in the area of Particulate Mechanics and Physics, especially on the bulk mechanical behaviour of particulate materials under industrial process conditions as a function of their single-particle scale properties. He performs a number of inter-disciplinary research projects including in the area of waste processing of materials. He uses a wide range of computational tools including DEM, FEM and advanced experimental procedures including photonic stress sensing. He has obtained the prestigious MIT Young Researcher Fellowship Award for Exemplary Research in Computational Mechanics. He published more than 140 international journal papers and conference proceedings. He is a member of editorial boards for many international journals in his filed including Journal of Nanotechnology. He serves as a regular referee for several international journals, including Physical Review Letters and Physical Review E. He has served as a guest editor for the Jl. Granular Matter and the lead editor of the book 'Granular Materials: Fundamentals and Applications', published by the Royal Society of Chemistry, London in 2004. He is a member of many professional bodies worldwide. His research sponsors include EPSRC, Royal Society, DTI, ICI, BNFL, P&G, Pfizer, Borax Hosakawa Micron, Bridon International Ltd, Merck Sharp & Dohme, DuPont and Kayser-Threde. His biography is included in the Edition of Marquis Who's Who in the World and the Directory of International Biography Centre, Cambridge.

Sk Arif Mohammad is a Ph.D. student at the Department of Chemistry, Indian Institute of Technology Bhilai, India. He received his B.Sc. degree in Chemistry (Hons) from Moulana Azad College (Kolkata) in 2014. He subsequently completed his M.Sc. (2016) from Indian Institute of Engineering Science and Technology (IIEST), Shibpur and M.Tech (2018) in Polymer Science and Technology from University of Calcutta. Subsequently, he qualified for GATE to start his research career. Currently, he is pursuing his doctoral research on the "Smart Polymer Materials: Design, Synthesis and Applications". He is a recipient MHRD fellowship to pursue his Ph.D.

Smita Mohanty is working as a senior scientist at LARPM-CIPET Bhubaneswar and has more than 12 years of research experience. She has also initiated advanced research at LARPM including E-waste recycling, biopolymers from natural resources and polymer nanocomposites.

Snehalata Ankaram is an Assistant Professor in Zoology at Vasantrao Naik College of Arts, Commerce and Science, Aurangabad, Maharashtra, India. She holds M.Sc in Zoology and Ph.D in Vermicomposting Biotechnology. Her research area extends in solid waste management and animal microbiology. She has authored 12 publications in national and international journals and one book chapter in Springer.

Sowjanya Makarla obtained her B.Tech in Mechanical Engineering from Nagarjuna University, M.Tech in Thermal Engineering from Jawaharlal Nehru Technological University Hyderabad (JNTUH). She obtained her PhD in Mechanical Engineering for her research work in 'Planar Flow Melt Spinning Process' a collaborative research between Jawaharlal Nehru Technological University Hyderabad and DRDO (DMRL, Hyderabad). She developed 5 numerical Models to simulate the 'Melt spinning' process. She is the recipient of 'Excellent Paper Award' (ICMO, 2013 held at New Delhi), 'Best Paper Award' (TIME, 2016 held at Hyderabad) and 'Speaker of the Day Award' (ThermaComp2018 held at IISc Bangalore). She presented Invited Talks at an International Conference on Production and Design Engineering held at Berlin, Germany (2016), M.G University, Kerala etc. Life Member of ISTE, ISHMT,IEI and ISME. Her research areas include Heat transfer analysis of real time problems, Non-Conventional Energy, Air Pollution Control. She has expertise in the application of Computational Fluid Dynamics (CFD) for Heat transfer, Rapid solidification and single or Multi phase flow simulations.

Sujatha Karuppiah is a Associate Professor in the Department of Physics at Vellalar College for Women, India. She had 14 years of Teaching, Research and Admin experience in an Arts & Science College in India. She completed her B.Sc. degree in the year 1997. She postgraduated in M.Sc. Physics from V V Vaniaperumal College for Women, India in the year 1999. She received his M.Phil. Degree in Physics in the year 2000 from Alagappa University, India. She disseminated his research results through regular publications in scholarly and peer reviewed Journals.

Sunanda Roy is PhD from Nanyang Technological University, Singapore under Singapore MIT-Alliance Programme. He is presently a faculty of Mechanical Engineering at Inha University. He has extensive working experiences on various materials and techniques such as energy from waste, synthesis and functionalizations of nanomaterials, cellulose based composites, polymer nanocomposites, and coatings, electrospinning process and so on.

Sunil S. Suresh earned M.Phil degree in Marine Chemistry from Cochin University of science and technology (India) in 2011. Presently, he is concluding PhD in Chemistry at Laboratory for Advanced Research in Polymeric Materials (LARPM), Central Institute of Plastic Engineering and Technology (CIPET), Bhubaneswar. His research interests include E-waste recycling, material recycling and value addition and environmental sustainability of the recycled plastics.

Dr. (Mrs.) Susmita Dutta, Professor, Chemical Engineering Department, National Institute of Technology, Durgapur, did her Bachelor of Technology (B. Tech.) in Chemical Engineering from University of Calcutta in 1997, Master of Technology (M. Tech.) in Chemical Engineering from Indian Institute of Technology, Kharagpur (IIT, Kharagpur) in 1999. She obtained her Ph. D. degree in Engineering from Jadavpur University in 2004. She started her professional career by joining Heritage Institute of Technology, Kolkata, India in 2003 in the position of lecturer in Chemical Engineering Department. In 2005 she joined the Department of Chemical Engineering, University of Calcutta as lecturer and worked there for two years. In November, 2007 she joined her present institution as Assistant Professor of Chemical Engineering Department. In 2010 she was promoted to Associate Professor and in October, 2018, she was promoted to Professor in the same department. She was awarded several scholarship viz., National Scholarship during B.Tech Program on the basis of the performance in B.Sc. (Hons. in

Chemistry) Examination, Institute scholarship during M.Tech. program. She got M. H. Shukla 1st Prize, 2008 for the Best Technical Paper presented in the 60th Annual Session (CHEMCON 2007) held in December 2007 at Kolkata. She was selected as a Member of the Editorial Board of the Journal of Environmental Engg. and Landscape Management (Taylor and Francis) in January, 2019. She has a proven track record in various research fields. She carried out a number of research projects funded by several agencies. She has supervised a number of B. Tech students and M. Tech students. She has guided five Ph. D. students. She has published 40 papers in highly refereed national and international (SCI) journals. She has more than fifty publications in different conference proceedings. She has contributed in number of book chapters.

Tahshina Begum completed her masters in chemistry from Dibrugarh University (2010). In 2017, she obtained her PhD from Dibrugarh University under the supervision of Prof. Pradip K. Gogoi and Dr. Utpal Bora in the field of organometallic chemistry and nanomaterials, with a fellowship of the CSIR, New Delhi. She has been working as a SERB-NPDF in CSIR-NEIST, Jorhat, India under the supervision of Dr. Sanjib Gogoi since 2017. Her research interests include synthetic methodologies and metal-catalyzed reactions.

Tarun Parangi is UGC-Dr. DS Kothari Post Doctoral Fellow at Department of Chemistry, Sardar Patel University, Vallabh Vidyanagar, Gujarat. He received his Master's degree in Organic Chemistry from the School of Sciences, Gujarat University, Ahmadabad (2008) and Ph.D. in Applied Chemistry from the Maharaja Sayajirao University of Baroda, Vadodara (2015). His prolific research involves the development of Titania-nanoparticles and related inorganic materials as photocatalysts (Heterogeneous Catalysis) for different applications such as environmental pollution control, carbon dioxide and hydrogen production etc. He also works on synthesis, characterization of metal phosphate & phosphonate as proton conducting materials to be used as solid electrolyte for fuel cell applications.

About the Editors

Jibin K P, Currently working as JRF under DST Nanomission Project at International and Inter University Center for Nanoscience and Nanotechnology, and PhD Fellow at School of Chemical Sciences, Mahatma Gandhi University, Kottayam.

Qualified UGC CSIR NET in 2016 (National level exam). Secured first rank in MSc Analytical Chemistry 2015–2016, Mahatma Gandhi University, Kottayam, Kerala. Participated and paper presented in international conferences ICMS2017, ICECM 2017, ICRM 2018, ICN 2018, ICMST 2018, ICRAMC 2019, ICSG 2019 etc. Participated in several national seminars on different areas of chemistry. It includes workshop for chemistry students and teachers on september 2014 jointly coordinated by JNCASR and NIIST at Trivandrum. Worked on a project entitled *"optical studies of samarium complex synthesised using curcumin isolated from turmeric extract as ligand* and also in the project *synthesis of Mg and Co codoped ZnO based diluted magnetic semi-conductor for optoelectronic applications.*

Dr. Nandakumar Kalarikkal is an Associate Professor at School of Pure and Applied Physics and Director of International and Inter University Centre for Nanoscience and Nanotechnology of Mahatma Gandhi University, Kottayam, Kerala, India. His research activities involve applications of nanostructured materials, laser plasma, phase transitions, etc. He is the recipient of research fellowships and associateships from prestigious government organizations such as the Department of Science and Technology and Council of Scientific and Industrial Research of Government of India. He has active collaboration with national and international scientific institutions in India, South Africa, Slovenia, Canada, France, Germany, Malaysia, Australia and US. He has more than 160 publications in peer reviewed journals. He has also co-edited 15 books of scientific interest and co-authored many book chapters.

Prof. Dr. Sabu Thomas is currently the **Vice Chancellor of Mahatma Gandhi University** and the Founder Director and Professor of the International and Interuniversity Centre for Nanoscience and Nanotechnology. He is also a full professor of Polymer Science and Engineering at the School of Chemical Sciences of Mahatma Gandhi University, Kottayam, Kerala, India. Prof. Thomas is an outstanding leader with sustained international acclaims for his work in Nanoscience, Polymer Science and Engineering, Polymer Nanocomposites, Elastomers, Polymer Blends, Interpenetrating Polymer Networks, Polymer Membranes, Green Composites and Nanocomposites, Nanomedicine and Green Nanotechnology. Dr. Thomas's ground breaking inventions in polymer nanocomposites, polymer blends, green bio-nanotechnological and nano-biomedical sciences, have made transformative differences in the development of new materials for automotive, space, housing and biomedical fields. In collaboration with India's premier tyre company, Apollo Tyres, Professor Thomas's group invented new high performance barrier rubber nanocomposite membranes for inner tubes and inner liners for tyres. Professor Thomas has received a number of national and international awards which include:

Fellowship of the Royal Society of Chemistry, London FRSC, Distinguished Professorship from Josef Stefan Institute, Slovenia, MRSI medal, Nano Tech Medal, CRSI medal, Distinguished Faculty Award, Dr. APJ Abdul Kalam Award for *Scientific Excellence – 2016, Mahatma Gandhi University-Award for Outstanding Contribution – Nov. 2016, Lifetime Achievement Award of the Malaysian Polymer Group, Indian Nano Biologists award 2017* and *Sukumar Maithy Award* for the best polymer researcher in the country. He is in the list of **most productive researchers in India** and holds a position of No. 5. Because of the outstanding contributions to the field of Nanoscience and Polymer Science and Engineering, Prof. Thomas has been conferred *Honoris Causa (D.Sc) Doctorate by the University of South Brittany,* Lorient, France and *University of Lorraine, Nancy, France*. Very recently, Prof. Thomas has been awarded *Senior Fulbright Fellowship* to visit 20 Universities in the US and most productive faculty award in the domain Materials Sciences. Very recently he was also awarded with *National Education Leadership Award – 2017* for Excellence in Education. Prof Thomas also won **6th** *contest of "mega-grants"* in the grant competition of the Government of the Russian Federation *(Ministry of Education and Science of the Russian Federation)* designed to support research projects implemented under the supervision of the world's leading scientists. He has been honoured with *Faculty Research Award* of India's brightest minds in the field of academic research in May 2018. Professor Thomas was awarded with *Trila – Academician* of The Year in June 2018 acknowledging his contribution to tyre industry. This year, Prof Thomas was also awarded with *H.G. Puthenkavu Mar Philoxenos Memorial Best Scientist Award.* In 2019 Prof. Thomas received prestigious *CNR Rao prize lecture Award.* Prof. Thomas has published over 900 peer reviewed research papers, reviews and book chapters. He has co-edited 117 books published by Royal Society, Wiley, Wood head, Elsevier, CRC Press, Springer, and Nova etc. He is the inventor of **5 patents.** The **H index** of Prof. Thomas is **92** and has more than **40,121 citations.** Prof. Thomas has delivered over 300 Plenary/Inaugural and Invited lectures in national/international meetings over 30 countries.

Prof. Ange Nzihou (M) is a Distinguished Professor of Chemical and Environmental Engineering since 2012.

Since 2000, he has developed outstanding expertise in research fields such as Energy and added value materials from Biomass and Waste; Elaboration, functionalisation phosphate based composites/hybrid materials (sorbents, fertilizers, catalysts, energy carriers, sensors) for energy, agriculture and depollution (liquid, gas, solid phases); Thermochemical processes (pyrolysis, gasification, reforming) for biochar, bioash and energy production; The behavior of pollutants such as heavy metals and aerosols (fine particles) and Kinetics and transfer phenomena are among its field of expertise.

Through these research and industrial projects, he has supervised 32 PhDs (21 graduated already, 10 co-advised with colleagues from top universities in USA, Canada, Ireland, China and India), 12 post-docs, published 130 papers in peer reviewed journals, 4 world patents (2 scaled-up at industrial level), 38 invited plenary and keynote lectures in international conferences, published about 170 proceedings papers at national and international conferences. He has chaired and co-chaired 10 conferences and 2 international summer schools. He has obtained about 5.3 M€ of research grants over the 20 last years. He is chair and principal Investigator of joint Laboratories with industry: Solvay (Belgium) (2002–) and Terreal (France) (2014–2019).

He is Visiting Professor in number of renown Universities (USA, Ireland, China, India). He is the Editor-in-Chief of the peer reviewed journal "Waste and Biomass Valorization" (www.springer.com/engineering/journal/12649), and the founding Chair of the WasteEng Conference Series dedicated to the organisation of conferences and seminars on Waste and Biomass Valorisation (www.wasteeng2016.org) since 2005. He is currently editing a Handbook on characterisation of biomass, biowaste and related by-products (Springer) to be published in June 2019.

He obtained some International and National significant scientific recognitions and awards such as the *Grand Prix 2018 of the Academy of Sciences* (France) in association with IMT (Institut Mines-Telecom) for his outstanding contribution in science and technology of Energy transition. He was also awarded with the *Erudite Professor of the Mahatma Ghandi University* (India) in 2018. The previous recipients of this professorship are well known and outstanding international scientists, academicians and Nobel Prize winners. He is also evaluator for Executive Committee of the World Bank (The PASET Scholarship in Innovation Fund programme). He was also appointed member of the scientific advisory board of the Worcester Polytechnic Institute WPI (USA) from 2017. He was also promoted *Knight of the National Order of Merit* (France) in 2017 on the besthood of the president of France. He has frequently worked as evaluator for EU projects (RTD, SME, Intereg) under the FP7 and H2020 programmes.

Printed and bound by CPI Group (UK) Ltd, Croydon, CR0 4YY

23/10/2024

01777696-0017